BOSTON STUDIES IN THE PHILOSOPHY OF SCIENCE

VOLUME XXVII

TOPICS IN THE PHILOSOPHY OF BIOLOGY

SYNTHESE LIBRARY

MONOGRAPHS ON EPISTEMOLOGY,

LOGIC, METHODOLOGY, PHILOSOPHY OF SCIENCE,

SOCIOLOGY OF SCIENCE AND OF KNOWLEDGE,

AND ON THE MATHEMATICAL METHODS OF

SOCIAL AND BEHAVIORAL SCIENCES

VOLUME 84

BOSTON STUDIES IN THE PHILOSOPHY OF SCIENCE

EDITED BY ROBERT S. COHEN AND MARX W. WARTOFSKY

VOLUME XXVII

TOPICS IN THE PHILOSOPHY OF BIOLOGY

Edited by

MARJORIE GRENE AND EVERETT MENDELSOHN

D. REIDEL PUBLISHING COMPANY

DORDRECHT-HOLLAND / BOSTON-U.S.A.

Library of Congress Cataloging in Publication Data

Main entry under title:

Topics in the philosophy of biology.

(Boston studies in the philosophy of science ; v. 27)
(Synthese library ; v. 84)
Includes bibliographies and index.
1. Biology–Philosophy. I. Grene, Marjorie Glicksman, 1910– II. Mendelsohn, Everett. III. Series.
[DNLM: 1. Biology. 2. Philosophy, Medical. QH 331 T674]
Q174.B67 vol. 27 [QH331] 501s [574'.01] 75-12875
ISBN 90–277–0595–X
ISBN 90–277–0596–8 pbk.

Published by D. Reidel Publishing Company,
P.O. Box 17, Dordrecht, Holland

Sold and distributed in the U.S.A., Canada and Mexico
by D. Reidel Publishing Company, Inc.
306 Dartmouth Street, Boston,
Mass. 02116, U.S.A.

Printed in The Netherlands by D. Reidel, Dordrecht

PREFACE

The philosophy of biology should move to the center of the philosophy of science – a place it has not been accorded since the time of Mach. Physics was the paradigm of science, and its shadow falls across contemporary philosophy of biology as well, in a variety of contexts: reduction, organization and system, biochemical mechanism, and the models of law and explanation which derive from the Duhem-Popper-Hempel tradition.

This volume, we think, offers ample evidence of how good contemporary work in the philosophical understanding of biology has become. Marjorie Grene and Everett Mendelsohn aptly combine a deep philosophical appreciation of conceptual issues in biology with an historical understanding of the radical changes in the science of biology since the 19th century. In this book, they present essays which probe such historical and methodological questions as reducibility, levels of organization, function and teleology, and the range of issues emerging from evolutionary theory and the species problem. In conjunction with Professor Grene's collection of essays on the philosophy of biology, *The Understanding of Nature* (Boston Studies in the Philosophy of Science, Vol. XXIII) and the occasional essays on these topics which we have published in other volumes (listed below), this volume contributes to bringing biology to the center of philosophical attention.

Everett Mendelsohn, 'Explanation in Nineteenth Century Biology' (Boston Studies, Vol. II, 1965).
David Hawkins, 'Taxonomy and Information', (Boston Studies, Vol. III, 1967).
Norman Geschwind, 'The Work and Influence of Wernicke', (Boston Studies, Vol. IV, 1969).
Carl Wernicke, 'The Symptom Complex of Aphasia: A Psychological Study on an Anatomical Basis', (Boston Studies, Vol. IV, 1969).
Norman Geschwind, 'Anatomy and the Higher Functions of the Brain', (Boston Studies, Vol. IV, 1969).
Milic Capek, 'Ernst Mach's Biological Theory of Knowledge', (Boston Studies, Vol. V, 1969).
June Goodfield, 'Theories and Hypotheses in Biology: Theoretical Entities and

Functional Explanation', (Boston Studies, Vol. V, 1969); with comments by Ernst Mayer and Joseph Agassi.

Floyd Ratliff, 'On Mach's Contributions to the Analysis of Sensations', (Boston Studies, Vol. VI, 1970).

Milic Capek, Part I, 'Bergson's Biological Theory of Knowledge', (Boston Studies, Vol. VII, 1971).

Edward Manier, 'Functionalism and the Negative Feedback Model in Biology', (Boston Studies, Vol. VIII, 1971).

William C. Wimsatt, 'Some Problems with the Concept of "Feedback",' (Boston Studies, Vol. VIII, 1971).

Eugene P. Wigner, 'Physics and the Explanation of Life', (Boston Studies, Vol. XI, 1974).

J. Bronowski, 'New Concepts in the Evolution of Complexity: Stratified Stability and Unbounded Plans', (Boston Studies, Vol. XI, 1974).

Kenneth F. Schaffner, 'The Unity of Science and Theory Construction in Molecular Biology', (Boston Studies, Vol. XI, 1974).

Huseyin Yilmaz, 'Perception and Philosophy of Science (Boston Studies, Vol. XIII, 1974).

Ernst Mayr, 'Teleological and Teleonomic, a New Analysis', (Boston Studies, Vol. XIV, 1974).

Norman Geschwind, *Selected Papers on Language and the Brain* (Boston Studies, Vol. XVI).

Stuart Kauffman, 'Elsasser, Generalized Complementarity, and Finite Classes: A Critique of His Anti-Reductionism', (Boston Studies, Vol. XX, 1974).

Center for the Philosophy and
History of Science,
Boston University

ROBERT S. COHEN
MARX W. WARTOFSKY

TABLE OF CONTENTS

PART IV / EVOLUTION

PART V / SPECIES PROBLEM

FOREWORD

Philosophy of science in the past has focussed largely on examples from physics; even since 'logical reconstruction' has yielded ground to a more historically oriented position, talk of 'scientific revolutions' and so on has dealt chiefly with cases like the shift from Newton to Einstein rather than, say, from Darwin and Mendel to Watson and Crick. At the same time, however, there has been increasing interest on the part, not only of students, but of philosophers and historians, in conceptual problems at the foundations of the biological, as distinct from the 'exact' sciences. There are, indeed, some books, as well as anthologies, available in the field; but we felt that there was room at this stage for a carefully selected, yet representative, collection that would exemplify the major problems and some of the typical approaches to their resolution.

"Science", James Franck is supposed to have remarked, "is either something people do or it is nothing at all." Both editors agree with this pronouncement; we have tried to implement our acceptance of it by placing a trio of essays in the history of biology at the head of our collection. The fact that two of these papers happen to be by us does not mean, however, that we consider our work more important than that of our other contributors. Nor are we alleging that philosophical and historical problems are identical; what we *are* alleging is that conceptual problems at the foundations of any discipline originate, like any other human problem, *within* an historical, social-political-intellectual-personal, situation – a situation that not only establishes necessary conditions for their emergence as conceptual problems, but shapes them as the problems they have become of and for those scientists, or, at one remove, philosophers and/or historians, for whom, in the apposite situation, they have arisen. 'Logical reconstruction' *in vacuo*, therefore, necessarily ignores the substantive conceptual issues to which, in their development, the sciences recurrently give rise. This has been a hard lesson for philosophers of science to learn, but they are learning it; we have tried to stress it, pedagogically, by putting history first on paper as it is in very truth.

Parts II to V deal with the issues usually raised in discussions of the philosophy of biology – each time, again, with a brief selection of papers standing for many more. The first question, of course, has been the question whether biology is anything at all, or only, in the last analysis, a part of physics. On this issue Michael Polanyi's paper has by now become a classic. True, his argument should be modified to take account of the fact that some of the four bases on the DNA chain do combine more readily than others, and so the improbability of the code is not as complete as it seemed to be when his paper first appeared. With that qualification, however, his fundamental thesis remains defensible. It should be emphasized, moreover, that there is no question here, as some critics have supposed, of an old-fashioned vitalism; no serious philosopher any longer espouses such a view, if any ever did. What is at issue is the hierarchically organized structure of living things, which permits their study by scientists on a number of levels. (On this, see *e.g.*, Pattee or Kauffman in Part III). Against this position, Professor Schaffner states the equally classic case for physicalism: for the view that science is physics and that's the end of it.

Supposing, however, that radical physicalism is untenable, and that there *are* problems unique to biology, it remains to pose them. First, there is the question of biological explanation, to which Part III of our anthology is devoted. There is a complex network of problems here, which we have tried to exemplify in three parts. First, the question of levels of organization, already raised in Polanyi's paper, is considered from the point of view of an embryologist by Grobstein, of a biophysicist by Pattee and of a philosopher by Wimsatt. Second, the question of teleological explanation, which used to be 'the' question about explanation in biology, is represented by two papers among many devoted to the subject. Beckner's, again, has become a classic in this field. Wright's is of interest as representing a serious consideration of teleology in the explanation of animal behavior, rather than in biological explanation as such. It thus opens the possibility, which the same author has developed elsewhere, of separating the problem of functional explanation ('What does the liver do?' and even 'How does it do it?') from teleological explanation in a narrower sense ('Why did the rabbit run into its burrow?'). Wright's paper is of interest also for its careful modification of Charles Taylor's formulation of teleological explanation in his important and influential work *The Explanation of Behaviour* (Routledge and Humanities Press, 1964). Kauffman's essay, presenting an original view of the uniqueness of biological explanation, seems to fall into a class by itself, and so we have placed it alone as III C.

In the field of evolutionary theory, or meta-theory, it has been difficult to select from an extensive literature. We might have introduced specific problems like those arising from the question of group selection; or, some may think, we should have included treatment of such grand topics as 'emergent evolution'. The former, however, though fascinating, would be too specialized for our purposes, and the latter, we agree, though it may engender lively metaphysical discussion, is no longer a live subject for debate in the philosophy of biology. By and large, evolutionary theory means the theory of natural selection. What supplementation it needs is ably suggested in Slobodkin's paper, and one of the debatable areas it still leaves open, the precise meaning and scope of 'adaptation,' is discussed by Munson. Lewontin's paper indicates the possible application of a mathematical technique, the theory of games, to evolutionary questions; this application seems to be bearing fruit, or at least to be stimulating further discussion at the present time. Ayala's contribution might perhaps have been included in Part II, or indeed III B, for it constitutes a programmatic pronouncement that evolutionary theory alone saves biology from reduction to physics, and that it does so because its explanatory power is teleological, while that of physics is not. This thesis can be, and has been, criticized, but, once more, the essay has already become a classic in the field; and, together with Slobodkin's, it presents contrary interpretations by evolutionists of the conceptual structure of their theory.

Our last section raises recent issues most hotly debated by contemporary biologists: problems of the foundations of taxonomy; and at the same time it introduces one of the most venerable problems of philosophy, the problem of universals. Are there real classes, and, if so, how are they to be characterized? Ernst Mayr speaks most authoritatively, and clearly, for the majority of taxonomists; the range of contemporary positions is summarized by Hull in his review paper, and Pratt contributes a recent discussion of the issues as a philosopher sees them. In this field, too, of course, there is a great deal more we might have included; the reading list, which in turn includes books with extensive bibliographies, will, we hope, guide the student in further exploration.

University of California, Davis MARJORIE GRENE
Harvard University EVERETT MENDELSOHN

ACKNOWLEDGEMENTS

The editors gratefully acknowledge the permission of the publishers and authors of the articles listed below to include these pieces in the current volume.

Marjorie Grene: 1972, 'Aristotle and Modern Biology', *J. History Ideas* **30**, 395–424.
Everett I. Mendelsohn: 1968, 'Philosophical Biology versus Experimental Biology', *Actes*, Tome I, B, Discours et Conferences Colloques, XIIe Congrès International d'Histoire des Sciences, Paris.
Stephen Jay Gould: 1971, 'D'Arcy Thompson and the Science of Form', *New Literary History* **II**, No. 2.
Kenneth F. Schaffner: 1969, 'The Watson-Crick Model and Reductionism', *British J. Phil. Sci.* **20**, (Camb. Univ. Press).
Michael Polanyi: 1968, 'Life's Irreducible Structure', *Science* **160**, (Am. Assoc. Adv. Sci.).
Clifford Grobstein: 1969, 'Organizational Levels and Explanation', *J. History Biology* **2**, No. 1.
H. H. Pattee: 1971, 'Physical Theories of Biological Co-ordination', *Quart. Rev. Biophys.* **4**, 2 and 3.
William C. Wimsatt: 1974, 'Complexity and Organization', *Philosophy of Science Association*, 1972, *Boston Studies in the Philosophy of Science*, D. Reidel, Dordrecht.
Morton Beckner: 1969, 'Function and Teleology', *J. History Biology* **2**, No. 1.
Larry Wright: 1973, 'Functions', *Phil. Rev.* **82**, No. 2.
Stuart A. Kauffman: 1971, 'Articulation of Parts Explanation in Biology and the Rational Search for Them', *Boston Studies in the Philosophy of Science* **VIII**, pp. 257–272, D. Reidel, Dordrecht.
Lawrence B. Slobodkin: 1964, 'The Strategy of Evolution', *Am. Scientist* **52**, (Sigma Xi).
R. C. Lewontin: 1961, 'Evolution and the Theory of Games', *J. Theor. Biol.* **I**, (Academic Press).
Francisco J. Ayala: 1968, 'Biology as an Autonomous Science', *Am. Scientist* **56**, No. 3, (Sigma Xi).
Ronald Munson: 1971, 'Biological Adaptation', *Phil. Sci.* **38**, No. 2.
Ernst Mayr: 1957, 'Species Concepts and Definitions', Publication No. 50 of the American Association for the Advancement of Science.
Vernon Pratt: 1972, 'Biological Classification', *Brit. J. Phil. Sci.* **23**, (Cambridge University Press).
David L. Hull: 1970, 'Contemporary Systematic Philosophies', *Ann. Rev. Ecol. Syst.* **I**.

The editors wish to give thanks to Heather Cole, Harvard College Library, for the preparation of the index, and Ruth Bartholomew, of the Department of the History of Science at Harvard, for help in preparation of the typescript.

PART I

HISTORY

MARJORIE GRENE

ARISTOTLE AND MODERN BIOLOGY

I

Science had its origin, if not in opposition to Aristotle, at least in opposition to Aristotelianism. But science in its most authoritative form was what came to be called physics, not biology. Faced with the recent crisis in biology, in which the life sciences have been threatened with reduction to microversions of themselves and ultimately to chemistry and physics, one wonders if the besieged biologists, or at any rate their philosophical defenders, might not after all learn something to their advantage by reflecting on the one great philosopher who was also a great biologist. And we *can* learn from Aristotle; not, however, in a simple or straightforward fashion. There is no use just contrasting, as some have been tempted to do, Democritean with Aristotelian science and putting physicists in the former class, biologists in the latter. Even if we reject Simpson's alleged reaffirmation of Roger Bacon and stoutly deny that 'the study of Aristotle increases ignorance',[1] we must nevertheless admit that in some important respects biology, like all modern science, really is, and must be, un-Aristotelian. This thesis could be defended in a number of ways; let me select four.[2]

First, the role of abstraction and the relation of scientific reasoning to everyday experience differ deeply in Aristotelian and in modern science. Only mathematics, Aristotle insists, abstracts *from* most of the ordinary perceptible properties (qualities, relations, states) of the things around us. Natural science, as he understands it, remains *within* the framework of everyday perception and makes more precise, within that framework, our formulation and understanding of the essential natures of quite ordinarily accessible entities. Modern science began and continues, on the contrary, precisely by closing its eyes to all but a few highly selected features of the world around us, abstracting from all but those variables which give promise of susceptibility to some sophisticated, usually quantitative, manipulation by the experimenter or the theorist.

Secondly, because he sticks so closely to the concrete, limited, and limiting physiognomy of things encountered in the everyday world, the Aristotelian scientist never confronts the teasing problem of induction which the modern scientist, or better, philosopher of science, necessarily has to face. The possible data of scientific calculations are infinite, the calculations themselves and their results are finite. However ingenious the arguments that have been used to comfort us before this gap, the gap remains. For Aristotle it does not exist. Admittedly even the most ingenious experimenter must rely, in Aristotelian fashion, on the stability of his surroundings: on his materials, his apparatus, as being reliably not simply this-here, but this-such.[3] Every time you pick up a handful of $CuSO_4$ crystals, Norman Campbell argued, you are in effect acknowledging a *law* of nature: you confidently expect these blue crystals to have the same chemical properties as they did yesterday and will tomorrow.[4] Confidence in such stabilities does indeed depend, if you like, on Aristotelian induction, where we move from a rough and ready perception of the character of a thing to a more precise knowledge of just what kind of thing it is. But – *pace* Hume – it is only after this everyday induction that the problem of induction in science first begins: that is, the problem evoked by the necessary disproportion between data and hypothesis, between evidence and conclusion. Just because it remains within the horizon of everyday things-to-hand, Aristotelian science evades altogether the problem of induction in this its modern form.

A third way to emphasize the same contrast is to stress the role of productive imagination in scientific discovery. Aristotelian *phantasia* is powerless to excel what Kant called reproductive imagination. The productive power of that faculty, not *a priori* and once-for-all, as in the Schematism of the *Critique of Pure Reason*, but advancing hazardously to and beyond ever new frontiers: that is the moving force behind the scientific adventure, a force wholly beyond the ken of the much cosier *Prinzipienforschung* of Aristotle, cradled as it is within the comforting embrace of the familiar everyday world.

All this concerns Aristotelian methodology. Cosmologically, too, and especially for biology, there are features of Aristotelian science which the modern mind radically rejects. Aristotle's world is finite, unique, eternal, consisting of a finite number of eternally existent species, 'endeavoring' in their re-production to simulate the eternal circling of the celestial spheres (so he argues at the close of the *De Generatione et Corruptione*[5]). For modern biology, this eternal frame is shattered. All things flow. For many modern biologists, indeed, the theory of evolution is comprehensive for their science; all biological research, they feel, somehow derives from and contributes to it. That claim may be exaggerated. But even research seemingly unconnected with evolution is nevertheless related to it in some degree, as figure to ground. Some of Aristotle's detailed biological work too was sound and still retains its validity despite the incorrectness of his cosmological theory, e.g., his description of chick development or his account of the life history of *Parasilurus aristotelis* or of *Mustelus laevus*. Much modern research, similarly, may reach correct conclusions independently of the general conception of life's development from the non-living or of the transmutation of species. Yet the overall thrust of modern biology has been radically altered by the idea of evolution, much as the Aristotelian science of nature was guided by the contrary view of a static and finite cosmos. In their overall implications, the two are incompatible.

Despite these contrarieties, however, there is much to be learned from Aristotle in relation to the philosophical problems of biology. I want to discuss in this context three Aristotelian concepts: τέλος or the οὗ ἕνεκα, that for the sake of which; εἶδος in contrast both to ὕλη and to γένος, that is *eidos* as form and *eidos* as species, and finally that most puzzling of Aristotelian phrases, τὸ τί ἦν εἶναι, the 'being-what-it-is' of each kind of thing.

II

First, *telos*. Again, it was Aristotelian teleology that seventeenth-century innovators were most emphatically determined to abolish from the study of nature. And again, as Aristotelian teleology had come to be understood, or misunderstood, this was a necessary move. Yet some sort of teleology, or teleonomy, as some modern biologists prefer to call it, keeps creeping back into biological language and thought. Let us consider, then, how and where something like Aristotle's *telos* occurs in modern biology and how modern usage compares with his.

Two misconceptions must first be set aside. (1) *Telos* is not in the first instance – and in the study of nature is not at all – 'purpose or plan'. In nature, 'that for the sake of which' a series of events takes place is the intrinsic endpoint in which, if nothing fatally interferes, that series normally culminates. What usually happens to a fertilized robin's egg, for instance? A baby robin hatches out of it; that is its *telos*. There is absolutely no question of any kind of 'purpose' here, either man's or God's. To suppose otherwise is to introduce a Judaeo-Christian confusion of which Aristotle must be entirely acquitted. (2) Nor is Aristotle interested primarily in one over-all cosmic *telos*. Despite the passage from *De Generatione et Corruptione* already referred to, and despite the 'teleological' causality of the unmoved mover, the kind of 'ends' that usually interest Aristotle are the determinate end-points of particular processes within the natural world. True, the stability of the universe is, for him, a necessary condition of the orderly processes of its components; but this is by no means the target of his primary interest. On the contrary, it was, again, the Judaeo-Christian God who (with the help of neo-Platonism) imposed the dominance of a cosmic teleology upon Aristotelian nature. Such sweeping purpose is the very contrary of Aristotelian.[6]

The concept of *telos* in exactly Aristotle's sense, however, does occur in exactly the area where he himself invokes it: in the study of ontogenesis. Here, indeed, the contrast of Aristotelian and Democritean science still appears valid at least to a first approximation. The embryologist must put questions to the living embryo, in terms to which, and in which, it can respond.[7] And this is impossible on principle in terms of a thoroughgoing atomism, since in those terms there *is* nothing alive. What might happen to a *really* Democritean scientist faced with a biological problem was suggested by Frank Baker in a paper I have quoted elsewhere. Imagine, he says, an observer looking in the field of a microscope at the filaments of a fungus. 'He witnesses', says Baker,

> that at the tips of the filaments are disposed a number of radiating branches more frequently segmented than those of the stalk to which they are attached; which adjointed elliptical segments are easily set free by pressure in the surrounding medium. But, supposing that he decides to investigate these segments, what kind of ideas are going to control his choice of further observations; how will be proceed, loosely speaking, to discover 'the nature' of these structures; and, in brief, in such a context, what does this notion of 'investigation' already imply?[8]

An old-fashioned chemist, Baker suggests, might throw these segments (which, unknown to him, we may call spores) into concentrated H_2SO_4. He would learn something; but would this 'lay an effective basis to the study of mycology'? And why are these not as good facts to start with as any other? But suppose our investigator places the segments on jam (where he first found them) or sugar, and in the warmth, and watches what happens. He may then discover 'their relation to the life cycle of the fungus'. Actually, Baker remarks, it would take a whole series of investigators 'animated by a single scientific impulse or tradition' to lay the foundation which would so much as show him where to start. When he gets this far he can then, but not before, undertake his chemical analysis. Only, in other words, when the concept of germination is understood and its designatum assumed to exist, do the details 'fall into order and acquire a significance', such that detailed analysis of some parts of the process of germination can be undertaken. The orderly development of the organism under investigation must have been assumed, Baker concludes, before the right 'facts' could be selected for further investigation and analysis.[9] An orderly development toward a normal end, therefore, is necessarily presupposed by the biological investigator before he can set out to make his investigation. A concept of 'that for the sake of which' the development is occurring, of its natural *telos*, is contained in the very question asked.

Such considerations place teleological (or teleonomic) thinking in the position of at least a *regulative* idea (in the Kantian sense) at the beginning of biological research. From this point of view *telos* is a signpost to the study of nature: a 'reflective concept' (*ein Reflexionsbegriff*) as Wieland argues.[10] Looking at the endpoint of the series helps us to start looking for its necessary antecedents; there is nothing 'unscientific' about this, not even anything very un-Democritean. But is that all there is to it? A Kantian regulative idea – say, the infinite divisibility of matter, or indeed natural teleology as Kant conceived it – is a pure *as-if*. And many modern thinkers would be content with this, with 'the appearance of end', as Waddington calls it.[11] It's a makeshift, they say, a crutch to lean on until we have mastered the necessary and sufficient conditions, or until we have constructed a machine to simulate an animal, or until we have synthesized life. Then we can throw away the crutch and walk alone.[12] Or perhaps, as Piaget argues, the very idea of a final cause, even in this seemingly harmless form, is based on a logical error and we don't need it at all. It rests, he maintains, on a confusion of three different notions: physical or physiological causality (cause *a* produces effect *b*), logical implication (the use of *A* implies the consequence *B*), and instrumentality (to get *B* 'we must' use *A*).[13]

Both these views are mistaken. To assert that a robin's egg hatches out a robin and not an oak tree is to state not a regulative idea but a fact of nature. Nor, *a fortiori*, does such an assertion represent any logical howler. At the least, it locates in the real world an orderly process, the details of which the biologist may undertake to study. It selects a certain segment of orderly temporal process *in* its orderliness as the locus of an inquiry. To this extent at least it locates real, not apparent, ends and suggests really, not seemingly or misguidedly, teleological questions.

But is that all? If there are real processes with natural endpoints, real τέλη in nature, are there not also teleological *answers* to the questions we put to nature? This seems to me a much more difficult problem. Professor C. P. Raven of Utrecht has written of the application of information theory to biology as 'the formalization of finality', and I had formerly taken him at his word.[14] Yet now I wonder. What is 'finalistic' about information theory? Admittedly, both teleological explanation and cybernetical explanation are complementary to classical causal explanation; in this sense, both resist a one-level, Democritean approach. But that does not make them equivalent. I prefer to leave this question open here, and postpone a consideration of multilevelled explanation in general until I come to the concept of *eidos*. For the moment we may take it that in the study of individual development, the concept of a normal endpoint of development helps methodologically to locate the place of the inquiry and to locate it really in nature, whether or not the concept of *telos* is also embedded in the solution to the embryologist's problem.

III

The most vexing problem with respect to Aristotelian *telos* and modern biology, however, concerns not ontogeny, but phylogeny. Granted, there is a definite, if perhaps limited, role for teleology (or teleonomy) in the study of ontogenesis; can one transfer the concept of end to the study of evolution, or can one at least discover in evolutionary explanation an analogue of teleonomic thinking? It has repeatedly been claimed both that Darwinian evolutionary theory rejects any cosmic *telos* and that it retains the concept of *telos* in some more acceptably 'scientific' sense. Indeed, it has even been argued that it is precisely by virtue of its teleological structure that evolutionary theory, and only evolutionary theory, rescues biology from reduction to physics and makes it 'an autonomous science'.[15] These claims may perhaps be clarified if we compare the 'teleonomic' thinking of modern evolutionists with Aristotelian teleology.

The fixed endpoint of a natural process, for Aristotle, is the mature form of the adult individual (strictly, of the adult male!) of the species in question. 'Nature is like a runner', he says, 'running her course from non-being to being and back again'.[16] The being in the case is the developed adult of such and such a kind. In modern evolutionary biology, however, there is no such fixed form; the *eidos* itself, which is the *telos* of individual development, is transitory. What remains? For Darwinism, the *telos* that remains when the eternal species is removed is simply survival. Survival of what? The individual perishes; what survives? In Darwinian terms: the descendants of the slightly more 'successful' members of a species. In neo-Darwinian terms: the alleles which made possible the development of phenotypes carrying the slightly more 'successful' characters, or rather, statistically, a higher ratio of those alleles in the gene pool of the next generation. A robin's egg is not, it seems, the way to make a robin, but a robin is a robin's egg's way to make more, and more probably surviving, robin's eggs. The locus of the goal of biological process is not, as it appeared to Aristotle, in the mature individual, who is, as such, mortal and of no concern to evolution, but in the future gene pool of the population of individuals of a potentially interbreeding population.

Take, e.g., Kettlewell's classic study of industrial melanism.[17] If tree trunks are blackened by industry, birds take more peppered moths, and hence more genes for peppered wings, than they do the *carbonaria* mutant of the species. Hence, fewer mutant genes are eaten and proportionally more survive. The *telos* of this process is the greater ratio of *carbonaria* genes in the next generation. Evolutionarily speaking, that is what the whole business of being a moth is *for*; not, indeed, just 'for' the survival of *this* gene, but if we had a complete count of all the genes in the population at time t_0 and time t_1 the differential ratio would give us the 'end' of the story: the differential survival of some genes rather than others.

But which genes? Whichever ones survive, of course. If we clean up industry and the tree trunks bleach again, more peppered and fewer *carbonaria* genes survive, and the endpoint goes the other way. Similarly in peacetime healthy human males are, other things being equal, better adapted – and that means of course in evolutionary terms more likely to leave descendants – than sickly ones, but in war time the contrary holds: the halt, the lame, and the blind are better adapted than the healthy. Biological process is first and last and always evolution; evolution is first and last and always a chronicle of survival, the survival of whatever survives.

A strange *telos*: we are told simply, what survives survives. But this, it has repeatedly been objected, looks like a mere tautology. And at first sight, at least, it has also been repeatedly objected, a tautology seems to have no explanatory power, let alone the explanatory power that would be characteristic of a teleological account. For a teleological account distinguishes, and sets out as aimed at, a goal to which it can then relate the antecedent steps. Here, however, goal and steps are collapsed into an empty identity.

Yet that identity, we are told, presides over a rich and precise elaboration of 'evolutionary mechanisms', and hence of teleonomic patterns of structure or behavior. The case for this view is argued in a thought-provoking book by George Williams, *Adaptation and Natural Selection.* Although he accepts, and celebrates, the tautological character of the principle of natural selection, as the survival of the fitter in the sense of the more probable survival of what will more probably survive, he insists nevertheless that this principle, correctly used, can preside over a vast range of teleonomic investigations:

A frequently helpful, but not infallible rule is to recognize adaptation in organic systems that show a clear analogy with human implements. There are convincing analogies between bird wings and airship wings, between bridge suspensions and skeletal suspensions, between the vascularization of a leaf and the water supply of a city. In all such examples, conscious human goals have an analogy *in the biological goal of survival,* and similar problems are often resolved by similar mechanisms. Such analogies may forcefully occur to a physiologist at the beginning of an investigation of a structure or process and provide a continuing source of fruitful hypotheses. At other times the purpose of a mechanism may not be apparent initially, and the search for the goal becomes a motivation for further study. Adaptation is assumed in such cases, not on the basis of a demonstrable appropriateness of the means to the end, but on the direct evidence of complexity and constancy.[18]

The study of the lateral line of fishes, he suggests, is a good example of this kind of reasoning:

The lateral line is a good illustration. This organ is a conspicuous morphological feature of the great majority of fishes. It shows a structural constancy within taxa and a high degree of histological complexity. In all these features it is analogous to clearly adaptive and demonstrably important structures. The only missing feature, to those who first concerned themselves with this organ, was a convincing story as to how it might make an efficient contribution to survival. Eventually painstaking morphological and physiological studies by many workers demonstrated that the lateral line is a sense organ related in basic mechanism to audition (Dijkgraaf, 1952, 1963). The fact that man does not have this sense organ himself, and had not perfected artificial receptors in any way analogous, was a handicap in the attempt to understand the organ. Its constancy and complexity, however, and the consequent conviction that it must be useful in some way, were incentives and guides in the studies that eventually elucidated the workings of an important sensory mechanism.[19]

How does this kind of teleonomic thinking compare with the use of *telos* in Aristotle? Aristotle presents his concept of 'that for the sake of which' as a guide to the study of nature in opposition to the thinking of Empedocles, who would elicit the phenomena of the living world, without ordered ends, out of a combination of chance and necessity. At one stage in cosmic history, Empedocles imagines, there were heads and trunks and limbs rolling about the world. Those that happened to come together in a viable combination survived; the others perished. This was a very crude theory of natural selection, to be sure, but a theory of natural selection, nevertheless. Aristotle as a practising biologist objected: ox-headed man progeny and vine-bearing olives, such as Empedocles envisages in his transitory world, are an absurdity. What we *always* have in nature is the ordered passage to a definite endpoint: man to man, cattle to cattle, grape to grape, olive to olive. Only where there are such functioning, ordered series does the study of life *begin*. Williams would agree. Where we can use only the concepts of chance and necessity, he insists, we should. Thus the descent of flying fishes can be explained in terms of physics alone; their flight, however, which is 'contrived', in analogy to human contrivance, needs, he argues, another and teleonomic principle of explanation, as any piece of machinery does.

So, as we saw with the case of ontogenesis, we need, it seems, a teleological approach to locate a biological problem. But is the explanation, in the case of selection theory, teleological as well? Have we even found, as in individual development, a directed process to describe – however we may eventually explain it? I think not; for explanation in terms of orthodox evolutionary theory collapses pretty quickly into pure Empedoclean chance-times-necessity. Had fishes not had the sensory mechanism of the lateral line they would not have 'heard' their predators coming and would not have survived. Or better: those whose 'hearing' was slightly more acute left descendants in the gene pool, those not so gifted left fewer and finally none. Chance mutations necessarily sorted out by the compulsions of environmental circumstances: that is a pure Empedoclean, anti-teleological process. The peppered moth case is a striking example. We have here, we are told, 'evolution at work': now we see the whole process in little. Extrapolate this 'mechanism' to the whole story of life and you have the vast panorama before you: no other principles are needed. But what is this story? The environment, for extraneous reasons (in this case the industrial revolution) changes; the gene pool is always changing; the changed environment necessitates changed predation (birds can't as easily see black moths on black trees as peppered ones); changed predation necessitates differential survival of some genes rather than their alleles. So we *necessarily* get more black moths than peppered ones. Extrapolate this process to the whole of evolution and you see a vast sequence of necessities. True, the sequence is triggered, and kept going, by a set of curious chances. These, however, are 'chances' only in the sense of being at a tangent to the 'normal' sequence of development. They are to be explicated, on principle, in terms of natural, that is, physico-chemical laws. Thus, given the nature of bituminous coal, the tree trunks had to be blackened. Given the chemical nature of DNA, one supposes, the 'errors' which occur in its replication will ultimately be explained also in physico-chemical terms. They happen by 'chance' – just as in Aristotelian chance – only in the sense that they are outside the usual sequence of events to which one has been attending, in this case the 'normal' development of the peppered moth. But they have, or will have, their physico-chemical explanation, which must ultimately exhibit their necessary occurrence. Again, extrapolate this reduction to the whole history of nature. Where is the teleology? It has served as a heuristic maxim to start us off on our inquiry, but in the sequence of survivals *it* does not survive even as a factor in the phenomena described, let alone as an explanatory principle.

Is there any other way to introduce teleology into evolution? Many philosophers and some biologists have tried to do this in terms of a theory of 'emergence'.[20] But these theories, compared either with the precise and limited teleology of Aristotle or with the vanishing teleology of Darwinism, are vague, and empty of explanatory power. If one says, for example, with Vandel (following Bergson) that life has moved toward an increase of 'le psychisme', with two high points, the insects (generic inventiveness) and man (individual inventiveness), two objections at once arise.[21] First, some of the diverse branches of evolution have gone that way, but not by any means all. What of parasitism, what of long stable forms like Lingula, what of the vast variety of birds or 'lower' mammals, what of the evolution of plants, etc., etc.? Second, even if we can see, very generally, some such tendency in the history of life on earth, *how* did it happen? What does such an assertion of the 'emergence' of psychic powers explain, and how? This appears an even stranger extrapolation than the Darwinian. For one *can* imagine the melanism story stretched back to the beginning of time. I suspect (indeed, I have argued elsewhere) that this extrapolation entails untenable pseudo-reductions of richer to poorer concepts; but still one can see how it's done. The 'emergence' extrapolation, however, I for one simply cannot follow at all. There is a goal, mind or thought or inventiveness, we are told, *for* which evolution happened. It is the achievement of this goal that we are studying when we look at evolution's course. But *whose* goal? *Whose* achievement? The giraffe, we know *contra* Lamarck, didn't get a longer neck by *trying*; and are we to believe that the brachiopods tried to achieve thought and left it to us to succeed, or tried to achieve social rituals and left it to the ants to carry them through? Achievements must be some *one*'s achievements. A goal, even if it is an Aristotelian *telos*, not a conscious purpose, must be the endpoint of some *entity's* becoming. Whose achievement is evolution? Whose goal, on an evolutionary scale, is thought? In any terms available to this writer at least the very question is nonsense. The concept of *telos* is intelligible and useful, I submit, only with reference to something already in existence. In the study of evolution, on the contrary, where we have no fixed individuals and therefore no fixed endpoint of process, we have, whether in the Darwinian view or in the efforts of 'emergence' theorists to revise it, only the appearance of teleology, not its flesh and blood.

IV

Still, that appearance keeps reappearing. Why? In emergence theory it is a case of metaphysical aspirations as yet unfulfilled. All honor to them; it may well be that this controversy will only come to rest once one has accepted a cosmology of some Whiteheadian kind. This, to most of us, has not yet happened, certainly not in such a fashion as to affect the practice or the thinking of biologists. But why does neo-Darwinism, as distinct from those broader and vaguer views, recurrently lay claim to being teleonomic in its structure?

The answer is not far to seek. To give it will permit one more comparison with the teleology of Aristotelian science.

Darwinian evolutionary theory *appears* teleological because it is first and last a theory of adaptation. Deriving from Paley, it views all organisms as adaptation machines, aggregates of devices for the adjustment of the organism to its environment. On this view it is, as Williams insists, thinking in terms of adaptation, and this alone, that distinguishes biology as a science from physics and chemistry. Yet adaptation in evolutionary terms is for survival and survival only. Everything non-trivial in specifically biological processes reduces to this one phenomenon. But explanation in these terms, as we have already seen, either collapses into tautology or is reduced to necessity, and so in either case fails to retain its alleged teleonomic structure.

How does this situation compare with that of Aristotelian teleology? First, in Aristotle we find for each kind of thing a given normal endpoint of development, and relate to it a set of what Aristotle calls 'hypothetical necessities'. Given, for instance, that a creature needs to hear – or be somehow sensitive to environmental vibrations – it will develop *some* kind of auditory organ, whether a vertebrate ear or a piscine lateral line. In the modern version, however, such hypothetical necessities become simple necessities. Since the endpoint to which one might refer them as means is not fixed in advance, it becomes simply the ineluctable issue of the preceding steps. Instead, therefore, of the necessary conditions being relative to the end, the end is the automatic product of its necessary (and sufficient) conditions. In Aristotle, secondly, the end being given, its achievement happens 'always or for the most part' – but it may fail. All along the way, there is room for abnormality, for chance. In modern theory, however, there is no such leeway. Even though, in terms of our present knowledge, most mutations may be 'chance' occurrences, that is so only in a sense analogous to Aristotelian *tyche*; that is, they are caused by some cause and effect sequence outside the usual pattern of development in the given case. But they are not – or cannot survive – as *to automaton* in Aristotle's sense, as sheer random happenings. Did they not prove to be adaptive as the environment changes, natural selection would eliminate them. And if they do prove adaptive, on the other hand, they have to survive. Depending on how you look at it, in other words, everything is teleological (adaptive) or everything is necessary. There is no middle ground for the merely contingent in the interstices of an otherwise orderly sequence.

It is just this two-edged comprehensiveness, finally, combined with the authority of an algebraic formulation, that lends to modern natural selection theory its great explanatory power. First, there is a formal, mathematical instrument, the algebra of Fisher or Haldane, which may be used to measure natural selection and which lends weight and precision to experimental results in this field. Such a formula, intrinsically tautological, is used to measure changing adaptive relations and therefore serves the appearance of teleology. But these relations in turn, when viewed as a series of organism-environment interactions, appear thoroughly necessitated, not teleological at all. When, however, such interactions are summed up, for long periods, in algebraic formulation, the results, neatly ordered, present apparent trends and thus once more give the appearance of teleology, an appearance once more reduced to necessity when we visualize the whole sequence of action/reaction from which they have eventuated. Thus if one accuses Darwinians of being 'mechanistic' they point to the 'trends' in evolution as they see it; if one accuses them of being 'teleological' in their thinking, they point to the necessity of the whole show in terms of environmental pressure and the consequent changes in gene ratios. And if we try to bring this to-and-fro to rest, what have we? Once more, tautology: well, after all, what survives survives. If you look at the process *a tergo* it appears teleological; if *a fronte*, necessitated. And *sub specie aeternitatis*, when the theory is summed up in a formula for measuring differential gene ratios, you have a theorem universally applicable because empty, totally comprehensive because it expresses a simple identity.[22]

Why do we keep going round this merry-go-round? 'Adaptation' is a means-end concept. Yet if all adaptations are for no specifiable end except survival, one keeps falling back into a universal necessity which is in turn reducible to the same old tautology. Stretch it out: it's teleology. Collapse it one level: it's necessity. Collapse it still further: it's tautology. What is lacking to stabilize this endless vacillation? This brings me at last to my second major Aristotelian concept: *eidos*. For what is lacking in the modern concept of adaptation is precisely the definite *telos*, which in Aristotle is the mature form of the species, of the type, the τοιοῦτον of the τόδε τι.

Nor, of course, do I mean by this some cosmic goal. Again, there are no such goals within Aristotelian biology. True, there are such in Aristotelian cosmology, as e.g. in respect to the proof of the unmoved mover. In the framework of biological investigation, however, we need not, indeed we may not, invoke such dialectical arguments. Modern biology, however, lacks even the more limited and concrete end-points of Aristotelian science. In short, when Darwinism evicted the watchmaker of Paley's famous watch, it threw out as well the *telos* of the watch itself. But without a *terminus ad quem* of development, without a *terminus ad quem* for our understanding of the organization of a living system, of an organism, of an organ, of an organelle, one has no univocal concept of adaptation, of the adjustment of these means to that end. True, the organism, the organ, the organelle is continuously adapting itself to its environment, both internal and external; but what for? To what end? In ontogenesis, to the end of maturation and self-maintenance of the organism, the organ or the organelle, that is, to the end of the origination and conservation of some *form*. The processes of adaptation, as distinct from their result and adaptedness, are thus related as dynamic processes to their goal, that is, to the actuality of the organized system which comes into being or maintains itself in being through those processes. And the result, adaptedness, as differentiated being-such-and-such of the parts of the organism, the organ or the organelle, is also subordinated, therefore, to the development or maintenance of the form, the *eidos*, of the whole. It is, then, precisely the Aristotelian concept of form, or some modern analogue thereof, which is lacking in the modern concept of adaptation, or better, of the organism as a pure aggregate of adaptive mechanisms.[23]

V

Let us look a bit more closely, then, at the relation between modern biology and Aristotle's concept of *eidos*.

Perceptions of form – the shape of an oak leaf, the walk of a cat, the metamorphosis of a butterfly – the grasp of such configurations and changes of configuration, are among the basic insights by which the subject-matter of biology is singled out. A certain freedom of form within form, Buytendijk has shown, is the criterion by which we see a shape as 'alive'.[24] Biological knowledge, the knowledge of men like Ray or Hooker, was a refinement of such elementary perceptions, a refinement to the point of genius, but not different in kind from its everyday counterpart. Modern biologists, however, at least the more theoretical, and more articulate, among them, sharply reject such old-fashioned connoisseurship. Form, and the recognition of form, are not only not (according to their own professions of faith) their central concern; they exhibit a positive dread of form. In the polemics that characterize contemporary taxonomy, for example, and in evolutionary controversy also, the favorite epithet of the combatants is 'typology'. To call a man a typologist is the worst insult you can bestow.[25] It is hard sometimes to tell quite what is meant by the term; it is not necessarily Platonic realism, perhaps something like Aristotelian realism. In any case, it's deadly. Perhaps, indeed, it is most of all this *eidophobia*, if one may so christen it, that makes biologists shudder at the very name of Aristotle. Yet the Aristotelian concept of *eidos* could teach reflective biologists much about the foundations of their discipline.

Eidos in Aristotle is used differently in different contexts, but when used technically it seems to represent a single concept, although it is already rendered in Latin (as in modern European languages) by two separate terms: *forma* and *species*. I have not found, however, any indication that Aristotle took the term *eidos* to be in any formal sense equivocal. Thus he includes *genos*, for example, in the philosophical dictionary (Delta 28), but uses *eidos* in that discussion as if there were no problem about its meaning.[26] Since he deals so carefully with the several meanings of equivocal terms, I can only conclude, therefore, that he thought of *eidos* as one comprehensive concept, with different applications in respect to different problems, and perhaps with two major applications which correspond, for us, to 'form' and 'species'. Or, if even this partial separation is incorrect, we may separate, for ourselves, the two aspects of his one concept when we try to see what it can tell us in the context of modern biological methods.[27]

First, then, form in contrast to matter. *Eidos* in this context functions in a number of striking respects in the same way as the concept of organization (or information) in modern biology.

A. For one thing, form and matter in Aristotle constitute a pair of concepts used relatively to one another and relatively to the problem at hand in a great variety of different contexts. The *eidos* of an entity or process is its organizing principle, the way it works to organize some substrate capable of such control. Though it is sometimes equivalent to *morphe* or shape, that is by no means always the case. Nor is form in nature a separate, self-subsistent 'absolute'; on the contrary, it must once more be emphatically affirmed, it exists in, and only in, that which it informs. In the context of the entity or process in question it exists *as* the organizing principle of that process, just as its matter exists as the potentiality of such (or other) organization. Thus in noses (one of Aristotle's favorite examples) snub is the form of the matter flesh and bone. In bone, however, boniness is the form of earth and fire or whatever elements compose it. Biological systems lend themselves *par excellence* to this dual – but not dualistic – analysis. It depends on the particular system one is studying what will be form and what matter in a given case; but the two-level analysis is always apposite. On the one hand, this matter as the matter of this form is by no means to be ignored, since natural form exists only as actualized in an appropriate matter. As the matter of this form, indeed, it can be studied with profit for itself – as modern biology studies, in much greater exactitude than Aristotle could dream of, the physico-chemical substrate of living systems. But the form too can be studied for its own sake, as it is by modern systems-theorists, even though it exists only as enmattered and depends for its existence and continuance on the laws governing the matter of which it is the form. And again, be it noted, if Platonists and scholastics generalized the Aristotelian concepts to form a cosmic hierarchy from the abstraction of prime matter to the Divine Mind, this was not Aristotle's primary intent. *Eidos* and *hyle* were for him a pair of analytical tools, to be applied in the study of nature relatively to one another and relatively to the particular inquiry.

Despite the simplicity of his examples and the crudity of his 'chemistry', Aristotle's methodological thesis is an important one. *Eidos* in the sense of organizing principle is indeed a definitive concept for biological method. True, in view of the advance of scientific knowledge since Aristotle's time, its modern counterpart is couched in very different terms. Thus, for example, G. L. Stebbins in his *Basis of Progressive Evolution* describes the principle characterizing living systems as that of relational order. 'In living organisms', he writes, 'the ordered arrangement of the basic parts or units of any compound structure is related to similar orders in other comparable structures of the same rank in the hierarchy, permitting the structures to co-operate in performing one or more specific functions'.[28] This is a much more precise statement, indeed, but it plays the same role relative to the chemistry and physics of living things as does Aristotle's concept of form in relation to matter. The colinearity of the DNA chain is a relatively simple example of such order. The concept of information may also play a similar part. In fact, it is even closer to Aristotelian form. For information can be found in any sort of system (and any system, in Aristotle's view, can, and should, be studied in form-matter terms), yet in living things the quantity of information is vastly greater – and more interesting – than in non-living systems.[29] From this point of view, indeed, Raven's 'Formalization of Finality' might better be called 'The Formalization of Form'. Just as, from the Aristotelian scientist's point of view, matter *is* the possibility of taking on one form or another, so the elements of an information-bearing system can assume any one of many equiprobable states, one of which, by virtue of its very improbability and in proportion to that improbability, becomes a bearer of information. In short, the relation between entropy and negentropy in biological processes expresses a quantitative equivalent of Aristotle's qualitative distinction between the material and formal aspects of a given system or subsystem, of an organism, organ system, tissue, etc.

B. As either of the instances just cited indicates, moreover, (that is, Stebbins on relational order or Raven on the information-theoretical aspect of biology), the role in biology of a concept of form, organization or information should demonstrate once for all the irreducibility of biology to physics and chemistry (at least in their classical, reduced, and one-level form). This was Aristotle's thesis also, against Democritus. Organized systems cannot be understood in terms of their least parts alone, but only in terms of those parts *as organized* in such systems. Organized systems are *doubly determinate*; they exist on at least two levels at once. True, the form-matter pair of concepts do not of themselves generate a stratified *cosmos*; but they do show us how to resist reduction to *one* single cosmic level, whether of Democritean atoms or of the fundamental particles of modern physics. There may or may not be a cosmic hierarchy; the fact remains that whenever we study living systems we are studying particular, limited systems that are hierarchically organized, organized on at least two levels. We are studying systems composed of elements obeying their own laws, but constrained at the same time by arrangements of those very elements which constitute, as such, laws of a higher level. The higher level – form, organizing principle, code, fixed action pattern or what you will – exists only *in* its elements, and depends on them for its continuance, yet the laws of the elements in themselves, corresponding to Aristotelian *hyle*, permitting any number of informing arrangements, do not as such account for the principle which in this case happens to constrain them.

If, moreover, the question of teleological explanation left us puzzled, here, it seems to me, the case for non-'mechanistic' explanation becomes much clearer. The concept of *eidos*, like that of *telos*, is indispensable to help locate a biological problem: it has *heuristic* value. If used with good judgment it locates, as *telos* does in limited areas, a real phenomenon, a structure or process, in nature; it has *descriptive* value. But much more clearly than *telos*, it also has *explanatory* power. To discover the working principle of an organized system, as in the specification of the DNA code, or in the functional explanation of the lateral line (apart from the question of its origin), is to *explain* the system just as truly as one explains it by analyzing its physico-chemical parts. To learn how an organized system operates is just as conducive to the understanding of it – the *scientific* understanding of it – as is the analysis of the same system into its elementary components. Indeed, the teleonomic study of biological systems is probably reducible, I would suggest, to the diachronic rather than the synchronic study of their form. Teleonomy sits uneasily on evolutionary theory, one would then suspect, because, since the sequence of living systems that have inhabited this planet does not itself constitute a living system, it has no *eidos*, and therefore no *telos*, to which the study of its necessary conditions could be referred.

C. Biological explanation, then, works in terms of form *or* matter, systems-theoretic study of wholes or part-analysis (ultimately physico-chemical) of their constituents, with the two kinds of explanation complementing one another. That complementarity, thirdly, however, is not symmetrical. Again, there is a striking resemblance here to the Aristotelian form/matter pair. All natural things, organized parts of such things and processes exhibited by them, are inherently informed matter and can be studied on both levels; but form is prior. In non-Aristotelian language: although the upper level of a doubly determinate system depends on the lower level, and the laws of the lower level, for its existence, and is inseparable from it, it is the upper level that makes the system the kind of system it is. We have to refer to the upper level, we have seen, to generate a problem, to describe the system we are studying, and in certain cases at least to explain the operation of the system as such. Thus in the code case or in physiological explanation, the problem-location, the description *and* the explanation all refer to form: in*form*ation. In some cases perhaps only the first and second obtain as explanatory principles. Research on the the lower level, physics in relation to molecular biology, or physiology in relation to behavior, or chemistry in relation to metabolism, may indeed go on indefinitely without explicit reference to the higher level, the structure of an enzyme, the typical course of a nesting behavior, the normal growth process of an embryo; but it will not be study of *this system* without some implicit reference at least to the organizing principle concerned. The higher level, though dependent on the lower, is both epistemologically and ontologically prior to it. For Aristotle, of course, it is also prior in time: man begets man eternally. That is what evolution has altered: for us potency is, in nature as a whole, prior to actuality. But given the existence of an organized being or process, form is then prior to matter, not only cognitively but ontologically as well. Prior cognitively: to know the system is to identify, describe, and understand it in terms of its operating principles, of the way it uniquely constrains its components to make this system of *this kind*. Prior ontologically: since the principles of 'matter' on their own could logically, and in terms of the laws of probability, take on any 'form', it is the existence of *this* form that makes the system what it is.

Once more, of course, we are talking about *relative* levels of a system or subsystem e.g., organic bases vs. their arrangement in a code, or the reactions entered into by chemical trace elements vs. the structure of the metabolic pathways in which they serve. Whatever system or subsystem we happen to be attending to in a given inquiry, however, this relative priority of the higher level obtains. One might call this relation, as I have suggested elsewhere, a principle of ordinal complementarity.[30] Aristotle understood it well.

D. Finally, there is a special area where the Aristotelian concept of form has proved strikingly parallel to some aspects of modern thought: that is, in his doctrine of soul, the form of organized bodies. We may leave the vexed question of active reason aside and note briefly two theses in Aristotle's 'psychology' to which modern reflection about biology seems to be slowly and painfully returning.

First, Aristotle's concept of psyche as such is functional. This is a concept of mind suggested by thinkers as different as Putnam, Ryle, and Polanyi.[31] Polanyi puts this thesis in his essay, 'Logic and Psychology', in terms of his distinction between focal and subsidiary awareness. We can formulate the distinction between mind and body, he says,

> as the disparity between the experience of a subject observing an external object like a cat, and a neurophysiologist observing the bodily mechanisms by which the subject sees the cat. The experience of the two is very different. The subject sees the cat, but does not see the mechanism he uses in seeing the cat, while, on the other hand, the neurophysiologist sees the mechanism used by the subject, but does not share the subject's sight of the cat ... to see a cat differs sharply from a knowledge of the mechanism of seeing a cat. They are a knowledge of quite different things. The perception of an external thing is a from-to knowledge. It is a subsidiary awareness of bodily responses evoked by external stimuli, seen with a bearing on their meaning situated at the focus of our attention. The neurophysiologist has no experience of this integration, he has an at-knowledge of the body with its bodily responses at the focus of his attention. These two experiences have a sharply different content, which represents the viable core of the traditional mind-body dualism. 'Dualism' thus becomes merely an instance of the change of subject matter due to shifting one's attention from the direction on which the subsidiaries bear and focusing instead on the subsidiaries themselves.[32]

Thus, mind is not a separate something but is what Ryle calls 'minding'. It is the higher-level, operating principle of a complex system:

Some principles, – for example, those of physics – apply in a variety of circumstances. These circumstances are not determined by the principles in question; they are its boundary conditions, and no principle can determine its own boundary conditions. When there is a principle controlling the boundary conditions of another principle, the two operate jointly. In this relation the first can be called the higher, the second the lower principle.
Mental principles and the principles of physiology form a pair of jointly operating principles. The mind relies for its working on the continued operation of physiological principles, but it controls the boundary conditions left undetermined by physiology.[33]

This approach is generalizable, moreover, to living things as such. In general the 'soul' of any living thing is its style of operating on and in its environment, no more, but also no less.

Secondly, the 'kinds' of 'soul' distinguished by Aristotle appear to correspond in general to the major divisions which, however we see the problem of 'higher' and 'lower' among organisms, do in fact seem to obtain. If we look, independently of any special theory of the *how* of evolution, at its general course, we find, I think, three and only three really 'surprising' 'advances'. First, there is the origin of life. Here we get what Aristotle calls 'nutritive soul'. Living things grow and reproduce. These are the minimal functions of all life. Secondly, living things acquire the principle of sentience and self-locomotion. This is 'sensitive soul'. We get organisms capable of behavior, centers of irritability, appetite, and self-motion. And thirdly we have the origin of man, of culture, of the human social world, of what Aristotle calls 'passive reason'. Surely these steps, distinguished by Aristotle, are just those that may well puzzle the evolutionary theorist.[34] These are very crude distinctions if you like, but it may be worth noting that with painful deviousness we are coming back to the simple divisions in the levels of organization in the world around us which Aristotle had recognized long ago, divisions which he made, as we are trying to do, not dualistically, like Plato, but in terms of function, of the inherent organizing and operating principles that mark off kinds of complex systems as unique.

VI

So much for *eidos* in contrast to *hyle*. What about *eidos* in contrast to *genos*? It is here that the opposition to Aristotle is centered. For if modern biologists in fact use concepts like organization and information in ways that resemble Aristotle's use of 'form' in the form-matter pair, they strongly object to the Aristotelian species concept, even though in Aristotle this appears to be, if not the very same concept, at least the same concept under another aspect. I want to make two points in this connection: first, to indicate how the use of the species concept in modern biology does still resemble the Aristotelian, and secondly, to explain, or at least to locate clearly, the modern resistance to Aristotelian thinking on this score. Again, however, let me reiterate briefly what I said at the outset about the different cosmologies associated with the ancient and the modern view. Aristotelian species are certainly eternal, modern species certainly are not. This meta-scientific contrast should not be underrated. A modern biologist can no more be a complete Aristotelian than he can be a complete Cartesian. Yet in the routine use of the species concept there is nevertheless a residual, though not a merely vestigial, similarity, and at the same time, in the epistemological foundation of that use, a very deep-seated contrast.

First, the similarity. *Eidos* and *hyle* form, we have seen, throughout the range of Aristotelian sciences, an analytical pair to be used relatively to one another and to the subject matter in question in a particular investigation. When it comes to the *eidos/genos* contrast, however, *eidos* assumes a different and less relational aspect. (This holds, I think, even if we admit that *genos means hyle*, as it sometimes does; see note 27.) Only individuals are real for Aristotle – the modern biologist would agree – but they are individuals of *such and such a kind.* The *infima species*, like any form, exists only in, and as form of, the individuals who exemplify it. The species is the sum total of its specimens, past, present, and future, and they are the individuals they are in virtue of their membership in that species. But there is nothing relative about this. *Eidos* interpreted as species takes on an absolute character which, in the natural world at least, eludes *eidos* as paired with *hyle*. This of course, in its explicit enunciation at least, is just what modern taxonomists, whether evolutionists or pheneticists, so stoutly object to. And yet practising taxonomists of whatever school do in fact continue to treat 'species' as having a special role, a role in some way less conventional and closer to the real ways of nature than the concepts designating 'higher categories'. True, as Simpson points out, all categories are 'objective' in that all taxa are collections of real organisms. And they are all 'subjective' insofar as they are all concepts in the minds of taxonomists. Nevertheless, he admits, 'species' is more clearly 'non-arbitrary' than other categories.[35]

Admittedly, some biologists, from Darwin himself to Ehrlich and Holm, have predicted that the species concept would wither away and we should be left with a classless aggregate of biological particulars.[36] Yet biologists still classify and argue about the foundations of such activities; and in the view of most of them the species concept has been refined, indeed, even transformed out of recognition, but not abolished. Perhaps this is correct. Certainly, the 'biological species concept', defined in terms of potentially interbreeding Mendelian populations, or the 'multi-dimensional species concept', tailored for the inclusion of non-sexually reproducing as well as of Mendelian populations, looks at first sight very unlike the traditional, originally Aristotelian, concept.[37] I want to point out only that it is not as *wholly* unlike as it is usually painted.

Aristotle is usually accused of tagging species by means of one single character selected *a priori* and abstractly.[38] This is unfair. He is certainly no *a priorist*. Again, modern science had to reject him at its outset because he was not 'a prioristic' enough: he stayed too close to the concrete pronouncements of everyday experience; he was too good an empiricist.[39] And even though he writes in the *Topics* of, and is perpetuated in the tradition as insisting on, definition *per genus et differentiam*, he himself suggests both in the *Post. Anal.* and in *De Part. Anim.* that the 'substance' of a thing (*ousia*) or its nature cannot be captured by specifying any one differentia alone.[40] Indeed, his opposition to Platonic 'division' is based at least in part on the insistence that we divide up natural things as nature demands, not by one character and its contrary, but by the cluster of characters which helps us to single out a natural kind within a larger group.[41]

In this piecemeal and empirical approach to classification he is not so different from modern taxonomists as it is now fashionable to consider him. And again, like modern biologists, he is driven, despite his insistence that only individuals are real, to grant to the species concept some kind of uniqueness, as the least and most 'real' of universals. In short, it is in *eidos* as species that the relativity of form is somehow or other anchored in reality.

Just how this anchoring comes about is hard to say. Just how is the relational concept which we render 'form' to be identified with the non-relational concept called 'species' in the inclusive but univocal concept *eidos*? I can give no satisfactory answer, but only suggest the location of an answer in the even more puzzling conception of the τί ἦν εἶναι, which I shall discuss briefly in the concluding sections of this paper. Even more difficult would be the task of specifying the lesson to be derived for modern biology from the plurality-in-unity of Aristotelian *eidos*. I can only register tentatively the suggestion that one might profitably reflect on the link, whatever it turns out to be, between organization or information on the one hand and species on the other.

Biologists study throughout the widely varying phenomena of living nature the organization of systems or subsystems at any number of levels. Living things, as information-bearing systems, have arisen gradually from non-living systems much poorer in information content, and, once evolved, they continued to vary continuously, to throw up, in correlation with their changing environments, myriad new patterns in every conceivable direction of novelty. At the same time, there are cuts in this continuum, not only the infinite number which we might make anywhere, but a few (in relation to the infinite possibilities) which present themselves as preeminently 'natural' or, in Simpson's term, 'non-arbitrary'. These we designate as cuts between 'species' that is, between carriers, for a time, of distinctive patterns of information. One can study the organization of any organized system or subsystem, of chloroplasts, cell membranes, muscle cells, populations of genes, populations of whole organisms, communities of populations of many species, etc., etc., but there are also some points at which the transfer of a stable pattern of organization (or information) from one living individual to another stops – stops 'really', not because we decide to stop analyzing just there, but because there is a gap, a real discontinuity. To populations confined by these plain discontinuities (plain at least in sexually reproducing organisms with a relatively long generation span),[42] we give the same name that Aristotle gave them; *eidos*, *species*, the very name he gave to organization, *eidos*, *forma*. But here we are not, as in multi-levelled analysis of forms and their matter, singling out such patterns as we discover by applying 'form' and 'matter' as shifting and relational tools for our own study of nature, tools that locate form-in-matter here, there, and everywhere. We are finding certain forms singled out for discontinuity within the continuity of the phenomena, as it were, by nature itself. It may of course be objected, as Simpson remarks about higher categories, that in a sense all forms discovered everywhere are equally objective. Every cell is really bounded by a real membrane: the cytologist studies in the parts of one kind of organism the structures of cells as such. Geneticists study drosophila not because the species of that genus are themselves of greater intrinsic interest than elephants or antelopes, but because they are good experimental subjects and so from them much can be learned about the organizing principles of all heredity. Yet there *is* something unique – even uniquely obtrusive – about species. The developmental biologist – as distinct from the old-fashioned zoologist or botanist – studies organized processes that range much farther than any given species, or even phylum or kingdom. Yet even he has to select, and learn to know, *some* species in order to study *in* it the universal life pattern that interests him.[43] Much as he would like to, he cannot evade these fundamental gaps. In practice he is still an Aristotelian in spite of himself.

VII

Yet in their *attitude* to taxonomy, ancients and moderns are very different. If forms are really pinned down into discontinuous species in a fashion not so very unlike that recognized by Aristotle, why should modern biologists so emphatically deny that any shadow of Aristotelian thinking lingers in their own methods? Partly because they take a crude and truncated 'Aristotelianism' as identical with the thought of Aristotle himself. But there is another and more deep-seated reason, and that concerns, in Aristotle, the relation of knowledge, especially the knowledge of species, to perception. Modern science, let us recall once more, began by rejecting the Aristotelian approach to nature, in part at least because it was too directly tied to everyday perception of natural entities and processes and so prevented the flights of abstractive thought and creative imagination on which, as we can now see, the development of science largely depends.[44] The ideal of 'scientific method' for many philosophers and scientists has become the correlation, not of perceptions, usually directed as they are to complex, concrete individuals, but of sheer particulars, of 'hard data', with abstract laws, whether universal or statistical: correlations peculiarly susceptible, it seems, to quantitative manipulation and experimental control. For Aristotle, on the contrary, knowledge, however theoretical, is rooted in the full, concrete, perceptual world; it analyzes that concrete world and gains new insight into it, but never leaves it as ultimate, as well as initial, dwelling place. Now such perceptual insight is indeed essential to certain kinds of biological practice: from the macroscopic recognition of specimens in the field to the recognition of structures in electronmicrographs. The late C. F. A. Pantin called such biological connoisseurship 'aesthetic recognition'.[45] The double meaning of 'aesthetic': informal or connoisseurlike on the one hand, and having to do with 'aesthesis', perception, on the other, should be kept in mind in considering what this means. Pantin recalls, for example, seeing a worm in the field and saying, 'Why, that's a *Rhynchodemus*, but it's not *bilineatus*, it's an entirely new species'. This, he points out, is not the yes-no procedure of the museum taxonomist, nor does it resemble at first sight the generalizing procedures of the exact scientist.[46] It is, precisely as for Aristotle, a case of seeing a *this-here* as a *such-and-such* – or in this case, not quite a *such-and-such* but a *somewhat-different*. But just such perceptual recognition of real kinds is what modern theorists profess to abhor. They seek to produce scientific knowledge in an abstracter and more completely specifiable way. Scientific knowledge has its pedigree, they claim, by mathematical thinking out of bare particulars, rather like Love in Diotima's story, who was begotten by Resource on Poverty.

This distrust of anything but bare particulars on the one hand and high flights of theory on the other comes clearly into view in the contemporary taxonomic controversy between the *phenetic* and *phylogenetic* schools.[47] The pheneticists profess to take all and any particulars, without prior weighting derived from taxonomic skill and experience, feed them into computers (those praiseworthy inorganic animals) and come out with classifications better (for what purpose?) than those derived from less restricted starting points and less quantitative manipulation. If they sometimes admit sadly to producing by this method something justifiably called 'types', they are at any rate, they claim, '*empirical* typologists', with no initial predilection for one cluster of characters rather than another. Phylogeneticists, on the other hand, hasten to cover their undoubted, but theoretically suspect, taxonomic insights, derived from the aesthetic recognitions of field experience, under the convenient bushel of evolutionary descent. Darwinian-Mendelian theory, in other words, serves them as an abstract and therefore scientifically respectable cover for their delicate perceptual discriminations: for the heritage of Ray, Hooker, and even, on the side of biological practice as against theory, Darwin himself. Particulars tied together by computer techniques, says the one side; particulars tied together, says the other, by lines of descent inferred though necessarily unobserved. Some writers, notably Gilmour, try to go between the horns of the phenetic-phylogenetic dilemma by espousing a pleasant pragmatism.[48] We classify as we need to for our uses, says he; as the use shifts, so does the classification. This easy way out, however, while correct in a way, since there are of course many possible classifications of anything for many possible uses, fails to still the controversy. For it neglects the fact that some classifications do seem to be, quite apart from our wants and uses, less arbitrary than others. To give due weight to the role of aesthetic recognition in taxonomy, I submit, and to acknowledge its rootedness in the perception of real *this-suches*, would permit taxonomists to make more sense than they have recently done of the real nature of their calling.

It should be duly noted, in passing, that the grounding of scientific knowledge, and especially of scientific discovery, in perception (rather than in sensation or the bare observation of bare particulars) is beginning at last to be acknowledged by philosophers of science. In a general way, perception as the paradigm of discovery was the *Leitmotif* of Hanson's writing.[49] The 'primacy of perception' as our chief path of access to reality was the central theme of Merleau-Ponty's work.[50] A similar theme dominates Straus's phenomenology.[51] And in Polanyi's *Personal Knowledge*, *Tacit Dimension*, and other essays, both earlier and more recent, one has, as distinct from those more general intuitions, a carefully articulated epistemology which explicitly makes of perception, understood in a Gestalt-cum-transactional fashion (not unlike Aristotelian aesthesis), the primordial and paradigm case of knowing, and explicitly makes the achievement of perception the primordial and paradigm case of discovery.[25] These lessons are beginning to have some impact on philosophers of science, especially on those who base their philosophy largely on physics. But biology, from which, through Aristotle's biological practice, the acknowledgement of the primacy of perception took its start, still (with a few honorable exceptions like that of Pantin) stubbornly resists this fundamental insight.

I am not, of course, alleging that perception in its newly-discovered role plays the *same* part as it did in Aristotelian science. It *was* the basis, and the home, of discovery and of knowledge; it *is* the primordial case of discovery and of knowledge, and all discovery and all knowledge are structured as it is. In Polanyi's terms: in all knowledge, as in perception, we rely on subsidiary clues within our bodies to attend focally to something in the real world outside. However 'abstract' that something be, both the bodily base and the from-to structure characteristic of sense perception persist. In the present context: there is no disgrace, therefore, in acknowledging the perceptual skill of field naturalists and taxonomists as part of science. It is not a 'primitive' survival, but a visible analogue of the achievement of knowledge, the paradigm case of our way of gaining contact with reality.

VIII

Let me return now to the question raised earlier: how does *eidos* as the correlate of *genos* escape the relativity of the *eidos/hyle* pair? *What* is unique about species that makes it the paradigm case of natural form? Here we come to that most idiosyncratic of Aristotelian terms: the τί ἦν εἶναι, what it is for a such-and-such to be a such-and-such. 'Essence', with its age-old accretions of misleading connotations, is a poor translation of Aristotle's phrase; perhaps 'being-what-it-is' is the best one can do.[53] When one speaks of 'form', this is what one's discourse is aiming at; the form of a given kind of thing is just what it is for that thing to be the kind of thing it is.

According to the *Topics*, the 'being-what-it-is' of a kind of thing is what its definition designates. Indeed, a definition is there defined as 'a phrase indicating the being-what-it-is'. To some interpreters, therefore, τί ἦν εἶναι appears to be primarily a logical term. Aristotelian science starts from first principles, including real definitions. These specify, among the properties of a thing, certain characters which are 'essential' to it, and that means characters which can be deduced from the initial predicating statement. Definitions, in other words, are premises of scientific demonstrations, and the τί ἦν εἶναι is of interest as the reservoir, so to speak, from which the predicate of such a premise can be drawn.[54] Such a rendering seems to bring the Aristotelian approach close, in form at least, to the so-called 'hypothetico-deductive method'. It must of course be noted, however, that Aristotelian demonstration is anchored, through *nous*, in the direct, 'intuitive' knowledge of first principles. It is not, like dialectical syllogism, merely hypothetical. Yet there is a parallel. For in both cases it is the deduction of some properties from others that is chiefly of interest, and the τί ἦν εἶναι is thought of in this context as ancillary to this primarily logical game. So far as the importance of deductive method goes, Aristotle himself, at least in his logical writing, certainly reinforces this impression.

Yet if one searches the corpus for scientific demonstration, one finds relatively little of it. Most of the arguments of most of the treatises do not look like assertions of defining phrases followed by deductions from these. Most of them appear to be not strictly demonstrative, but inductive, dialectical, or aporetic. They move from common experience or common opinions, weighing the views of others, analyzing difficulties, in the hope, it seems, of arriving at (not starting from) an insight into some specific nature. This *searching* nature of much of Aristotle's writing leads W. Wieland to a different interpretation. For him the τί ἦν εἶναι is not so much a guide to definition, and therefore to demonstration, as it is a heuristic tool, a *topos* or path along which the thinker may seek insight into some special problem.[55]

For form and matter this heuristic or methodological interpretation is indeed fruitful, as I have emphasized. Yet *via* form as species we have seen that there is also a resting place for form-matter analysis, a place at which form becomes uniquely non-relational. And it is here that form exhibits its *ontological* foundation in 'the being-what-it-is' of each kind of thing, the very foundation which, according to Aristotle, his predecessors lacked, the ignorance of which prevented them from discovering the right method for the investigation of nature. Neither the logical nor the methodological approach to Aristotelian science makes sense without this frankly ontological foundation. The τί ἦν εἶναι is expressly what definition is about, both its target and its presupposition. Indeed, without the τί ἦν εἶναι as the referent of definition, the real being-what-it-is of this kind of thing which the defining phrase designates, the demonstrations that follow on definition would be simply 'hypothetico-deductive' in the sense of positivism or phenomenalism. They would not be rooted, as Aristotelian science was certainly meant to be, in the natures of the things themselves, and in our understanding of these natures. Without the τί ἦν εἶναι as end-point and foundation of inquiry, moreover, the inquirer would be confined to an endless and directionless groping; but that is not the case. On the contrary, the real being-what-it-is of the kind of entity under investigation is constantly guiding his search, and directs it to its successful issue. Despite the apparent formal correctness of the logical interpretation, therefore, and despite the importance of the methodological aspect, the traditional ontological interpretation of the τί ἦν εἶναι is still fundamentally sound.[56]

What does it teach us? We would not link the being-what-it-is of things, as Aristotle did, to eternal kinds, nor would we restrict these 'kinds' to kinds of substance. Everything becomes, including species; and *what* becomes may as well be events or processes as more literally 'things'. But despite these deep differences, there are, I believe, two important lessons to be learned from the Aristotelian concept of being-what-it-is as the designatum of definition and the target of inquiry. Aristotle understood, as most modern philosophers of science until recently have not, that the investigation of nature arises out of puzzlement about some particular problem in some limited area: nobody investigates, or can investigate, everything at once. And in such an investigation, further, it is the real nature of the real entity or process that the investigator seeks, and sometimes finds. Science is *pluralistic* and *realistic*, not uniform and phenomenalist, as modern orthodoxy has supposed.

IX

If we put these brief remarks together with our earlier reflections about form and matter, we find, in conclusion, three important methodological lessons to be derived from the study of Aristotle. Through the concept of form as an analytical tool, correlative with matter, Aristotle can remind us of the many-levelled structure both of inquiries into complex systems and of the systems themselves, and thus of the inadequacy of a one-levelled atomism for the understanding of such systems. In conjunction with the grounding of form in the τί ἦν εἶναι of each kind of thing, further, he can remind us of the falsity of two other modern misconceptions: the unity of science concept on the one hand, the claim that the subject-matter and method of science are everywhere the same, and, on the other, the insistence that science must renounce any claim to seeking contact with reality: that theories float, as pure constructs, on the surface of the phenomena, with no mooring in the real nature of the real events or things.

The first of these reminders is plainly related to the concept of organization or information and hence (as I have already argued) to the subject-matter of biology, and to the question of its reducibility or irreducibility to chemistry and physics. The second reminder, of the plurality of science, a reminder of the good Kantian principle[57] that we can have no systematic knowledge of the whole of nature, should help also to liberate biology, or thinking about biology, from the overabstract and reductive demands imposed by taking one science, classical physics, as the ideal of all. And lastly, the acknowledgement of scientific realism should release the biologist to admit the insights into the concrete manifold of his subject-matter, from which his work originates and in which, however abstract and sophisticated it may become, it still is anchored. Nor, finally, as I emphasized at the outset, is this a plea for a return to Aristotelianism. It is a plea for us to listen, despite our fundamental differences of metaphysic and of method, to some of the tenets that Aristotle, as a biologist-philosopher, advocated long ago, and to try to interpret them in ways that could be useful to us as we attempt to articulate and revise our conception of what the investigation and the knowledge of living nature are.

NOTES

[1] G. G. Simpson, *Principles of Animal Taxonomy*, New York and London, 1961, 36n: 'I tend to agree with Roger Bacon that the study of Aristotle increases ignorance.' In fact Bacon was objecting to the current translations of Aristotle, not to Aristotle's teachings themselves. His statement reads as follows: 'Si enim haberem potestatem super libros Aristotelis ego facerem omnes cremari qui non est nisi temporis amissio studere in illis et causa erroris et multiplicatio ignorantiae ultra id quod valeat explicari. Et quoniam labores Aristotelis sunt fundamenta totius sapientiae, ideo nemo potest aestimare quantum dispendium accidit Latinis quia malas translationes receperunt philosophi.' ('Compendium Studii Philosophiae', *Bacon Opera Inedita*, Rolls Series number 15, 469). I am grateful to my colleague, Professor John Malcolm, for finding this passage.

[2] See my *Portrait of Aristotle*, Chicago and London, 1964, esp. Ch. VII.

[3] Wolfgang Wieland (*Die Aristotelische Physik*, Göttingen, 1962, 95n.) argues that the fashion in which modern physicists take a given experiment as general is still Aristotelian: 'Auch die neuzeitliche Physik liest an einem speziellen experimentellen Fall eine allgemeine Regel ab und prüft dann, ob es sich um die wahre Allgemeinheit handelt. Erst dann nämlich kann sie sehen, ob sie von einem speziellen Fall ausgegangen ist oder nicht; d. h. ob sie diesen Fall von seinen besonderen oder von seinen allgemeinen Merkmalen her gedeutet hat.' It is, in his view, modern *theories* of induction that mislead us here. But those theories *have* recognized a logical gap which Aristotle failed, and for his purposes did not need, to recognize.

[4] Norman Campbell, *What is Science?*, New York and London, 1921, Ch. II.
[5] *De generatione et corruptione*, II, 11, 338b7ff.
[6] See the exposition of Wieland in the work referred to, note 3, above. For a close study of Aristotle's use of *telos* in the explanation of generation, as well as, and in relation to, *eidos* and *hyle*, see the excellent paper by Anthony Preus, 'Science and Philosophy in Aristotle's *Generation of Animals*', *J. Hist. Biol.* **3** (1970), 1–52; see also his *Science and Philosophy in Aristotle's Biology*, Darmstadt, in press.
[7] Cf. M. Grene, *The Knower and the Known*, New York and London 1966, p. 237.
[8] Quoted *ibid.*, pp. 235–236, from A. F. Baker, 'Purpose and Natural Selection: A Defense of Teleology', *Scientific Journal of the Royal Coll. of Science* **4** (1934), 106–19, 107–108.
[9] Baker, *op. cit.*, 108–10.
[10] W. Wieland, *op. cit.*, 254–77.
[11] C. H. Waddington, *The Strategy of the Genes*, London 1957, p. 190.
[12] See the (by now classic) interpretation of organicism in Ernest Nagel, *The Structure of Science*, New York 1961, pp. 428–46.
[13] Jean Piaget, *Biologie et connaissance*, Paris, 1967, pp. 225–226. Piaget refers here to an argument by J.-B. Grize: 'J.-B. Grize, qui a étudié ces trois relations au point de vue du calcul logistique, montre de même que la relation de 'cause finale' est mal formée logiquement, du fait qu'elle mêle les relations réelles de la 'langue' (instrumentalité et causalité) avec les relations d'isomorphisme appartenant à la 'méta-langue' et utilisées pour mettre en correspondance cette causalité $a \rightarrow b$ et l'instrumentalité $B \rightarrow A$.' (*ibid.*, 226n.)
[14] *The Knower and the Known*, Ch. IX, 238.
[15] F. Ayala, 'Biology as an Autonomous Science', *Amer. Scientist* **56** (1968), 207–21.
[16] *De Gen. Anim.* 741b 21ff.
[17] H. B. D. Kettlewell, 'Selection experiments on Industrial Melanism in the Lepidoptera', *Heredity* **9** (1955), 323ff. 'A résumé of investigations on the evolution of Melanism in the Lepidoptera', *Proc. Roy. Soc. Lond. B.* **145** (1956), 297ff.
[18] George Williams, *Adaptation and Natural Selection*, Princeton, 1966, p. 10.
[19] *Ibid.*, 10–11.
[20] Among biologists critical of the neo-Darwinian synthesis, see for example E. S. Russell, 'The Diversity of Animals', *Acta Biotheor.* **13** (Suppl. 1) (1962), 1–151. Cf. A. Vandel, *L'Homme et L'Evolution*, Paris, 1949. Among recent philosophers, see M. Polanyi, *Personal Knowledge*, Chicago and London, 1958 (Torchbook edition: New York, 1962), Ch. 13.
[21] A. Vandel, *op. cit.*
[22] Cf. my analysis of R. A. Fisher's 'Genetical Theoretical Theory of Natural Selection', in *The Knower and the Known*, pp. 253–66; Chapter VIII of this volume.
[23] Cf. T. Dobzhansky, 'On Some Fundamental Concepts of Darwinian Biology', *Evol. Biol.* **2** (1968), 1–33, where efforts are made, not wholly successfully, to disentangle some of these concepts.
[24] See my account in *Approaches to a Philosophical Biology*, New York 1969, Ch. 2, pp. 74–75, and this volume, Ch. XVIII.
[25] R. R. Sokal, 'Typology and Empiricism in Taxonomy', *J. Theoret. Biol.*, **3** (1962), 230–67.
[26] For a detailed account of Aristotle's usage in his own biological writings the reader should consult Professor David Balme's definitive treatment, as well as A. L. Peck's notes in the Introduction to his edition of the *Hist. Anim.* D M. Balme, 'ΓΕΝΟΣ and

ΕΙΔΟΣ in Aristotle's Biology', *Class. Quart.* **12** (1962), 81–98; A. L. Peck, Introduction, in Aristotle, *Hist. Anim.*, volume I, Cambridge, Mass., 1965, esp. notes 5–11. Cf. also D. M. Balme, 'Aristotle's Use of Differentiae in Zoology', in *Aristote et les Problèmes de Méthode*, Louvain 1960, pp. 195–212.

[27] The two aspects I am separating are, admittedly, brought very close together by Aristotle himself, not only in *Post Anal.* (94 a 20ff.), where *genos* is given a place corresponding to that of 'material cause' in the *Physics*, but also in *Metaphysics* Δ 1024b 8, Z 1038 a 6, H 1045a 23 f. and I 1058 a 23, where *genos* is identified with *hyle*. I am grateful to Professor Balme for calling my attention to the latter passages, and confess that I might not have made the distinction between the two pairs of terms as flatly as I tried to do, had I read his papers (just referred to) before writing this essay. No one interested in Aristotle's biology and its relation to his philosophy of science can afford to neglect Professor Balme's careful and illuminating work.

[28] G. L. Stebbins, *Basis of Progressive Evolution*, Chapel Hill, N.C., 1969, pp. , 5–6.

[29] The *locus classicus* for the application of information theory in science is *Science and Information Theory*, Leon Brillouin, New York, 1956. The application of information theory to biology has been discussed in a number of places, notably by Henry Quastler, see *The Emergence of Biological Organization*, New Haven, 1964. The point that the distinction between living and non-living systems with respect to information is quantitative – though great enough to appear qualitative – I owe to Dr. Thomas Ragland of the University of California, Davis, who has lectured on this subject to my class in the philosophy of biology. Michael Polanyi, in 'Life's Irreducible Structure' (in *Knowing and Being*, Chicago and London, 1969, pp. 225–39), argues, on the contrary, that the distinction between the two kinds of systems is logical and qualitative; yet he admits, in terms of evolution, a continuous transition from one to the other.

[30] *The Knower and the Known*, p. 233.

[31] Hilary Putnam, 'The Mental Life of Some Machines', *Intentionality, Minds and Perception*, Detroit, 1967, pp. 177–200; Gilbert Ryle, *The Concept of Mind*, London 1949; Michael Polanyi, *op. cit.*

[32] M. Polanyi, 'Logic and Psychology', *Amer. Psychologist*, **23** (1968), 39–40.

[33] *Loc. cit.*

[34] Cf. for example, the obscure but thought-provoking argument of David Hawkins in *The Language of Nature*, New York, 1967.

[35] G. G. Simpson, *op. cit.*, 114.

[36] Darwin, *Origin of Species*, Ch. XV ('species are only well-marked varieties'); Paul R. Ehrlich and Richard W. Holm, 'Patterns & Populations', *Science* **137** (1962), 652–57.

[37] See Ernst Mayr, *Animal Species and Evolution*, Cambridge, Mass., 1969, pp. 18–20.

[38] See e.g., Simpson, *op. cit.*

[39] See *Portrait of Aristotle*, esp. Ch. III, and Wieland, *op. cit.*

[40] *Post. Anal.* II, 13, 96a33ff. and *Part. Anim.* 634b 29ff.

[41] *Part. Anim.*, *loc. cit.*

[42] For the complexities, e.g., of bacterial taxonomy, see Mortimer P. Starr and Helen Heise, 'Discussion', *Systematic Biology, Nat. Acad. Sci.* publication 1962 (1969), 92–99.

[43] Professor Dennis Barrett of the University of California, Davis Zoology department, while denying that he is a 'zoologist', admits sadly that he has to know 'his' organism, the sea urchin, in order to study in it the development of the fertilization membrane.

[44] Such founders of modern science as Harvey and Newton, indeed, *thought* they

derived their great discoveries very directly from experience; we, with three centuries of hindsight, know they were more daringly imaginative than they believed.

[45] C. F. A. Pantin, 'The Recognition of Species', *Science Progress* **42** (1954), 587-98; cf. his posthumous Tarner lectures, *The Relations between the Sciences*, Cambridge 1965.

[46] 'The Recognition of Species', p. 587.

[47] D. Hull, 'Contemporary Systematic Philosophies'. *Annual Review of Ecology and Systematics*, I (1970), pp. 19–54 and M. Starr and H. Heise, *op. cit.*

[48] J. S. L. Gilmour, 'Taxonomy', *Modern Botanical Thinking*, Edinburgh 1961, pp. 27–45.

[49] N. R. Hanson, *Patterns of Discovery*, Cambridge, 1965. Cf. also his posthumously published *Perception and Discovery*, San Francisco, 1969.

[50] M. Merleau-Ponty, *La Phénoménologie de la Perception*, Paris 1945; *The Phenomenology of Perception* (transl. by Colin Smith), New York and London, 1962.

[51] See E. W. Straus, *The Primary World of Senses* (trans. by J. Needleman), New York 1963 and *Phenomenological Psychology* (trans. in part by Erling Eng), New York 1966, and this volume, Ch. 17.

[52] M. Polanyi, *op. cit.*, also *The Tacit Dimension*, New York, 1966. Cf. William T. Scott, 'Tacit Knowing and The Concept of Mind', *Phil. Quart.* **21** (1971), 22–35.

[53] Cf. 'What-is-being', the rendering of Joseph Owens in his *Doctrine of Being in the Aristotelian Metaphysics*, Toronto, 1957.

[54] The 'logical' is one of the aspects of the τί ἦν εἶναι distinguished by C. Arpe in his dissertation, *Das* τί ἦν εἶναι *bei Aristoteles* (Hamburg 1937). Cf. E. Tugendhat, ΤΙ ΚΑΤΑ ΤΙΝΟΣ (Freiburg/München 1958), 18, n. 18: 'Erst Arpe hat, anknüpfend an Natorp (*Platons Ideenlehre*, S.2) gezeigt, dass die Form des Ausdrucks nur aus der Definitionssituation verstanden werden kann und sich, wie am besten aus der Topik zu ersehen ist, auf jede beliebige Kategorie bezieht'. But cf. also Wieland, *op. cit.*, 174; the 'logical' here is perhaps closer to what I am calling the 'methodological' interpretation. It is closer to heuristics – the search for principles – than to demonstration *from* them. The deductive aspect is stressed by Prof. Moravcsik in his reading of the τηε (oral communication).

[55] Wieland, *op. cit.*, pp. 174–75. Wieland emphasizes the methodological function of the τηε (which he identifies, by implication, with Arpe's 'logical' function) to the exclusion of its other aspects, esp. what Arpe calls the 'physical' or 'teleological' and the 'ontological'.

[56] See Arpe, *op. cit.* For a clear summary of the traditional view, see Bernard J. Lonergan, S. J., *Verbum – World and Idea in Aquinas*, Notre Dame 1967, pp. 16–25.

[57] In a discussion of a similar argument, Professor Günther Patzig has pointed out to me that the principle of the plurality of science is non-, even anti-Kantian, if by it we mean to espouse a plurality of scientific methods. For Kant the method of science was indeed one. What *is* Kantian, however, is the denial that we can have one unified, finished system of knowledge for the *whole* of nature. If we could have such a system, we could *not* have diversified sciences with diversified methods. If we cannot have such a system, on the other hand, then a plurality of fields, and of methods, is at least logically possible, and on Kantian grounds.

EVERETT I. MENDELSOHN

PHILOSOPHICAL BIOLOGY VS EXPERIMENTAL BIOLOGY: SPONTANEOUS GENERATION IN THE SEVENTEENTH CENTURY*

I. SPONTANEOUS GENERATION AND ITS HISTORIANS

By the time John Harris published his *Lexicon Technicum, or an Universal English Dictionary of Arts and Sciences* (1704) he was able to dispose of spontaneous (or equivocal) generation in the following manner:

> Equivocal Generation, is the Production of Plants without Seed, Insects or Animals without Parents, in the Natural way of Coition between Male and Female.
>
> The Learned World begins now to be satisfied, that there is nothing like this in Nature; and since the use of Microscopes, and a more particular Application to Enquiries of this kind, a prodigious Number of Plants have been discovered to have seeds; and of Animals (Insects) have been found to be produced Univocally, or in the ordinary way of Generation, which before we thought to be equivocally produced.[1]

The historiography of the problem was thus well set by the turn of the century; the focus was to be upon observations. When he returned to the problem in volume two of the *Lexicon Technicum* (1710), Harris noted one or two hold outs which had not yet yielded to the experiential moves. The production of worms, especially intestinal worms, had proven to be very hard to understand but even these now could be explained if one accepted the words of Nicolas Andry who had concluded that even these worms could be shown to breed from a seed which entered the body, and thence gave forth the worm in normal manner.[2]

> He judges that the Eggs of small Insects enter our Bodies by the air we breathe, and in our Food, and that they are hatched there, only when they find a agreeable Heat and Disposition of the Humours and Parts.[3]

Harris closes his entry (as any good scientific paper should) with a series of references to the literature: Dr. George Garden's microscopic observation of insect eggs in fruit trees indicated that insects did indeed breed *ex ovis*, with previous observers being fooled by not being able to recognize the manner of propagation; John Ray's claim that there is no such thing as equivocal generation since all insects were the natural issue of Parents of the same species; the other Englishmen, Dr. Martin Lister and

Mr. Francis Willoughby are joined on the list by Francesco Redi who "cleared up this point in his *de Generatione de g'l Insetti*".

At first glance it appears that there is no problem left by the turn of the century and indeed that there had not really been much of a problem previously; if only naturalists had given trust to their senses and verified their conjectures with observation. It is at just this point that I would contend the whole problem is set. In turning back to biological studies of the Seventeenth Century, it is important to know just what expectations one had of science and also important to recognize the range of views held as to what constituted a 'proper', and 'acceptable', explanation in science in general and in biology in particular. Later historians, it turns out, have been no less prone than Harris to look backward and declare, or at least imply, that spontaneous generation was a 'non-subject' in the history of biology.

Several trends have marked writings in the history of biology and these have in turn didacted an attitude toward spontaneous generation. One common practice has been to adopt an orthogenetic view of the evolution of ideas and to write histories based upon the 'single correct path' to the present. Unfortunate deviations are appropriately labelled and observations, concepts and theories are read from the present backward. For example Arthur W. Meyer in his often informative volume *The Rise of Embryology* evinces great praise for Louis Pasteur whose

> experiments afforded a complete explanation for phenomena which had baffled the ablest investigators for centuries, he not only finally disposed of the theory of spontaneous or equivocal generation but also opened broad vistas for a better understanding of living things.

As though this were not enough Meyer provides an insight into his conception of history:

> Indeed, it is not improbable that had Pasteur followed Leeuwenhoek immediately the whole course of biology might have been changed.[4]

In today's argot this sounds like a plea for 'instant history'.

Another part of this same approach to history would strip the context of an idea or operation to the absolute minimum so that a 'correct' or 'prophetic' observation or speculation might be applauded even when it exists within the context of a 'mistaken' theory. For example, Joseph Needham in his truly outstanding *History of Embryology* exhibits one source of the interests that led to his writing of the study (originally cast as an introduction to a three volume study of chemical embryology) by singling out some observations made by the illustrious Englishman Thomas Browne.[5] Needham cites a number of Browne's discussions of what we might identify as chemical properties of embryos and credits Browne with taking encouraging steps toward the beginning of chemical embryology. Needham felt, however, that he could safely bypass the fact that Browne's *Pseudodoxia Epidemica* (the source of the 'biochemical' remarks) is filled with detailed and favorable discussions of spontaneous generation. The question we would pose is whether there was any relationship between the anticipation of chemical embryology, a 'modern view', and Browne's partiality toward spontaneous generation, considered the holdover from the past.

One other trend would recognize that there might be benign or even beneficial effects in the 'errors' of the past. In her stimulating recent book, *Investigations into Generation*, Elizabeth Gasking indicates a special role for the theories of spontaneous generation and preformationism.[6] These mistaken views of the past led to detailed research on insects and insect life cycles which in turn served to undermine the theory of spontaneous generation. Preformationism was thus accorded its moment in history as the means of overturning spontaneous generation. By and large historical writings have adopted the attitude that spontaneous generation was something to be gotten out of the way.[7]

Perhaps the strongest move in the positivist construction of the history of embryology was that made by Arthur W. Meyer, when he moved to deny a place in history to those undeserving scientists who have been misleading. René Descartes was the target of his wrath and after citing one of Descartes' more extravagant explanations of generation he fairly explodes:

> Since these words are not an exception but truly indicative of the nature of the discussion both as to disregard of facts and its fantastic nature, Descartes deserves no place in the history of embryology. He set a bad example, indeed, for men of science... he contributed nothing to our *objective knowledge* of development... he had *assumed* the existence of things which he had never seen and had *formulated* his ideas on embryology in the easy manner of the philosopher or philosophizing scientist.[8]

Of course from the point of view of my study it is just such distinctions as how to 'do' biology (and embryology) and how to construct or formulate a biological explanation that are the critically important factors to understand in the work and writings of the scientist. Clearly there is a relationship between the underlying assumptions of the scientist and his experimentation as well as his theory building; this is almost certainly true in every case and examination of this relationship should serve as one of the major foci of writing the history of science.

A trend not unrelated to the last mentioned and which has conditioned the historical treatment of spontaneous generation (and is widely used in other areas as well) is the search for 'turning points', 'revolutions', and more recently 'paradigm shifts'. Obviously we are interested in charting or measuring change and the fixation on some new conceptual scheme certainly has marked scientists at many periods. An old problem has been solved and a new one is now in central focus. It is surprising, though, how often competing explanatory schemes kept bodies of scientists working, often side by side, and sometimes the competition is found even within the work of a single man.

What is curious about spontaneous generation is that its overthrow as a paradigm has been celebrated at least three times: once each in the seventeenth, eighteenth and nineteenth centuries. The heroes of each of these scientific revolutions can be readily identified: Redi and Leeuwenhoek working primarily on insect life cycles in the seventeenth century; Spallanzani who was concerned with the Infusoria in the eighteenth century; and Pasteur who once again 'disproved' spontaneous generation through his work on bacteria in the nineteenth century. By putting the problem in this form it is clear that spontaneous generation, as a paradigmatic concept, did not die each time, but rather in descending order of size of organism its usefulness as an explanatory model for the generation of organisms was replaced. In light of current biological research, especially that on molecular genetics and the replication processes of DNA and RNA, some form of a concept of abiogenic formation of living 'stuff' may well still be a very strong explanatory model. Certainly a concept with this many lives deserves the type of careful scrutiny accorded it by this symposium.

The relationship of method to explanatory model has always loomed large in the biological sciences where empiricism so marked the attitude of its practitioners. For today's scientist-historian looking backwards a very strong link is expected between a proper (the more 'modern' the better) experimental technique and an acceptable explanation. Speculation has regularly been decried and as we noted above speculators, like the atomists, have suffered at the historian's hands. Indeed we find the historian willing to applaud 'modern' method even when the results are 'mistaken'. A. W. Meyer in his discussion of John Turberville Needham, the eighteenth century proponent of spontaneous generation, describes one of Needham's attempts to demonstrate the spontaneous generation of eels and is led to comment: "Although sadly misled, Needham deserves much credit for his experimental approach."[9] Today's historian finds it difficult to understand how a 'modern' method did not lead to a correct concept.

One of the major and most persistent questions in the philosophy of biology concerns the reducibility of biology to the concepts and laws of the physical sciences. In many important ways the approach of the physicist and his mode of action is different from the biologist. The physicist today, and often in the past, has shown a greater willingness to idealize, to construct abstract models, than has the biologist. The latter has often displayed a degree of distrust for what appears to be a reductionist move in biology, a distrust that has at times spilled over to include all analogical thinking where the physical sciences are the analogue.

The seventeenth century controversy between William Harvey and René Descartes gave a good preview of the conflicting attitudes which marked later biological thought. Harvey firmly rejected mechanical, corpuscular and atomistic explanations for biology and indeed contrasted to them the empiricism of the anatomists:

> It is a common mistake with those who pursue philosophical studies in these times, to seek for the cause of diversity of parts in diversity of the matter where they arise.... Nor do they err less who, with Democritus, compose all things of atoms; or with Empedocles, of elements. As if *generation* were nothing more than a separation or aggregation, or disposition of things.[10]

Harvey and the empirical tradition of which he is part, and upon which he had great influence, had a very clear idea of the 'proper' manner in which to study and explain generation. It is not that Harvey was without metaphysical commitments, for surely the strong Aristotelianism displayed in his embryological work cannot be discounted, nonetheless he argues the empirical case against the 'philosophers'. As he opens his *Treatise on Generation* he declares that he will set before his readers only "what I have observed on this subject from anatomical dissections." It is the method of the anatomists that Harvey praises most fulsomely. Errors, "like phantoms of darkness ... suddenly vanish before the light of anatomical inquiry. Nor is any long refutation necessary where the truth can be seen with one's proper eyes ... and ... of how base a thing it is to receive instruction from others' comments without examination of the objects themselves ... as the book of nature lies so open and is so easy of consultation."[11] This is the tradition which comes to guide many seventeenth century biologists and causes them to give much more attention to each supposed instance of spontaneous generation and its particular characteristics and have less interest in formulating any general explanatory model, especially one which has as a major goal transcending the boundary between the animate and the inanimate.

Looking backward we can have little doubt that mechanical and atomistic explanations of living phenomena were often naïve. Yet within the context of the seventeenth century and its successful mechanization of the general world view it is critically important to understand the attempts made to mechanize biology. In the hands of the mechanistic-atomistic biologists of the period, biology had to be recast and the nature of its explanatory models made acceptable to the more general program of mechanization. After all, they might argue, if the universe is mechanical and can be likened to a great clock work, should the organism be exempted? If not, if you opt for a unified science with a simplicity of law as a continued guide for theory building, then the task of the biologist is to search for, and to construct, mechanical and/or corpuscular explanations for the biological world. On the ideal level these explanations give no greater difficulty than analogous ones in the physical sciences. A century later when Charles Bonnet argued in favor of a preformist theory of generation (including spontaneous generation) he caught the spirit of these moves and praised the victory of rational science over sensate knowledge.

Finally in this historiographical section I wish to point to the importance of *context*, fitting ideas into their contemporaneous surrounds, as absolutely requisite to an examination of spontaneous generation. Without proper context spontaneous generation, especially when viewed through scientists[1] eyes, may well look like a 'non-problem'. Walter Pagel has understood this point well in his studies of Harvey, Paracelsus, and van Helmont; and has provided a rich fabric of religio-philosophical background.[12] Similarly, Jacques Roger in his fine study of French biological thought of the eighteenth century (with a good deal of seventeenth century background) places the important biological concepts, spontaneous generation included, in their proper intellectual milieu.[13] He makes the perceptive point that the atomists-corpuscularians – Highmore, Gassendi, and Descartes – all began their treatises on generation with a lengthy discussion of spontaneous generation. Why? The question they were interested in, the context of their examination of generation, is the question of matter, its organization, its motion, and its transformations. Spontaneous generation to them emerged as the simplest subspecies of the general case.

What we found then in our examination is that historians who came to the study concentrating upon the development of modern embryology and modern concepts of development avoid, or treat with circumspection, the theory of spontaneous generation. Those interested in the underlying assumptions or commitments of science during the seventeenth century were able to place spontaneous generation within the matrix of ideas, concepts, and policies which guided the scientist or student of generation both in his empirical studies and in his construction of explanatory models.

II. SPONTANEUS GENERATION IN SEVENTEENTH CENTURY THOUGHT

One need only scan the biological literature of the seventeenth century to realize that for many authors spontaneous generation (equivocal generation) was a real issue. It was treated as a real issue by its proponents and its critics, although there was a third group for whom it did not really matter. The lines of commitment were often strangely drawn; numbered among the staunch advocates of spontaneous generation were the orthodox Aristotelians *and* the philosophical atomists! Among the opponents were those who based their views on the new experimental tradition and others who objected on religious grounds.

The question we are immediately led to ask is whether the theories of spontaneous generation talked about in each of the general categories are the same? In many ways it turns out that they are not; even though the operational end product, some sort of generation of a recognized form of life, may be the same. After all, is the concept 'living thing', 'life', the same for an Aristotelian and an atomist?

In this paper I will not try to outline a complete history of spontaneous generation in the seventeenth century, for this is a task beyond my ability and outside my interest.[14] I will turn instead to a number of, hopefully representative, examples and examine the nature of the arguments proposed, the structure of the explanatory models outlined and the context of the theory in relation to other portions of the biologist's work.

That spontaneous generation was widely discussed and argued during the seventeenth century can be seen in the published works of numerous authors but also in the letters of learned men of the period. For example Henry Oldenburg, whose correspondence is now appearing in a multi-volume edited edition,[15] wrote from Paris in June, 1659 of a meeting held at M. Montmor's house[16] during which spontaneous generation was one of the topics. At this date, 1659, prior to the work of Redi, Leeuwenhoek and the other experimental critics of spontaneous generation he is able to report what seemed to be a prevailing attitude:

> The consensus of opinion was that no animals whatever are formed by any other means than by generation from semen fermented in the natural womb, or from some equivalent....[17]

This claim, he reported, was based upon experience and was to include insects where the same sort are always engendered in the same matter. Clearly, therefore, some determinate 'seminal spirit' was producing insects of the same kind and never any others. The criticism of spontaneous generation was important for explaining examples of generation that were irregular or haphazard. In response to the argument put forward for the equivocal generation of intestinal worms or worms found in fruit and in vinegar the local 'consensus' that Oldenburg reported believed that air was impregnated with a vivifying spirit when drawn into some sorts of matter in the proper situation. The effect was seen to be exactly the same as semen in the womb. Just whose arguments were being relayed is not apparent but the point of view was characteristic of the many arguments used to counter claims of spontaneous generation during the mid-century. They did not rely on experiential evidence of any great weight yet they did refer quite often to the specific organisms under discussion.

Although it would really be inaccurate to characterize William Harvey as an opponent of spontaneous generation many later authors point to the frontispiece slogan of the *De Generatione Animalium, 'ex ovo omnia'* and identify it with a theory of generation and development which had little room for spontaneous generation.[18] A reading of Harvey's treatise, however, gives no instance in which he denies spontaneous generation, indeed, several examples can be found in which he accepts it, but when it is discussed it is evident that it has its special place within a total theory of generation. Generation for Harvey, as it had been for Aristotle, was a process wherein form is given to matter and spontaneous generation, as generation in general, is dependent in Harvey's work upon his theory of matter and how matter becomes organized. Thus, when Harvey talks about the generation of an insect out of putrescent material (and similar examples) he refers to the process as *metamorphosis*. In this case, heat or moisture are the accidental factors which serve to uncover a form "due to the potency of a pre-existing material" as contrasted with the red-blooded more perfect animals which are made by epigenesis or the "super addition of parts." There is a scale of generative power in which the less perfect animals are less preservative of their own race whereas the contrary is seen to be the case with the lion and the cock which owe their existence to an "operative faculty of a divine quality and require for their propagation an identity of species, rather, than is the case with imperfect animals, a supply of fitting material."[19]

Harvey was intent upon developing a theory of epigenetic development, a theory based, as Pagel pointedly claims, upon an anti-materialist position[20]; but, also upon an empirical basis laying great store in observation of the stages of development. It is within this context as a key part of his epigenetic theory, that Harvey developed the concept *egg*. This is not the ovum of later biology and is definitely not a simple morphologic unit. In one passage the *egg* concept emerges as that from which anything which goes through generation has been generated:

All living things do derive their Original from something, which doth contain in it both the matter, an efficient virtue and power: which therefore is that thing, both out of which, and by which, whatsoever is born, doth deduce its beginning. And such an Original or Rudiment in animals... is a certain humour, which is concluded in some certain coat, or shell; namely a similar body, having life actually in it, or *in potentia*; and this, in case it be generated within an Animal, and do there remain, untill it have produced an Univocal Animal, is commonly called a Conception: but if it be exposed without, by being born, or else assume its beginning elsewhere, it is called either an

> Egg, or a Worm. But I conceive that both ought alike to be called *Primordium*, the first Rudiment from which an Animal doth spring; as Plants assume their nativity from the Seed: and all these *Primordia* are of one kinde, namely Vital.[21]

But as Keynes wisely comments, in his biographical study, multiplying quoted passages from Harvey will yield little because there are many passages which would seem to permit of spontaneous generation and many others which would seem to deny it.[22]

The question we might ask, however, is how important was the concept of spontaneous generation for Harvey? My answer, now, would be that it was not important at all. The evidence he had at hand would not resolve the problem for Harvey, although he does imply that he had collected a good deal of material on insect studies.[23] We can further conclude that readers of the *De Generatione Animalium* would, however, be left with a view generally unfavorable to spontaneous generation. We can go one step beyond this, for considering the source from which the strongest arguments for spontaneous generation were coming – that is the atomistic and mechanical biologists – Harvey leaves his reader with a vigorous anti-atomist, anti-corpuscular, anti-materialist view; he denies the role of the four classical elements, "the so-called elements, therefore, are not prior to those things that are engendered, or that originate, but are posterior rather – they are relics or remainders rather than principles."[24] And he is no less critical of the followers of Democritus and Epicurus who believed all things to be composed of congregations of atoms of diverse figure: "it is a vulgar error at the present day, to believe that all similar bodies are engendered from diverse or heterogenous matters." Then we come to the core of Harvey's critique. "For on this footing, nothing even to the lynx's eye would be similar, one, the same, and continuous; the unity would be apparent only, a kind of congeries or heap – a congregation or collection of extremely small bodies; nor would generation differ in any respect from a [mechanical] aggregation and arrangement of particles." This last, of course, is the exact point that the atomists, Gassendi, Highmore *et al.* were claiming. When we recall that Highmore's *History of Generation* was also published in 1651 (the same year as Harvey's *De Generatione Animalium*) and that Harvey undoubtedly was familiar with this treatise and its atomistic proposals it is clear that his remarks are not at random but rather that his aim is good and his intent clear. Harvey's approach to spontaneous generation, then, is understandable only within the context of his full theory of generation and development, and his concept of matter and living stuff.

Before turning to the atomistic proponents of spontaneous generation I will examine other critics and the framework of their criticism and rejection.

John Ray went right to the heart of many of the objections when he claimed:

Spontaneous Generation of Animals and Plants upon due Examination will be found to be nothing less than a Creation of them.[25]

The concern he had for any tampering with the scriptural account of the Original Creation is clear. After all how was God able to rest on the Seventh day? If spontaneous generation is a daily occurrence then nothing was accomplished at the original creation which was different. Creation, Ray confidently claims, is the work of omnipotency and is not granted to any creature; it is beyond the power of nature or natural agents, all things produced since have been produced out of the original seed. This is exactly the same concern with which Francesco Redi opened his famous treatise on the *Generation of Insects*. Once the earth, at the command of the creator had brought forth the first plants and animals it has never since produced plants and animals, whether perfect or imperfect, except from the true seeds of those originally produced.[26]

Redi was aware that daily observation seemed to imply that an infinite number of insects and worms were being generated out of decayed material, but he countered this view with his own:

I [am] inclined to believe that these worms are all generated by insemination and that the putrefied matter in which they are found has no other office than that of serving as... a nest.[27]

Shortly later in a letter to Oldenburg published in the *Philosophical Transactions*, John Ray gave voice to a very similar view, indeed he credits Redi as having done an excellent job of discrediting the theory of spontaneous generation. He noted only two possible exceptions, the generation of insects from galls and from other animals.[28] In the first edition of the *Wisdom of God*, Ray had denied spontaneous generation and in response to an attack upon his views he expanded the critique of spontaneous generation in the second edition (1692) to a lengthy section detailing all the evidence and argument he could marshal.[29]

As his biographer Charles Raven notes the criticism of spontaneous generation was one of John Ray's strongest convictions.[30]

Ray's reasons for rejecting spontaneous generation have been summarized in the following fashion:

(1) The production of an animal from matter is an act of creation and God stopped creating on the sixth day.
(2) If spontaneous creation can occur why has all the reproductive apparatus, including the two sexes, sexual organs, etc. been provided?
(3) It is absurd to believe that lower forms can produce higher ones or that lifeless matter can produce the living.
(4) The experimental evidence is strong; Redi has provided it for animals and Malpighi for plants.
(5) If insects can be spontaneously produced why not elephants and man?
(6) Why have there been no new species produced but only the same types as are born of natural propagation?
(7) Finally, the opinions of the experts confirm opposition to spontaneous generation: Redi, Malpighi, Swammerdam, Lister, Leeuwenhoek.[31]

Both Ray and Redi made religio-philosophical arguments an important part of their critique of spontaneous generation also, of course, relying upon the new experimental and observational evidence to substantiate their claim. When Ray pushed his point about no new species being generated he was clearly aiming his opposition at the atomistic proponents of spontaneous generation. For new species to have come into existence heat, from the sun, would have had to serve as the generative principle acting to put passive matter (particles) into motion. These particles would differ only in figure, magnitude and gravity. And Ray finds it almost inconceivable that heat, putting particles in motion, can in any way sort them and place them in such a manner as to form an animal body. It is even harder for him to understand how a process, seemingly random, goes on to fashion animals of already extant forms rather than to produce some new unknown creatures.[32] A similar, if less developed, anti-atomism slipped into the opening pages of Redi's treatise when he criticized that "class of wise persons who hold it to be true that generation proceeds from certain minute agglomerations of atoms, which contain the seed of all things."[33]

There is little doubt that the strongest thrust against spontaneous generation came during the second half of the seventeenth century from the experimentalists and the observers. I have already provided some indication of what it was that motivated their work. There was that underlying religio-philosophical conviction concerning creation and generation. There was a firm doubt of any continuity between the organic and the inorganic realms which manifested itself in anti-atomistic and anti-mechanistic views. There was, of course, the strong, new commitment to empiricism as the operating mode for biological investigation. At its best this was good experimentalism, at its worst it was rampant, uncritical observationism.

The focus for empirical research on spontaneous generation was insects; even Harvey had turned his attention to them for a period, although as he tells us his notes of these researches were lost during the Civil War.[34] Most of the work involved macroscopical observations although the various magnifying devises were utilized. The problem, as we have come to know it, was the determination of the various stages of the life cycle of the organisms – the hidden stages, parasitic stages and others which caused so much trouble and provided the exceptions to the growing consensus in opposition to spontaneous generation. Guiding concepts were not unimportant and early in his work, Redi indicates the critical role of Harvey's dictum concerning the *egg* for his own investigations.[35]

While the generalization that no spontaneous generation occurred was applied to plants, it was held with less than full conviction. John Ray for example qualified his claim:

> In my denial of the Spontaneous Generation of Plants, I am not so confident and premptory; but yet there are the same Objections and Arguments against it as against that of animals [36]

Yet, he went on to use the commonsense argument (teleological as well) that if spontaneous generation took place why were all the apparatus of vessels, preparation of the seed and probable masculine-feminine distinction present. Even the strongly held convictions of Francesco Redi were compatible with an exception or two and he cited the production of insects from galls on trees and worms in the liver of a sheep. He even rationalized the production of insects (worms) by plants noting that the same natural principle that produces the fruits of the plant could well be responsible for the insects or worms. But in spite of these 'lapses' the strong inference of the empirical attack upon spontaneous generation was that careful scrutiny would yield the natural parents and the 'egg or seed' from which *any* generation proceeded (at least this held for organisms the size of insects). A new rule of simplicity was applied – assume that generation proceeds from an egg or seed unless there is good evidence to the contrary.

Support for this type of inference had come from many sources; one additional bit having been added by Robert Hooke in his *Micrographia* of 1665. In the course of his description of the mite, for which a good microscope was judged a necessity, he described its occurrence in putrifying material and emmeshed in mold. While Hooke claims that he "cannot positively from any Experiment, or Observation, I have yet made determine ..." whether or not the mite is spontaneously generated it seems probable to him that wandering mites have laid their eggs in places that Nature has instructed them as being convenient for hatching and therefore, "I am apt to think, the most sorts of Animals, generally accounted spontaneous, have their origination ..." in this manner.[37] Thus Hooke was committed to the same inference against spontaneous generation even before Redi had published *Observations* in 1688.

It comes as no surprise that Antoni van Leeuwenhoek addressed himself to the problem of spontaneous generation and that having taken up the problem his treatment was almost devoid of metaphysical commitment, except that to the power of the senses as magnified by a good lens; and the added strong belief, aided by his microscopic examination of the minutest animals, that the male semen contained animalcules (spermatozoa) responsible for generation. Thus when Leeuwenhoek turned to Hooke's mite, in 1680, he confirmed the English microscopist's earlier speculation:

> We can now be assured sufficiently that no animals, however small they may be, take their origin in putrefaction, but exclusively in procreation.[38]

The basis for Leeuwenhoek's judgement was his theory of generation based on the preformed animalcules which he found in the semen of all animals, "from the largest down to the little despised animal, the flea." He also observed structural similarities up and down the range of the animal kingdom. Firm believer that he was in the magnifying lens, Leeuwenhoek was willing to extend his views just as far as the power of the microscope would take him, even "the smallest animals, smaller than the flea, nay even the very smallest animalcules, have the perfection that we find in a flea."[39] He is referring in this case to infusoria and implied that there existed no such thing as a primitive animal, the kind so often claimed as the product of spontaneous generation. Leeuwenhoek's published 'Letters' are replete with numerous other observations, repetition of the observations and experiments of Redi and others on lice, decaying veal, plant tubercules, maggots, etc., all reaching the same conclusion, there is no evidence for spontaneous generation.[40]

In the course of one of his numerous refutations of spontaneous generation, Leeuwenhoek focused his attention on the familiar problem of creation. The view he expressed fits the orthodoxy, scriptural and biological, found already in Ray and Redi. No creature, he claims, now originates except through propagation, "for however small it may be its first production depends upon the beginning of Creation". Any thought to the contrary, any view that an immobile substance such as stone, wood, earth, plants or seeds bring forth an animal, "would be a Miracle and its production once again be dependent upon the Great Almighty Creator."[41]

But even as he provided new evidence contradicting the spontaneous generation of insects, worms, maggots, etc. he came upon another problem which would not be resolved for at least two centuries. In response to the work of what he describes as 'a certain Gentleman' (almost certainly Francesco Redi) who believed that "no living being can originate in a firmly closed bottle previously filled with meat-juice or meat," Leeuwenhoek undertook a series of experiments. After a carefully set-up (but obviously non-sterile) preparation had been introduced into sealed glass vessels and allowed to sit for several days he examined the contents for animalcules; and he found them in several varieties "so small they were hardly discernable." The bacteria which he was later to describe in greater detail had violated the principle of Redi. Leeuwenhoek's conclusion set the limits beyond which the denial of spontaneous generation could not go in the seventeenth century:

> I think that when that Gentleman speaks of living little animals, he only means worms or maggots, such as we commonly see in rotten meat....[42]

What kind of theory was it that brought out this rash of observation and experiment? Just what were the bounds claimed for spontaneous generation in the seventeenth century? And why, in the final analysis, was the theory as advocated in the seventeenth century such an easy target, so easily overthrown?

As I indicated earlier, there were two sets of proponents and defenders of spontaneous generation. One group was made up of the orthodox seventeenth century Aristotelians, men like Alexander Ross who would not accept even the slightest deviation from doctrine. The full vigor of Ross's defense is seen in his reaction to a passage in Dr. Thomas Browne's *Pseudodoxia Epidemica* (1646). At the end of a volume filled with favorable discussion of spontaneous generation Browne questioned one of the more extravagant claims of the Paracelsians.[43] Ross replies with sharpness and conviction:

He doubts *whether mice can be procreated of putrifaction*. So he may doubt whether in cheese and timber worms are generated; or if butterflies, locusts, grass koppers, sel-fish, snails, eels, and such are procreated of putrified matter.... To question this is to question Reason, Sense, and Experience.[44]

Ross would have the doubter travel to Egypt and examine the muddy banks of the Nile and there find fields swarming with mice "begot of the mud of *Nylus*." The full title of Ross's volume gives indication of the range of his polemics: *Arcana Microcosmi; or, the hid secrets of man's body discovered; in an anatomical duel between Aristotle and Galen concerning the Parts thereof;... with a refutation of Dr. Brown's Vulgar Errors, the Lord Bacon's Natural History, and Dr. Harvey's book De Generatione, Comenius and others....* Sir Kenelm Digby also felt the wrath of Alexander Ross in the latter's *Philosophicall Touchstone... in which errors are refuted and Aristotelian Philosophy Vindicated* (1645). In this instance Ross challenged Digby's atomistic leanings and played the role of the loyal Aristotelian.[45]

Paracelsus is another important source for the belief in spontaneous generation in the early seventeenth century. Walter Pagel has analyzed this material with great care in his splendid monograph and reaches the conclusion that Paracelsus believed that *all* generation was basically the same and that all examples could be understood as a process of putrefraction.[46] In the first instance a seed putrefies in the ground, forming a *prima materia* for that which will ultimately grow from it. This is the source from which the new organism derives its form. The process is very much in the Aristotelian tradition involving a cyclical alternation of putrefraction of one thing and growth of another. And as in the Aristotelian view original heat serves to provide the basis for the differing putrefying processes. Paracelsus, thus, finds no difficulty in understanding how putrefaction serves as the basis for spontaneous generation, which for him are the instances of animals *not* begotten by their likes e.g. snakes, toads, scorpions, spiders, wild bees, ants, etc.[47] Thus within the Paracelsian tradition spontaneous generation takes its place as part of a broader scheme of generation as well as reflecting the Paracelsian theory of matter. Generation is a very broadly conceived process which includes not only natural procreation but artificial recreation or reproduction as it might be understood by the alchemist. Digestion, for example, is viewed as a form of generation in which new parts come to replace those which are worn away.

The Paracelsian tradition can be traced through a number of the early seventeenth century defenders and users of spontaneous generation. Almost a century after Paracelsus, J. B. van Helmont was still willing to claim that a mouse could be generated by fermenting wheat with water for twenty one days. Both seeds and ferments were seen to have equal vital strength and therefore the capacity for generation.[48] For van Helmont, also, spontaneous generation was understood in the context of the full theory of generation and also the manner in which matter was organized. Generation was understood as a chemical phenomenon carried out by a ferment and directed by an Archée. In the limited case of spontaneous generation the process occurs without the direction of the Archée and relies instead upon a material spirit. Spontaneous generation, though irregular, is by no means incomprehensible.[49]

Daniel Sennert was described by Pagel as playing the role of reconciler between the new iatro-chemical school of Paracelsus and the traditional philosophical views of the ancients. This task emerges clearly in his treatise: *On the Consensus and Discord Between the Chemists and the Followers of Aristotle and Galen.* Two other lengthy works treat generation and spontaneous generation. The latter concept, which is embraced by Sennert emerges with a very special meaning in his writings:

> Truly when generation is self made, or rather the formation of an organical Body, the generation of a Spontaneous living Thing differs not from that of a non-spontaneous. For in both the motion and the formation of the Body is made by an agent which lies concealed in the matter.[50]

One agent that supplies the direction in both cases is a seminal faculty; in the case of a spontaneously generated organism, it does "not proceed from an external equivocal agent, but from an internal principle; which if any man will call a Seed, or a Seminal principle... he shall have my consent."[51] No real distinction exists between the two forms of generation for in Sennert's manner of interpreting the process, "nothing can properly be said to be Spontaneously generated; but everything is bred of its own Seed, though hidden and not discernable by our senses."[52] There is little mystery why theories of this sort could be challenged by the experimentalists. As long as no search was made for the instruments of generation this sort of rationalized process might be widely acceptable as a form of explanation, particularly in as much as it was in harmony with an explanatory model and a general metaphysics which had much wider appliability. But as soon as experience was seen as a legitimate challenge, the Aristotelian and Paracelsian models of generation lost their strength, except for the committed.

If the Aristotelians and Paracelsians found support for spontaneous generation in one ancient tradition, the atomists, in their advocacy of spontaneous generation, called upon another and oftimes antagonistic tradition, from antiquity. By way of transition we might examine the views of Sir Kenelm Digby whose mixture of Aristotelianism and corpuscularism provided an interesting focus for discussion in the fourth decade of the seventeenth century.[53] While Digby developed a mechanical approach to physiology, his theory of generation emerges, one part mechanical and one part metaphysical. It might well be said of Digby that he made a virtue of compromise in his attempts to explain generation. But rather than spend time with this 'ornament of England' or his equally interesting compatriot Sir Thomas Browne, let me turn directly to the major protagonists of mechanical and atomistic embryology, René Descartes, Pierre Gassendi, and Nathaniel Highmore.

So much has been written about Descartes that one dispairs at having anything new to add; his basic physiological view that man and animal are fundamentally machines is well known and has been fully adumbrated elsewhere.[54] It is with this very point of view that Descartes opens his treatise on the formation of the fetus, only the last section of which actually comes to deal with the formation of animals. Joseph Needham has characterized this as Descartes' attempt to construct an embryology *more geometrico demonstrata*. By way of commentary on its value, Needham cites these words of the contemporary naturalist George Garden:

We see how wretchedly Descartes came off when he began to apply the laws of motion to the forming of an animal.[55]

Descartes had wished to explain generation, as he had physiology as a whole, in the only way he believed to be proper – in a rigorous mathematical-mechanical fashion. The formation of an organism is due to a process analogous to fermentation, as that process was understood in the first half of the seventeenth century, that is involving pressure, collision, and particles in motion. In just what manner the particles became ordered and got to the proper place and come into combination with the other proper particles in the correct proportions is nowhere made clear in Descartes' rendition. He, like so many other naturalists of the period, relied upon the heat within the seed to account for the movement and the forms which emerged. Descartes never sets out the complete mechanism by which generation occurs, as he tried to do for other parts of his physiological system, but the strength of his mechanical commitment comes through quite clearly:

> If one knows well what are all the parts of the seed of a certain species of animal in particular, for example of *man*, one could deduce from that alone, by the reasons entirely mathematical and certain, the whole figure and conformation of each of its members, as also, reciprocally, in knowing many particulars of this conformation, one can deduce from that what the seed is.[56]

Thus in the terms of matter and its motion and following the laws of mechanics the seed is in essence 'preformed'. Descartes knew what mode of explanation he wished to provide but in the case of generation, at least, he was unable to manage the empirical system.

Spontaneous generation is briefly discussed by Descartes in the *Treatise on the Generation of Animals*, that contested fragment found among his manuscripts at his death.[57] It should be clear at once that given his views of generation among sexual animals spontaneous generation is not a difficult process for him to conceive:

> For spontaneous production of living things outside a uterus it is necessary only that two subjects not very distant from one another be aroused in different ways by the same force of heat, so that it compels subtile parts (vital spirits) to come forth from the one, and thicker parts (blood or vital humor) from the other....[58]

Descartes explains that this is the very same process, which involving substances from the lung and liver, ultimately form the heart.

If Descartes has seemed unsatisfactory because of his failure to provide a full explanation of generation while only proposing the proper form that the explanation should take, Gassendi and Highmore are much more thorough. Gassendi, who is certainly best known for works other than biological ones, was treating the problems of generation at about the same time that Harvey, Digby, Highmore and Descartes were similarly engaged. Furthermore there is strong evidence suggesting that he was in contact with all these figures, even if only indirectly.[59] Gassendi was a thorough atomist and atoms are at the basis of his theories of physiology (animal functions) and generation. Atoms were important for him in explaining nutrition, digestion and sensation. As with previous commentators, Gassendi was careful to separate generation from creation, the latter involving the formation of new matter or the production of a natural body from nothing; this act belonged to God alone. Generation on the other hand is a production from already extant matter, probably involving only a change in the disposition of atoms, that is a change of form, as for example occurs when a sculptor produces a statue from clay.[60]

The key to the process of generation as Gassendi described it, is the formation of the seed and it is at this point, the provision of direction, that he introduces the hand of God, indeed God working at the first moment of creation; God created the "seeds of all that which was made then and which must be engendered afterwards."[61] These seeds are in the form of atoms with various mass, shape and motion which enter into combinations to form molecules, and other small structures from which the actual seeds are thence fashioned.[62] Generation, of any sort then, can be understood simply as the rearrangement of pre-formed matter, which is then set out in some determined order.[63] In Gassendi's view this can be understood as a process analogous to nutrition since the small animal formed out of atoms within the seed becomes visible through the addition of atoms to the basic formation. In carrying this process one step further Gassendi enters the ranks of those later considered preformationists and thus joins the argument against the Harveyan (and Aristotelian) theory of epigenesis.

Although he did not dwell at length on spontaneous generation, it is quite clear that the concept very readily fits into his full theory of generation, and just as importantly into the theory of matter upon which generation rests.[64] In the course of his discussion of plants he explains:

The corpuscles which change into a seed within a plant are also attracted, from the earth itself, and the only difference [between seeds spontaneously produced and these] is that the fashioning of seeds can be more easily accomplished within a plant because a more select supply of corpuscles or similar principles has already been produced and is now gathering.[65]

Since generation is based upon a combination of atoms and molecules, as Gassendi understands these terms, which are in a continuous state of motion,

nothing forbids the creation of those molecules, which because of the heat variously enclosed by them, become the seeds of various. Hence it can also be said that *even now the seeds of animals are being formed* either from atoms or from other principles which God created in the beginning and willed to be endowed with these forms and motions....[66]

Thus Gassendi is able to speak about two kinds of animals, one, which are sexually wrought and require a longer period for their formation and therefore are brought into being in a closed receptable. The second type require a shorter time period and thus a more open receptacle is sufficient. The function of the enclosure is the preservation of the "spirituous and fiery substance of the seed". The first kind of animal requires that it be formed within the body of a similar animal and these therefore are the sort that are always born from similar animals. For the second kind of animal, however, the seed does not need a similar body within which to mature and therefore may spring forth from the most dissimilar things. This second kind of animal may arise from putrid or other materials and its seed is hidden in foreign and unexpected matter; these may be said to arise spontaneously or as if by chance and are referred to as animals which arise of their own accord. Within Gassendi's theory of generation both sorts of animal have an internal cause *but*,

usage has nevertheless prevailed, so that the cause of the first is said to be univocal, that of the second equivocal, because only the external and apparent cause is customarily taken into account, not the internal, hidden one and consequently the external cause of the first and the seed containing it are observed to be of the same sort, whereas in the second they are observed to be different.[67]

Faced with the question of form and resemblence (what we would call heredity) Gassendi seems to adopt an idea very reminiscent of pangenes.[68] The blood gathers together atomic parts from all segments of the body whence they are collected in the genitals. But the collection, for Gassendi, is not a wholly mechanical one as he implies would have been the case for Epicurus, for the collection "yields to the vessels that which Nature has destined."[69] Since God's hand is visible in the original creation the

Epicurean view, that the atoms are mixed and arranged by chance and that the observed world is the last possible disposition of all the atoms, is not a wholly acceptable one for Gassendi; indeed Epicurus is thought to be impious.[70]

Throughout the whole of Gassendi's attempt to construct an atomic-mechanical biology there runs an ambivalence. He is willing to claim that the body works like a machine, but it was not formed by chance. The body reflects, in its construction, reason, foresight, and design and quite clearly was destined for a particular use.[71] This is a familiar argument for the seventeenth century; it is the story of the clock work designed and constructed originally by that Master Craftsman, but now it becomes the natural philosophers task to explain its varied operation in mechanical terms.

While Descartes and Gassendi were mechanical and atomistic philosophers attempting to apply their general principles to the generation of living forms, Nathaniel Highmore was a student of the living organism, a physician, in search of an explanation for a most complex biological phenomenon. His *History of Generation* appeared in 1651, the same year in which William Harvey brought out his own treatise, *De Generatione Animalium*. While discussing many of the same specific problems, no two texts could be further apart in basic outlook. Nathaniel Highmore has written what is certainly the first English work treating embryology from a fully atomistic stand point and probably the earliest work in any language expressing that point of view; and Highmore is a thorough going atomist believing in atoms of heat, cold, color and smell which come to the sense and thus provide for our ability to believe in their existence.[72]

Highmore's treatise, while ostensibly written as a critique of the theory of generation briefly propounded by Sir Kenelm Digby, became in turn an alternate proposal. Although he attempts a more mechanical theory than Gassendi and relegates God even further into the background after the initial creation, Highmore still indicates that in generation, at least, some agent of direction or design is necessary. This stands in contrast to his physiological theories which are almost totally mechanical.

The seed, from which all things are generated, is formed by processes almost identical to nutrition. While he objects to the pangenism implied in Digby's theory, the seed, as Highmore describes it, also contains the substance and properties of all parts of the body; but in this case in the form of atoms. Generation, then,

> is preformed by parts selected from the generators, retaining in them the substance, forms, properties, and operations of the parts of the generators, from whence they were extracted: and this *Quintessence or Magistery* is called the seed.[73]

By using this very process Highmore is able to provide an explanation of spontaneous generation. All creatures take their beginning in this fashion with "some laying up the like matter for further procreation of the same species." In other cases atoms of the original extract become lost, or diffused "and there sleep, till by a discerning corruption they are set at liberty" and begin a process of generation. This explains why swarms of living creatures are found to be generated in the putrefying bodies of others.[74] Seminal principles of this sort are found to generate in wood, from other animals and even from our own flesh. The irony Highmore sees in that having been trapped or imprisoned, when they obtain their freedom, these seeds, by a kind of revenge feed on their prison and devour that very thing which had preserved them from being even further scattered.

The concept of the seed and its role in generation add another interesting insight to the position of Aristotelian thought in the seventeenth century. If I interpret him correctly, Highmore seems to propose an atomic version of Aristotle's system of generation, in which the female was responsible for the material and the male contributed the form. The seed that Highmore describes has two parts, 'material atomes' which in turn are "animated and directed by a spiritual form."[75] The form is always the one proper to the species of whose seed it is a part. This form seems to play the same role that the animal soul did for Gassendi and the part corporeal and part incorporeal soul in Harvey's theory of generation. Both the material atoms and the natural forms remain indestructible through mutations or corruption, according to Highmore. They may separate, with the atoms of form becoming scattered or even linked with other material atoms. But continuity is there and the preservation of organisation through the generations is assured.[76] It is the reunion of the material with the formal atoms which accounts for the generation of plants and animals. But it is exactly this system which also accounts for anomalies and of course for spontaneous generation.

Highmore refers directly to the case of animals which arise from the corruption of other creatures, eels from mud, flies and worms from beasts, scarabeus from oxen and lice from filth of most creatures:

These I say grow up upon the mutual juncture of such Atomes, which before lay scattered in the bowels of some other compound.[77]

These dispersed 'atoms' in coming together are able to form 'another living thing' which may differ significantly from the creature wherein it was born. Thus Highmore explains the emergence of the Mistletoe on other sorts of trees. Seminal atoms, he points out can be taken up in nourishment and only later gather together and form the new organism, within, or in union with a host organism. But there is a rationale to the emergence of these imperfect organisms, they do not just spring up haphazardly. For example one plant might retain the seminal atoms of one type of insect, while other plants retained the seeds of a different insect. The same, he believed, to be the case with the spontaneous generation of insects and other small creatures from the bodies of animals.

There is implied in Highmore's theory of generation a type of preformism and it is this notion which is of great help in accounting for spontaneous generation and metamorphosis. But this is preformationism of a special type relying on the formal atoms, as the preformed entities, growing to visible size through a nutritive process. Thus in the work of Highmore, as in Gassendi, an atomistic theory of generation avoids the chaos of reliance upon chance processes, by being closely linked with a concept of preformation. This link seems to be the important one for building a satisfactory explanation for the equivocal or spontaneous generation of new organisms.

A peculiar irony emerges in Highmore's work. Not only was he the first to treat embryology from an atomistic point of view he was also among the earliest to make use of the microscope in the study of generation.[78] His *History of Generation* is studded with remarks about what could be seen through the microscope and magnifying lens. Yet it was probably the microscope which did more to undermine the atomistic interpretation of generation than any set of alternative theoretical proposals. Atomism was a suitable and indeed for some, a necessary form of explanation for the activities at the 'micro' level, only so long as the naked eye was the major tool of observation. When the microscope came to the aid of the eye, revealing a whole new level of organization, the fascinations of atomic explanations were surplanted by the seeming unlimited possibilities of empirical exploration. What might survive as a philosophical system within biology was severely challenged by an experimentally oriented alternate system of explanation.

But the atomic-mechanical explanations of living phenomena suffered for other reasons, as well, during the second half of the seventeenth century. While they might serve to illuminate some physiological activities, generation seemed to offer too many problems, not only of complexity but of organization as well, to yield to the fairly simple models of atomic-mechanical theories. These theories had rescued spontaneous generation, for a brief period, by explaining it as little more than an extreme case of the normal process. Indeed, it provided an explanation completely compatible with the theory of matter and its organization which for the atomist was at the basis of all phenomena.

Rejection of the atomic-mechanical explanation came ultimately for two reasons. First, because the mysteries surrounding many of the supposed examples of spontaneous generation were resolved by the detailed studies of insect life cycles which occupied many late seventeenth century biologists. If the generation of an ever enlarging group of insects could be explained in 'normal' terms it became less and less necessary to provide an often cumbersome alternative theory. Secondly, and perhaps most tellingly for a period of increased empiricism in the biological sciences, the atomic-mechanical theories of generation proved unfruitful. They led neither to new experimentation nor to the provision of useful new explanations of the very complex events surrounding procreation and morphogenesis. The physical sciences of the late seventeenth century, the very successful atomic and mechanical models of physics and chemistry, were not nearly subtle enough to match the problems that active embryological research was posing. Embryology yielded much more readily to morphological study than it did to functional or physiological study, and in truth it remained very much a study of form rather than function well into the twentieth century.

Harvard University

NOTES

* The author acknowledges the help given in a grant from the National Science Foundation.

[1] John Harris, *Lexicon Technicum, or an Universal English Dictionary of Arts and Sciences*, London, 1704.

[2] Nicolas Andry, *De la Génération des Vers dans le Corps de l'Homme*, Paris, 1700.
[3] John Harris, *Lexicon Technicum*, vol. II, London, 1710.
[4] Arthur W. Meyer, *The Rise of Embryology*, Stanford, 1939, p. 53.
[5] Joseph Needham, *A History of Embryology*, Cambridge, 1959, 2nd edition revised with the assistance of Arthur Hughes, pp. 131–133. Originally published as 'Part II, The Origins of Chemical Embryology', pp. 41–227 of *Chemical Embryology*, Cambridge, 1931, vol. 1. It was first published as a separate book in 1934.
[6] Elizabeth B. Gasking, *Investigations into Generation, 1651–1828*, London, 1967, p. 63.
[7] The major exceptions to this trend are the several historians who focused directly on the problem of spontaneous generation. Certainly the most thorough study of this sort is by Edmund O. von Lippmann, *Urzeugung und Lebenskraft. Zur Geschichte dieser Probleme von den ältesten Zeiten an bis zu den Anfängen des 20 Jahrhunderts*, Berlin, 1933. See also Sten Lindroth, 'Uralstringen. Ett Kapitel ur Biologiens Äldre Historia'. *Lychnos*, 1939, pp. 159–192.
[8] Meyer, *Rise of Embryology*, p. 293. Italics added.
[9] *Ibid.*, p. 51.
[10] William Harvey, *Anatomical Exercises on the Generation of Animals* (1651) in *The Works of William Harvey, M.D.* translated by Robert Willis, London: Printed for the Sydenham Society, 1847, pp. 143–586, p. 207. Referred to hereafter as Harvey. *On Generation*. See the important and provocative study by Walter Pagel, *William Harvey's Biological Ideas: Selected Aspects and Historical Background*, Basel/New York, 1967, in which Harvey's intent is subjected to thorough scrutiny. See also my earlier comments in: Everett Mendelsohn, 'The Changing Nature of Physiological Explanation in the Seventeenth Century', in *Mélanges Alexandre Koyré*, vol. 1, *L'Aventure de la Science*, Paris, 1964, pp. 367–386. An interesting discussion of Harvey's methodology and its clash with the Cartesian view will be found in J. A. Passmore, 'William Harvey and the Philosophy of Science', *Australian Journal of Philosophy* **36**, pp. 85–94, 1958.
[11] Harvey, *On Generation*, pp. 151–2.
[12] Walter Pagel, *William Harvey's Biological Ideas: Selected Aspects and Historical Background*, Basel, New York, 1967; *Paracelsus, An Introduction of Philosophical Medicine in the Era of the Renaissance*, Basel/New York, 1958; *The Religious and Philosophical Aspects of van Helmont's Science and Medicine*, Baltimore, 1944.
[13] Jacques Roger, *Les Sciences de la Vie dans la Pensée Française du XVIIIe siècle*. Paris, 1963. See p. 138.
[14] E. O. von Lippmann, *Urzeugung und Lebenskraft*, mentions a good many names in his encyclopedic monograph, although there is not much critical analysis.
[15] *The Correspondence of Henry Oldenburg*, edited and translated by A. Rupert Hall and Marie Boas Hall, vols. I–IV, 1965–1967.
[16] The role of the 'Montmor Academy' as the direct precursor of the Académie royale des Sciences has been examined in some detail by Harcourt Brown, *Scientific Organizations in Seventeenth Century France (1620–1680)*, Baltimore, 1934, pp. 64–134, especially.
[17] *Oldenburg Correspondence*, vol. 1, p. 262, Oldenburg to Saporta, 18 June, 1659.
[18] The motto is found on the frontispiece of the first edition, printed by Pulleyn in London, 1651. The changes in title page and frontispiece can be followed in Geoffrey Keynes, *A Bibliography of the Writings of Dr. William Harvey, 1573–1657*, Cambridge, 1953, pp. 52 ff. An analytical study of Harvey's treatise is found in Arthur W. Meyer, *An Analysis of the 'De Generatione Animalium' of William Harvey*, Stanford, 1936. The place of the *De Generatione Animalium* within the context of Harvey's other work is

discussed by C. Webster, 'Harvey's *De Generatione*: Its Origins and Relevance to The Theory of Circulation', *Brit. Jour. Hist. Sci.* **3**, pp. 262–274, 1967.
[19] Harvey, *On Generation*, pp. 334–5.
[20] Pagel, *Harvey*, see pp. 234f., 251–282.
[21] Cited in Geoffrey L. Keynes, *The Life of William Harvey*, Oxford, 1966, pp. 351–2, from the first English translation of the *De Generatione*, published, 1653, pp. 514–5.
[22] Keynes, *Life of Harvey*, p. 352.
[23] Needham, *History of Embryology*, p. 143.
[24] Harvey, *On Generation*, p. 517.
[25] John Ray, *The Wisdom of God Manifested in the Works of the Creation*, London, 1691. I have taken most of my citations from the seventh and final edition of 1717. See p. 300. An attack upon atomism and mechanism begins in the first edition and is elaborated upon thereafter. Ancient atomists and mechanists are rejected along with Gassendi, Descartes and other seventeenth century figures.
[26] Francesco Redi, *Esperienze Intorno all Generazione deg l'Insetti*, Florence, 1688. I have used the translation prepared by Mab Bigelow, Chicago, 1909, for my citations.
[27] Redi, *Generation of Insects*, p. 27.
[28] Charles E. Raven, *John Ray, Naturalist*, Cambridge, 1942, p. 375 cites the letter from *Phil. Trans. Roy. Soc.* (London) **VI**, no. 74, 2219, 'The Extract of a Letter Written by Mr. John Ray.... July 2, 1671, concerning Spontaneous Generation'.
[29] John Ray, *Wisdom of God....* London, 1692, 2nd edition, Part II, pp. 73–143.
[30] Raven, *John Ray*, p. 469.
[31] *Ibid.*, p. 376.
[32] Ray, *Wisdom of God*, 7th ed., p. 310.
[33] Redi, *Generation of Insects*, p. 24.
[34] Harvey, *On Generation*, p. 321. See also Redi, *Generation of Insects*, p. 25, who comments *a propos* Harvey: "Perhaps, however, he would have stated his opinion with greater clearness and precision if the notes which he had collected on this subject, [spontaneous generation], had not been dispersed during the tumult of civil war, to the deplorable loss of the republic of philosophy".
[35] Redi, *Generation of Insects*, pp. 24–25.
[36] Ray, *Wisdom of God*, 7th ed., p. 325.
[37] Robert Hooke, *Micrographia*: *Or some Physiological Descriptions of Minute Bodies Made by Magnifying Glasses with Observations and Inquiries Thereupon*, London, 1665, p. 214.
[38] Anthony van Leeuwenhoek, *Collected Letters*, edited, illustrated and annotated by a Committee of Dutch Scientists, Amsterdam, vols. 1–8, 1939–1967, vol. 3, p. 329.
[39] *Ibid.*, vol. 3, p. 329.
[40] See *Collected Letters*, vol. 6, pp. 61–65, vol. 7, pp. 35, 99–101.
[41] *Ibid.*, vol. 7, p. 35.
[42] *Ibid.*, vol. 3, pp. 261–265.
[43] Cited by E. S. Merton, *Science and Imagination in Sir Thomas Browne*, New York, 1949, p. 55. See Sir Thomas Browne, *Pseudodoxia Epidemica, or, Enquiries into Very Many Received Tennents and Commonly Presumed Truths* (1646), in *The Works of Sir Thomas Browne*, edited by Geoffrey Keynes, vols. 2 and 3, London, 1928. See the discussion on pp. 162–3 of volume 2 as an example of Browne's treatment of spontaneous generation.
[44] See Alexander Ross, *Arcana Microcosmi*, London, 1652, p. 155.
[45] Pagel, *Harvey*, p. 347.

[46] Pagel, *Paracelsus*, p. 155ff.
[47] *Ibid.*, pp. 115–6.
[48] Jan Baptiste van Helmont, *Oriatrike, or, Physik Refined*, trans. J(ohn) C (handler), London, 1662, p. 112. See Meyer, *Rise of Embryology*, pp. 38–9.
[49] Roger, *Les Sciences de la Vie*, pp. 100–1.
[50] Daniel Sennert, 'Discourse 5, Concerning the Spontaneous Generation of Live Things', in *Thirteen Books of Natural Philosophy*, by Daniel Sennert..., London, 1661. The original latin was contained in *Hypomnemata Physica*, Frankfurt, 1636.
[51] *Ibid.*, p. 173.
[52] *Ibid.*, p. 223. See also Pagel, *Paracelsus*, p. 341.
[53] Kenelme Digby, *Two Treatises. In the One of Which, The Nature of Bodies*; *in the Other, the Nature of Man's Soule*; *is Looked into*: *In Way of Discovery, of the Immortality of Reasonable Soules*, Paris, 1644. See pp. 209, 215, 219–220.
[54] See Auguste Georges-Berthier, 'Le Mécanisme Cartésien et la Physiologie au XVIIe siècle', *Isis* **2**, 37–89, 1914; **3**, 21–58, 1920; also, H. Dreyfus-Le Foyer, 'Les Conceptions Médicales de Descartes', *Rev. Métaphys. Morale* **44**, 237–86, 1937, A. C. Crombie, 'Descartes on Method and Physiology', *Cambridge Journ.* **5**, 178–86, 1951, and Georges Canguilhem, *La Formation du Concept de Réflexe au XVIIe et XVIIIe siècles*, Paris, 1955.
[55] Needham, *History of Embryology*, p. 156.
[56] René Descartes, *La Description du Corps Humain, De la Formation de l'Animal* (1648) in *Œuvres de Descartes*, edited by Charles Adam and Paul Tannery, Paris, 1909, vol. XI, pp. 276–277. See the study by George F. Davis, 'Metaphysics and Mechanics in Seventeenth Century Embryology', Senior Thesis, Harvard University, 1968, typescript.
[57] René Descartes, *Primae Cogitationes circa Generationem Animalium in Œuvres*, Adam and Tannery, vol. XI, pp. 505–538. For a discussion of authenticity see the 'Avertissement', by the editors, vol. XI, pp. 501–504.
[58] Descartes, *Œuvres XI*, *Primae Cogitationes*, pp. 505–6. I have used the translation given by Howard B. Adelmann, *Marcello Malpighi and the Evolution of Embryology*, 5 vols., Ithaca, N.Y., 1966, p. 725.
[59] See Robert H. Kargon, *Atomism in England from Hariot to Newton*, Oxford, 1966, p. 66.
[60] Pierre Gassendi, *Abrégé de la Philosophie de Gassendi en VIII Tomes*, Lyon, 1678, vol. 3, pp. 327–9. Gassendi's biological thought has not received much attention. See Paul Hoffmann, 'Atomisme et Génétique: Étude de la pensée philosophique et physiologique de Gassendi, appliquée à la définition de la Notion de Féminité', *Rev. de Synthèse*, **87**, 21–44, 1966; also Georges Martin-Charpenel, 'Gassendi Physiologiste', *Actes du Congrès du Tricentenaire de Pierre Gassendi*, Paris, 1955, pp. 207–215.
[61] Gassendi, *Abrégé*, vol. 3, p. 402.
[62] *Ibid.*, vol. 3, pp. 356ff.
[63] *Ibid.*, vol. 5, pp. 509–10, 535.
[64] See Roger, *Les Sciences de la Vie*, pp. 139–40.
[65] Translated by Adelmann, *Marcello Malpighi*, p. 798, from *De Plantis*.
[66] Translated by Adelmann, *Marcello Malpighi*, p. 803, from *De Generatione Animalium*; Italics added.
[67] *Ibid.*, pp. 804–5.
[68] Cf. Adelmann, *Marcello Malpighi*, p. 876.
[69] Gassendi, *Abrégé*, vol. 5, p. 527.
[70] *Ibid.*, vol. 3, pp. 347–9.

[71] *Ibid.*, vol. 5, pp. 435–6.
[72] Nathaniel Highmore, *The History of Generation*, London, 1651, p. 119.
[73] *Ibid.*, p. 26.
[74] *Ibid.*, p. 27.
[75] *Ibid.*, p. 27.
[76] *Ibid.*, pp. 43–4.
[77] *Ibid.*, p. 58.
[78] See Needham, *History of Embryology*, p. 125.

STEPHEN JAY GOULD

D'ARCY THOMPSON AND THE SCIENCE OF FORM

> Our own study of organic form, which we call by Goethe's name of Morphology, is but a portion of that wider Science of Form which deals with the forms assumed by matter under all aspects and conditions, and, in a still wider sense, with forms which are theoretically imaginable.
>
> *Growth and Form*, p. 1026

PREFACE

In 1945, the Public Orator of Oxford lauded D'Arcy Thompson as *unicum disciplinae liberalioris exemplar*[1]; in 1969, the Whole Earth Catalog called his major work 'a paradigm classic'. Few men can list such diverse distinctions in their compendium of honors. But then, few men have displayed so wide a range of talent. D'Arcy Wentworth Thompson (1860–1948), Professor of Natural History at Dundee and St. Andrews,[2] translated Aristotle's *Historia Animalium*, wrote glossaries of Greek birds and fishes, compiled statistics for the Fishery Board of Scotland and contributed the article on pycnogonids[3] to the Cambridge Natural History.[4] He also wrote a book of one thousand pages, revered by artists and architects as well as by engineers and biologists – the 'paradigm classic', *On Growth and Form* (1917, 2nd edition, 1942). To P. B. Medawar it is "beyond comparison the finest work of literature in all the annals of science that have been recorded in the English tongue."[5] To G. Evelyn Hutchinson, it is "one of the very few books on a scientific matter written in this century which will, one may be confident, last as long as our too fragile culture."[6] In it D'Arcy Thompson displayed his thoughts on organic form; these are curious in places, almost visionary in others and always profound. Almost thirty years after the second edition, and more than half a century after the first, they have gained new impact in a science that only now has the technology to deal with his insights.

I. INELUCTABLE MODALITY OF THE VISIBLE

"Extension, figure, number, and motion...," wrote John Locke, "may be properly called real, original or primary qualities, because they are in the things themselves, whether they are perceived or no." [7] Stephen Daedalus, walking along the Irish sea side, reviewed the dilemmas of epistemology and even performed the crucial experiment: "Ineluctable modality of the visible.... Then he was aware of them bodies before of them coloured. How? By knocking his sconce against them.... Open your eyes now. I will. One moment. Has all vanished since?... See now. There all the time without you: and ever shall be, world without end." [8]

Form pervades the material world. Aristotle was willing to consider form without matter for prime movers and demiurges, but not matter without form.[9] In studying nature ever since, most biologists have agreed with Needham that "the central problem of biology is the form problem." [10] To anyone who views form as an outdated concern, suited only for the few surviving taxonomists of a world dominated by the chemistry and energetics of molecular biology,[11] I recommend a book that stands second only to D'Arcy Thompson's as a 20th century paean to form: J. D. Watson's *The Double Helix*. For this book describes the pursuit of a shape "too pretty not to be true," [12] a molecular shape that would underlie and explain the phenomena of heredity. Watson's success emerged directly from a concern with form, from a methodology that prescribed the construction of models in preference to the search for a more subtle chemistry. He describes a lesson imparted by Francis Crick: "I soon was taught that... the key to Linus' [Pauling] success was his reliance on the simple laws of structural chemistry.... The essential trick... was to ask which atoms like to sit next to each other. In place of pencil and paper, the main working tools were a set of molecular models superficially resembling the toys of preschool children." [13]

All biologists must deal with form, but it does not follow that they treat it adequately. In my own field of evolutionary biology, I detect three approaches that seem especially insufficient when compared with the insights of D'Arcy Thompson:

(1) Theories that attempt to render form in such non-morphological terms as motion, flow, and energy: Darwin's success and Lamarck's failure is no simple consequence of their differing positions on specific

points; it also reflects their opposing approaches to form. For Lamarck, motion and becoming were primary; organic matter, ever in flux, mounted the scale of being, impelled by "the force which tends incessantly to complicate organization." As Gillispie[14] writes: "Lamarck's theory of evolution was the last attempt to make a science out of the instinct... that the world is flux and process, and that science is to study, not the configurations of matter, nor the categories of form, but the manifestations of that activity which is ontologically fundamental as bodies in motion and species of being are not." To Darwin, paradoxical as it may seem, form was primary; we sense this in his almost poetical admiration for the elaborate forms of orchids, exquisitely designed to insure pollination by insects.[15] Gillispie continues: "What [Darwin] did was to treat that whole range of nature which had been relegated to becoming, as a problem of being, an infinite set of objective situations reaching back through time.... The Darwinian theory of evolution turned the problem of becoming into a problem of being and permitted the eventual mathematization of that vast area of nature which until Darwin had been protected from logos in the wrappings of process."[16]

The triumph of Darwinism, however, did not assure an adequate treatment of form. The analysis of biological form must emphasize the concept of adaptation – the fitness of a structure to perform functions beneficial to an organism. Good design is an attribute of animals that fit comfortably into the metaphysic of a Paley or a Cuvier. It should, of course, have been just as congenial to Darwinism (as it was to Darwin himself), for evolution had merely substituted natural selection for God as the efficient cause of adaptation. Evolutionists and creationists were equally happy to find good design in nature, the former because it illustrated the effects of natural selection, the latter because it manifested God's plan for a world worthy of his creative interference. Yet modern evolutionary theory has tended to deemphasize form, "to dissolve, despite its great *verbal* emphasis on function, genuine adaptation into the non-morphological concepts of gene-pool, genetical 'fitness,' adaptive zones, etc."[17] In part, this is the legacy of early evolutionists who chose to avoid the concept of design (which represented the old biology, though it fit as well with the new) and to use form only as a clue to the tracing of lineages. It has led to some curious enigmas, most notably to the charge that Spencer's catch-phrase, "the survival of the fittest," is circular because fitness is measured as the capacity for survival. Indeed, the modern geneticist does so define 'fitness'; but Darwin did not. To him, fitness was a property of form, a measure of good design that did not entail survival, *a priori*. In many anti-Darwinian theories of evolution, the fit, in this sense, do not necessarily survive (notions of racial old-age or orthogenesis propelling a lineage on straight courses beyond the point of adaptation). Spencer's phrase, therefore, expresses the primary prediction of Darwin's system and embodies the test by which it can be compared with other theories and accepted.

(2) Theories that deal with form directly but do not attempt to explain it: Among taxonomists who pretend to engage in pure description, there is a mystique that exalts the unsullied objectivity of this most humble service to Nature, this display of Her forms, free from the intrusion of human speculation and the vanity of theory. The irony of this attempt lies in the impossibility of its attainment, for no human being can see the *Ding-an-sich*. 'Pure' description, the piece-by-piece compendium of an organism, is as firmly rooted in theory as the most abstract and mathematical approach to form; the problem is not only that its theory is hidden, but also that it is wrong. A standard species description catalogs the organism part by part; this implies, in a way that is subtle and captivating because it is inexplicit and even unintended, that an animal is merely a framework for its separate parts and that its complexity is irreducible. In purely heuristic terms, this theory is sterile and we must hope that its correspondence to reality is slim; for once it is stated we can do little with form but catalog it in wonder.

(3) Theories that try to explain form, but do so incompletely: In the 1890's D'Arcy Thompson developed an explanation for the regular form of sponge spicules. He ascribed them to adsorption at cell contacts; their simple geometry (Figure 1) merely reflects the regular form, assumed under surface tension, of an aggregate of cells. Michael Foster, an eminent evolutionary physiologist, was displeased: "I confess I am not very much attracted by the line of work.... Does your result wholly destroy the diagnostic value of the spicules? If the form is constant in a group – it does not matter how the form is brought about."[18] This statement epitomizes the approach to form that characterized early evolutionary thought and offended D'Arcy Thompson. This is the 'sign' theory of morphology – the idea that form is to be used only as a guide to the tracing of lineages;

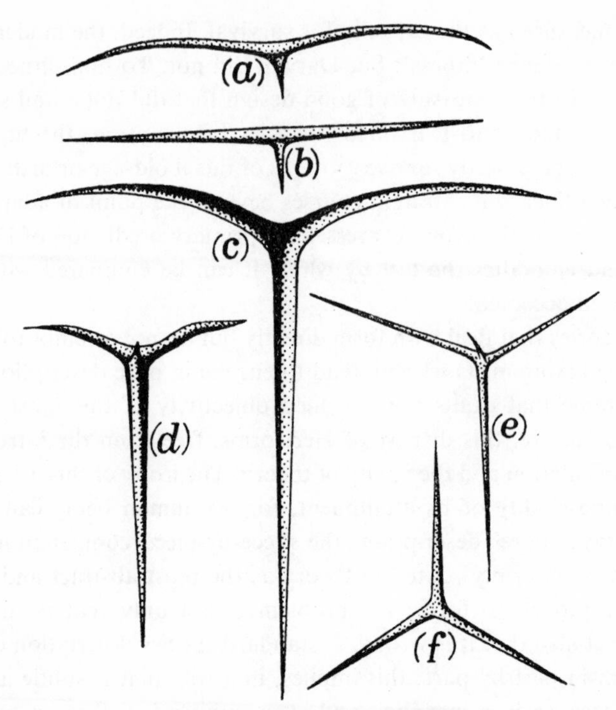

Fig. 1. D'Arcy Thompson's illustration of the regular form of sponge spicules. From his chapter ix, 'On Concretions, Spicules and Spicular Skeletons'.

that structures are to be analyzed as signs of ancestry, not as designs for modern existence; and that form is explained once its antecedents in ontogeny and phylogeny are identified. Thus, under the theory of recapitulation, juvenile structures are studied to identify the ancestral adult that they represent, not to determine how they function in the young animal. D'Arcy Thompson did not gainsay the use of form to trace lineages; he did decry a morphology that considered its work completed when it had done this – for explanation, as the translator of Aristotle understood,[19] requires the identification of many 'causes'. We try, for example, to determine the function of a well-designed structure, to identify its final cause. This remains insufficient still. In an early article on the shapes of eggs, he mentions a variety of final causes (prevention of rolling out of the nest, close packing within the nest to permit more eggs in a given space) and comments: "Whatever truth there be in these apparent adaptations to existing circumstances, it is only by a very hasty logic that we can accept them as a *vera causa* or adequate explanation of the facts."[20] We must state the efficient cause as well: "In Aristotle's parable, the house is there that men may live in it; but it is also there because the builders have laid one stone upon another.... All the while, like warp and woof, mechanism and teleology are interwoven together, and we must not cleave to the one nor despise the other; for their union is rooted in the very nature of totality" (p. 7).[21]

D'Arcy Thompson anchored his view of nature to the idea of organic form (*contra 1*); he tried to explain form by reducing its complexity to simpler elements that could be identified as causes (2); he recognized that explanation is multifaceted; that the identification only of a purpose, a mechanism, or a phyletic ancestor, provides a pitifully incompletc analysis of form (3).

II. D'ARCY THOMPSON'S THEORY OF FORM

In the epilogue to *Growth and Form*, D'Arcy Thompson stated his aim: "to shew that a certain mathematical aspect of morphology is...helpful, nay essential, to the proper study and comprehension of Growth and Form" (p. 1096). In the preface, he wrote: "Numerical precision is the very soul of science, and its attainment affords the best, perhaps the only criterion of the truth of theories and the correctness of experiments" (p. 2). In the opening sentence, he invoked Authority with his customary erudition: "Of the chemistry of his day and generation, Kant declared that it was a science, but not Science – *eine Wissenschaft, aber nicht Wissenschaft* – for that the criterion of true science lay in its relation to mathematics. This was an old story: for Roger Bacon had called mathematics *porta et clavis scientiarum*, and Leonardo da Vinci had said much the same...." (p. 1 – there follows an untranslated Italian footnote).

Yet D'Arcy Thompson's mathematics has a curious ring. We find none of the differential equations and mathematical statistics that adorn modern work in ecology or population genetics; we read, instead, of the partitioning of space, the tetrakaidekahedron, the Miraldi angle, the logarithmic spiral and the golden ratio. Numbers rarely enter equations; rather, they exemplify geometry. For D'Arcy Thompson was a Greek mathematician with 20th century material and insights. *Growth and Form* is the synthesis of his two lives: eminent classicist and eminent zoologist. As he stated in a Presidential Address to the Classical Association (1929): "Science and the Classics is my theme today; it could hardly be otherwise. For all I know, and do, and well nigh all I love and care for (outside of home and friends) lies within one or the other; and the fact that I have loved them both has colored all my life, and enlarged my curiosity and multiplied my inlets to happiness."[22]

Those who recognize the classical roots of *Growth and Form* generally link that work to Aristotle. Yet Aristotle represented only one of the two classical inputs to D'Arcy Thompson's science – and the one of lesser importance for *Growth and Form.* D'Arcy Thompson revered Aristotle as a descriptive naturalist: "In his exhaustive accumulation and treatment of facts, his method is that of the observer, of the scientific student, and is in the main inductive."[23] Moreover, he identified as Aristotle's weaknesses what most critics consider the twin strengths of *Growth and Form*: aesthetic style[24] and mathematical skill.[25]

D'Arcy Thompson's father, a classicist himself, had compared Aristotle's teachings "to the invigorating properties of sea-bathing, open air, regular exercise, wholesome diet and quinine" and the "preaching" of Plato to "opium or hachish."[26] It is to Pythagoras and the later Plato of the *Timaeus* that D'Arcy Thompson owes his vision; for he tried, as did Bertrand Russell, "to apprehend the Pythagorean power by which number holds sway above the flux."[27] D'Arcy Thompson accepted neither Pythagoras' doctrine that 'things are numbers',[28] nor Plato's vision of a realm of ideal numbers existing apart from the physical bodies that are but their fleeting and imperfect representation. But he did share their general attitudes: that solutions to the world's mysteries should be sought in "the geometrical aspect of number"[29]; that simplicity, regularity, symmetry, harmony and truth would be found conjoined[30]; that the world of 'science' could be approached and comprehended by poetic vision. As he wrote in reviewing a book on the Pythagoreans: "The Pythagorean, or Platonic, or Jewish concept of Number is a hard saying to the unpoetic, non-mystical modern and Western world; and many a way is found to show that Plato and Pythagoras meant something prosy and commonplace after all. But to some it is still as plain as ever that Number is the clue to the greatest of earthly mysteries, and that what we call beauty, whether of sound or form, is but its resultant expression."[31] In *Growth and Form*, Plato and Pythagoras are linked as "those great philosophic dreamers" (p. 2). "Not only the movements of the heavenly host must be determined by observations and elucidated by mathematics, but whatsoever else can be expressed by number and defined by natural law. This is the teaching of Plato and Pythagoras, and the message of Greek wisdom to mankind" (p. 1097).

Growth and Form is not a text in anachronistic Greek biology. It uses classical concepts to criticize modern theories (Darwinism in particular), and it shows a man as familiar with the problems of engineering in bridge design as with the tetractys of Pythagoras. It expresses the tensions and conflicts that must inhere in any philosophy compounded from such disparate sources. Of these tensions, two are particularly important in *Growth and Form*. (1) His view of Plato and Pythagoras (mathematics, generality and deduction) versus Aristotle (description and induction). (2) His Greek commitment to a pure and abstract understanding of form versus his Baconian idea that knowledge is power, as expressed by the engineer's love for a good design because it works.

As Hutchinson showed in his perceptive essay,[32] D'Arcy Thompson's theory of form encompasses three themes: an old idea to which he brought new light and grace, a theory on the production of form through growth and a mathematical method for representing the causes of differences in shape among related organisms.

(1) That organisms are well designed is among the oldest ideas of biology. What D'Arcy Thompson did, following Galileo and Borelli,[33] was to conduct this observation away from passive wonder at nature's marvels toward the analytic techniques of physics, mathematics and engineering. When this is done, the study of form becomes a science: predictions can be made and tested; actual forms can be compared with the ideal configurations that engineers prescribe; the superiority of one form over another can be assessed and even measured.

As one illustration of this theme, the pages of *Growth and Form* abound with examples demonstrating that organic shapes conform to the physical forces prevailing at their scale. The pervasive effect of size upon form is a simple consequence of physical laws and geometric arguments. As an animal (or any object) grows, its volume will increase as the cube of its length while its surface, if it maintains the same shape, increases only as the square; the larger the animal, the greater the ratio of its volume to its surface. Thus, a large animal lives in a world ruled by forces, primarily gravitation, that work upon its volume (i.e. its mass). A smaller animal is influenced little by gravity; the forces that act upon its surface hold sway because that surface is so large relative to the mass that gravity affects. Still other forces, the random shocks of the Brownian motion for example, come into play at bacterial dimensions.[34] This physical influence of size explains some of the most confusing of obvious statements: why any fly can walk up a wall, but only Jesus could walk on water[35]; why elephants have thicker legs than gazelles[36]; and why the giant ants of *Them* really couldn't have reached the Los Angeles sewers.[37] Kirby and Spence, extrapolating ant power to elephant size and calculating the prodigious powers of such a leviathan, thanked God in His wisdom for weakening the elephant, lest it should cause "the early desolation of the world" (quoted on p. 36). D'Arcy Thompson showed that if God had had any hand in the matter, he had worked in rather mundane fashion through the physical laws of size.[38]

Following the general introduction, D'Arcy Thompson began his work with the eloquent chapter 'On Magnitude' – a testimony to the significance of size and its physical consequences. The chapter ends thus (p. 77):

Life has a range of magnitude narrow indeed compared to that with which physical science deals; but it is wide enough to include three such discrepant conditions as those in which a man, an insect, and a bacillus have their being and play their several roles. Man is ruled by gravitation, and rests on mother earth. A water-beetle finds the surface of a pool a matter of life and death, a perilous entanglement or an indispensable support. In a third world, where the bacillus lives, gravitation is forgotten, and the viscosity of the liquid, the resistance defined by Stokes' law, the molecular shocks of the Brownian movement, doubtless also the electric charges of the ionized medium, make up the physical environment and have their potent and immediate influence on the organism. The predominant factors are no longer those of our scale; we have come to the edge of a world of which we have no experience, and where all our preconceptions must be recast.

(2) The relationship of size and shape links the first theme to the second – that physical forces exert a direct and immediate influence in shaping organisms as they grow. This is the guiding concept of *Growth and Form*; all but three or four of its seventeen chapters are devoted to the citation of correspondences between organic forms and the shapes that physical forces produce in acting upon non-living material of similar

size, density, viscosity, and rigidity. From these correspondences, D'Arcy Thompson inferred that organic forms had been fashioned in the same way – by the direct action of physical forces.

Many critics have failed to grasp this view of action of physical forces; they dismiss *Growth and Form* as a book filled with curious and ingenious analogies that describe some aspects of form but explain nothing. To D'Arcy Thompson, however, a correspondence between organic and inorganic is more than an analogy; it is a demonstration that two objects probably have the same efficient cause (p. 10):

> The waves of the sea, the little ripples on the shore, the sweeping curve of the sandy bay between the headlands, the outline of the hills, the shape of the clouds, all these are so many riddles of form, so many problems of morphology, and all of them the physicist can more or less easily read and adequately solve: solving them by reference to their antecedent phenomena, in the material system of mechanical forces to which they belong, and to which we interpret them as being due.... Nor is it otherwise with the material forms of living things. Cell and tissue, shell and bone, leaf and flower, are so many portions of matter, and it is in obedience to the laws of physics that their particles have been moved, moulded and conformed.

D'Arcy Thompson's analysis begins at small sizes, in the realm of surface forces. He compares protozoans with Plateau's surfaces of revolution (surfaces of minimal area radially symmetrical about an axis), and infers from the correspondence (Figure 2) that surface tension shapes the simple, single-celled animal. As a liquid cylinder is stretched beyond the limits of stability, it breaks up into "a series of equal and regularly interspaced beads, often with little beads regularly interspaced between the larger ones" (p. 386). D'Arcy Thompson finds these same beads in the 'dew-drops' of a spider's web and provides a mechanical explanation far more simple and satisfactory than that of earlier naturalists (p. 387):

> The same phenomenon is repeated on a grosser scale when the web is bespangled with dew, and its threads bestrung with pearls innumerable. To the older naturalists, these regularly arranged and beautifully formed globules on the spider's web were a frequent source of wonderment. Blackwall, counting some twenty globules in a tenth of an inch, calculated that a large garden-spider's web should comprise about 120,000 globules; the net was spun and finished in about forty minutes, and Blackwall was filled with admiration of the skill and quickness with which the spider manufactured these little beads. And no wonder, for according to the above estimate they had to be made at the rate of about 50 per second.

At larger sizes, gravity comes into play. In the medusae of coelenterates, 'jellyfish' and their relatives, D'Arcy Thompson was "able to discover

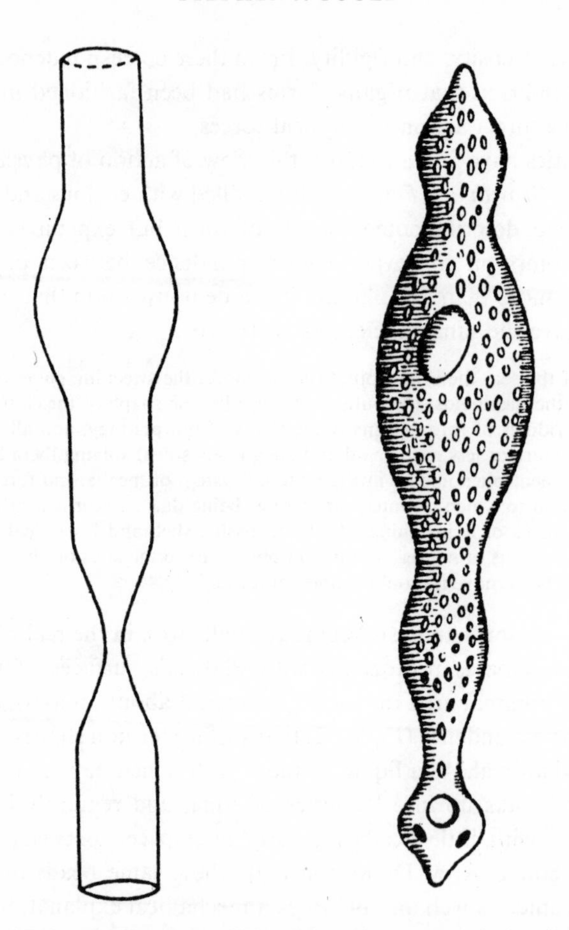

Fig. 2. Comparison between one of Plateau's surfaces of revolution (left – unduloid with positive and negative curvature) and "the flagellate 'monad' *Distigma proteus*". From Chapter v, 'The Forms of Cells'.

various actual phases of the splash or drop [Figure 3].... It is hard indeed to say how much or little all these analogies imply. But they indicate, at the very least, how certain simple organic forms might be naturally assumed by one fluid mass within another when gravity, surface tension and fluid friction play their part" (p. 397). At still larger sizes, surface tension becomes so negligible that rigid hard parts are needed to maintain shape, lest gravity make a world of pancakes. In the internal trabeculae

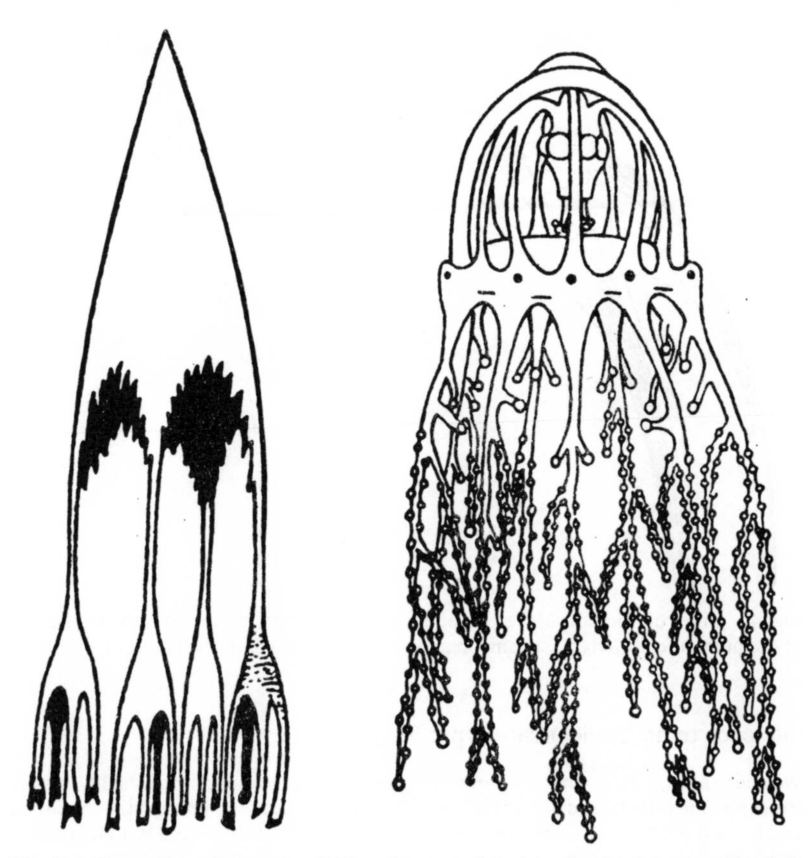

Fig. 3. Comparison between a falling drop produced by fusel oil in paraffin (left) and the medusoid 'jellyfish' *Syncoryme* (right). From chapter v, 'The Forms of cells'.

of vertebrate bones D'Arcy Thompson found patterns that mirror the stresses imposed upon them by the body's weight:

A great engineer ... happened (in the year 1866) to come into his colleague Meyer's dissecting-room, where the anatomist was contemplating the section of a bone. The engineer, who had been busy designing a new and powerful crane, saw in a moment that the arrangement of the bony trabeculae was nothing more nor less than a diagram of the lines of stress, or directions of tension and compression, in the loaded structure: in short, that Nature was strengthening the bone in precisely the manner and direction in which strength was required; and he is said to have cried out, "That's my crane" (pp. 976–977 and Figure 4 of this work).

From this observation of good design, D'Arcy Thompson made his usual

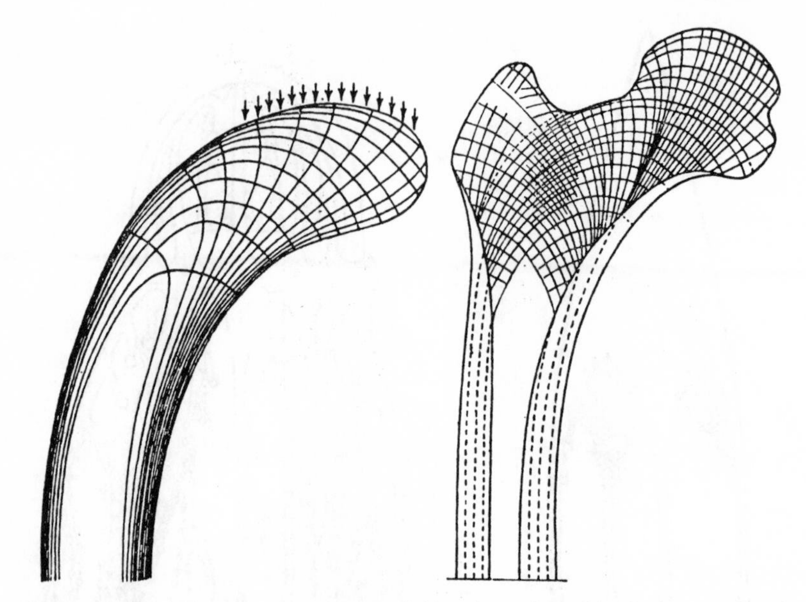

Fig. 4. Lines of force in Meyer's crane (left) and as reflected in the bony trabeculae of the human femur (right). From Chapter xvi, 'On Form and Mechanical Efficiency'.

inference to an efficient cause (pp. 984–985):

> We must always remember that our bone is not only a living but a highly plastic structure; the little trabeculae are constantly being formed and deformed, demolished and formed anew. Here, for once, it is safe to say that 'heredity' need not and cannot be invoked to account for the configuration and arrangement of the trabeculae: for we can see them at any time of life in the making, under the direct action and control of the forces to which the system is exposed.... Herein then lies, so far as we can discern it, a great part at least of the physical causation of what at first sight strikes us as a purely functional adaptation: as a phenomenon, in other words, whose physical cause is as obscure as its final cause or end is apparently manifest.

For sheer ingenuity, nothing in *Growth and Form* matches D'Arcy Thompson's famous, and likely correct, explanation for the narwhal's horn (actually a tooth). This horn may project eight or nine feet beyond the creature's head; "it never curves nor bends, but grows as straight as straight can be" (p. 907); winding about this straight axis is a screw of several low-pitched threads (Figure 5); in rare cases, when two horns are formed, the threads run the same way in each: they are not mirror images (an extraordinary situation for bilaterally symmetrical animals). D'Arcy

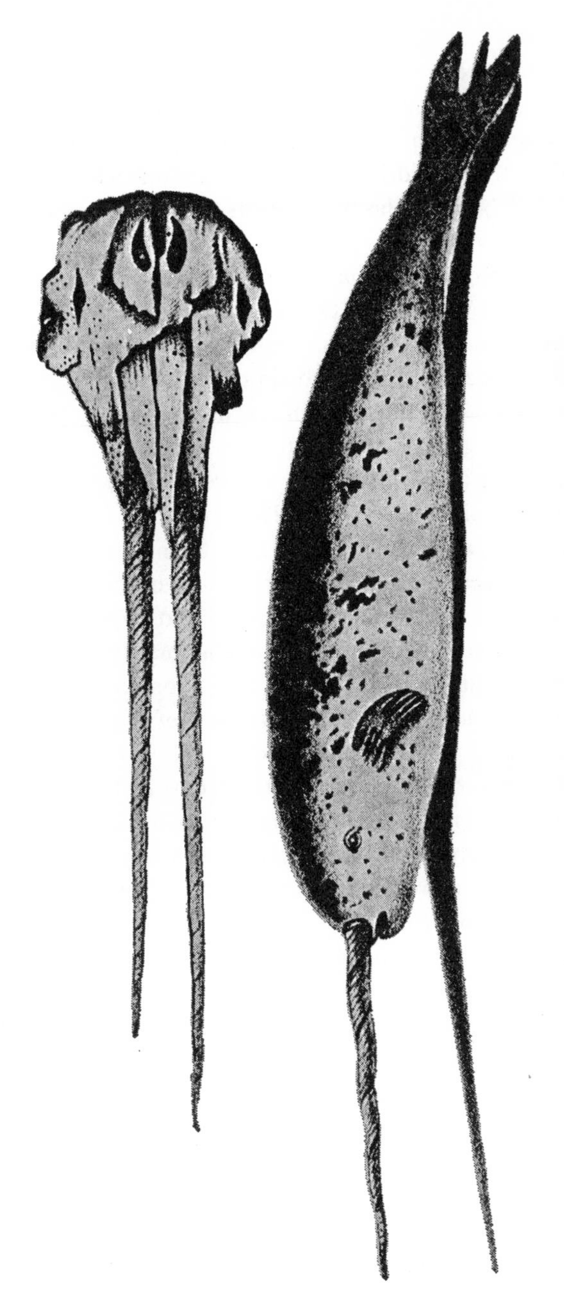

Fig. 5. Below: narwhal showing horn with straight axis and low-pitched threads. Above: skull of rare two-horned narwhal; threads run in the same direction on both horns. This interesting illustration is from the 1826 edition of the Comte de Lacépède's *Histoire naturelle des cétacées* (originally published in 1804 as one of the completing volumes of Buffon's *Histoire naturelle*). Lacépède, a colleague of Cuvier and Lamarck, wrote the last volumes of Buffon's monumental work.

Thompson then notes that screw-threads are made by combining forward and rotatory motion. Now the propulsion of closely-related dolphins contains a rotatory component originating at the tail. To this component, the flexible body responds more actively than the horn which is both rigid and located far from the driving impulse of the tail. The horn's helix, therefore, is formed because the narwhal, as it swims forward during growth, slowly rotates about its own horn!

(3) The last chapter of *Growth and Form* seems, at first, curiously out of place in a deductive work. In it D'Arcy Thompson imposes a net of rectangular coordinates upon various animals and generates series of related species by subjecting that net to simple deformations (Fig. 6 for a set of crab carapaces). Some have seen this as an exercise in empirical description, others as a game, a scherzo to a book that never received its final movement. But it is the finale to a coherent work. For his critics have missed a central point: that D'Arcy Thompson was interested in the deformed net, not primarily in the animal that it generated. He saw that net as a diagram of forces; and just as the trabeculae of the stressed femur reflected the forces responsible for their deposition, so would the deformed net depict the forces that could transform one animal to another. Since these forces might produce a form directly, the deformed net is no mere framework for description; it may be a display of efficient causes. If "diverse and dissimilar fishes can be referred as a whole to identical functions of very different co-ordinate systems, this fact will of itself constitute a proof that variation has proceeded on definite and orderly lines, that a comprehensive 'law of growth' has pervaded the whole structure in its integrity, and that some more or less simple and recognizable system of forces has been in control" (p. 1037).

The method of transformed coordinates is D'Arcy Thompson's provisional mathematics for complex structures. Though he might render a cell or even a jellyfish by forces reflecting a simple physical law, he knew that a complex vertebrate could not be encompassed so easily. Yet he felt that mathematical representation must be sought for the complex as well, even when the physical laws behind that representation could not be specified, and even if this meant accepting one form as given and achieving the simplification and causal insight of mathematics only for the generation of related forms. Thus, as Hutchinson wrote, D'Arcy Thompson constructed "what may be called by analogy with the floating

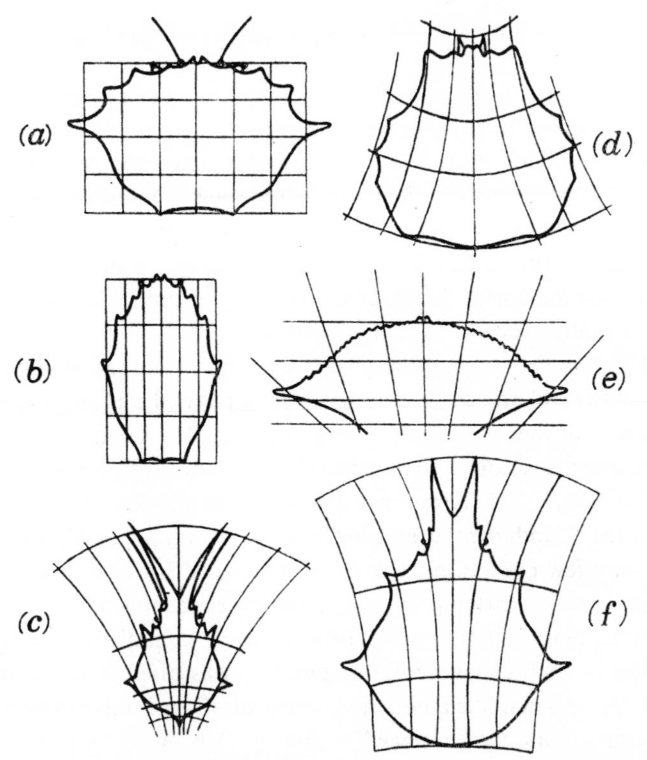

Carapaces of various crabs. (*a*) *Geryon*; (*b*) *Corystes*; (*c*) *Scyramathia*; (*d*) *Paralomis*; (*e*) *Lupa*; (*f*) *Chorinus*.

Fig. 6. Carapaces of various crabs displayed as transformations of a coordinate net imposed upon *Geryon* (a). From Chapter xvii, 'On the Theory of Transformations.'

chronology of the archaeologists, a floating mathematics for morphology, unanchored for the time being to physical science, but capable of valid generalization on its own level."

The study of form may be descriptive merely, or it may become analytical. We begin by describing the shape of an object in the simple words of common speech: we end by defining it in the precise language of mathematics ... The mathematical definition of a 'form' has a quality of precision which was quite lacking in our earlier stage of mere description; it is expressed in a few words or in still briefer symbols, and these words or symbols are so pregnant with meaning that thought itself is economized; we are brought by means of it in touch with Galileo's aphorism (as old as Plato, as old as Pythagoras, as old perhaps as the wisdom of the Egyptians) that 'the Book of Nature is written in characters of Geometry' (p. 1026).

III. D'ARCY THOMPSON AND HIS CRITICS

The critics of *Growth and Form* are numerous. Their major objections can be condensed to three categories.

(1) D'Arcy Thompson was a relic of past times and a perpetrator of their inadequate science. I detect three separate arguments here:

(i) His writing is so overblown, so pompous in its Victorian meter, so ostentatious in its Greek allusions, that harried professionals cannot grasp its essentials with the requisite economy of time. I do not deem this brand of philistinism worthy of comment.

(ii) His non-experimental approach belongs to an earlier natural history, not to modern science; he never manipulated nature in controlled situations, but merely described and interpreted what nature presented. This criticism is serious and justified, but its roots are with Aristotle, not with Linnaeus. For D'Arcy Thompson's inadequacy is what Sambursky cited as the "fundamental and decisive"[39] limitation of Greek science: "With very few exceptions, the Ancient Greeks throughout a period of 800 years made no attempt at systematic experimentation.... The consequence was that induction was limited to the systematic observation and collection of such experimental material as was offered by the study of natural phenomena. Such induction was naturally primitive in terms of the conception of modern science."[40] But if D'Arcy Thompson did not provide his insights with the verification of experiment they are not falsified thereby. And he did not, at least, subscribe to Sambursky's second limitation of Greek science: "It does not aim at the conquest and control of nature, but is motivated by purely intellectual curiosity.... For this reason technology finds no place in it; and it suffers from the lack of that synthesis of pure knowledge and practical application which is the strength of modern science."[41]

(iii) He gave scant notice to modern work that vied with his basic premises and even neglected to update material contested in the interim between editions. The latter charge is true only in part. In the 1917 edition (pp. 48–49), for example, he speaks well of Arrhenius' fantastic notion that the minutest of organisms might escape the earth's atmosphere on the electrical force of the Aurora, thence to be propelled by the radiant energy of light, "like Uriel gliding on a sunbeam" (p. 49 of 1917 ed.), to Jupiter in 80 days and to Alpha Centauri in 3000 years – thus to disseminate our life throughout the universe (or to seek the origin of ours in dispersion from other worlds). In 1942, this finale to 'On Magnitude' was quietly dropped in favor of the statement quoted on page 76. But it could be excised because its implausibility did not threaten any premise of D'Arcy Thompson's system: it had merely served as a dubious illustration of an incontestable truth.[42]

D'Arcy Thompson's stand was this: he did not alter his basic statements when someone challenged their application to particular cases. After all, a good theory is the best guide for separating correct from incorrect 'facts' – especially in natural history where 'pure' data are often the passive and subjective observations of men subtly prejudiced with unacknowledged *a prioris*. When confronted with contradictory evidence, D'Arcy Thompson trusted his theory and his aesthetic vision.[43] As a strategy in science, this is a dangerous gamble, one that can be recommended only for the great. The winner becomes a prophetic genius, the loser a blind dogmatist.

The view of D'Arcy Thompson as a relic will be embraced or dismissed depending upon one's conception of the history of science. In the Comptian perspective of inexorable progress – rendered by many scientists as a march to truth mediated by the continuing accumulation of facts – D'Arcy Thompson's hearkening to the Ancients, his opposition to modern facts in conflict with Greek Truths, must be decried. But if historians of science have any message for the practicing scientist, it is that great men fashion great theories with the elusive qualities denoting 'creativity' and 'genius': a sense (often aesthetic) that something is amiss with old ideas, an ability to bring new ways of thinking to an old problem, and an 'intuitive' feeling for the relative importance and reliability of conflicting information. And *nihil sub sole novum* or no, new ways are often forgotten old ways.

(2) D'Arcy Thompson was a stubborn opponent of modern ideas on evolution. Many have taken D'Arcy Thompson's antagonism to natural selection as yet another sign of his cantankerous obsolescence. In fact, his objections reflect no general stand against modernity, but a specific conflict between his vision and Darwin's. Seen in this light, his opposition is both intelligible and judicious. Yet it is also incorrect, or so we judge today – and here his vision fails in its strictest implication (but triumphs, we shall see, in a reinterpretation). He had three major objections[44]

to Darwinism: the first methodological, the last two substantive.

(i) To most Darwinians of his time, morphology was "an endless search after the blood-relationships of things living and the pedigrees of things dead and gone" (p. 3). D'Arcy Thompson, as we have already seen (p. 5), decried this limited view that sought only final causes or antecedent states. Of complex organisms, he wrote that we must both "look upon the coordinated parts as related and *fitted to the end* or function of the whole *and* as *related to* or *resulting from the physical causes* inherent in the entire system of forces to which the whole had been exposed, and under whose influence it has come into being" (p. 1020; my emphasis). D'Arcy Thompson must have enjoyed the dialogue of King Lear and the Fool as a prophetic satire upon evolutionists who would consider only final causes, and trivial ones at that:

> Fool: Canst tell how an oyster makes his shell?
> Lear: No.
> Fool: Nor I neither: but I can tell why a snail has a house.
> Lear: Why?
> Fool: Why, to put's head in.

(ii) In relating organic form to abstract geometry, D'Arcy Thompson approached the Platonic notion of a real realm of pure form; this realm would specify limits for the imperfect representation of its forms among earthly animals. If mathematical shapes cannot be transformed one into the other through insensible gradations, neither can major discontinuities in organic form be bridged by the imperceptible chain of intermediates that Darwinian theory requires. Thus, D'Arcy Thompson held that major transitions in evolution often occur suddenly, by some form of 'macromutation'.[45]

> An algebraic curve has its fundamental formula, which defines the family to which it belongs.... We never think of 'transforming' a helicoid into an ellipsoid, or a circle into a frequency-curve. So it is with the forms of animals. We cannot transform an invertebrate into a vertebrate, nor a coelenterate into a worm,[46] by any simple and legitimate deformation. ... Nature proceeds from one type to another ... and these types vary according to their own parameters, and are defined by physico-mathematical conditions of possibility. Cuvier's 'types' may not be perfectly chosen nor numerous enough, but types they are; and to seek for stepping-stones across the gaps between is to seek in vain, forever (pp. 1094–1095).

(iii) D'Arcy Thompson's belief in the direct molding of form by physical forces led him to distrust a primary guide to the tracing of lineages – that degree of similarity is a rough measure of evolutionary affinity (recency of common ancestry). It is something of a trade secret that most of our evolutionary trees are based not upon the direct evidence of fossils, but upon inferences that equate reasonable series of modern forms with actual affiliation in time. But if the forms of these series are fashioned directly, the sequence reflects no history, but only a gradation of physical influences obeying timeless laws. Of the single-celled Foraminifera he wrote (pp. 869–870):

While we can trace in the most complete and beautiful manner the passage of one form into another among these little shells ... the question stares us in the face whether this be an 'evolution' which we have any right to correlate with historic time. The mathematician can trace one conic section into another and 'evolve' for example, through innumerable graded ellipses, the circle from the straight line: which tracing of continuous steps is a true 'evolution', though time has no part therein. It was after this fashion that Hegel, and for that matter Aristotle himself, was an evolutionist – to whom evolution was a mental concept, involving order and continuity in thought but not an actual sequence of events in time. Such a conception of evolution is not easy for the modern biologist to grasp, and it is harder still to appreciate.

For these simple forms, subject by their small size to surficial and molecular forces, D'Arcy Thompson may well be right. At least his suggestion has the "audacity of imagination" that John Dewey found in "every great advance in science"; at least it might free the students of these animals from automatic allegiance to theories developed for vertebrates (and not wholly correct even for them). However, when he applies it to large and complex forms, I begin to doubt its propriety while continuing to admire its sheer ingenuity. Molluscan shells, for example, are spread neither evenly nor randomly across the range of form potentially available to them: certain shapes tend to occur again and again. To a modern evolutionist, these recurrent shapes are selected because they are functionally superior to rare or non-existent ones; to D'Arcy Thompson, they are simply the shapes that controlling physical forces produce most easily (p. 849):

It is hard indeed (to my mind) to see in such a case as this where Natural Selection necessarily enters in, or to admit that it has had any share whatsoever in the production of these varied conformations. Unless indeed we use the term Natural Selection in a sense so wide as to deprive it of any purely biological significance; and so recognize as a sort of natural selection whatsoever nexus of causes suffices to differentiate between the likely and the unlikely, the scarce and the frequent, the easy and the hard: and leads

... one type of crystal, one form of cloud, one chemical compound, to be of frequent occurrence and another to be rare.

(3) D'Arcy Thompson's central idea – that form is fashioned by the direct action of physical forces operating during growth – is applicable to very few of the cases he cites. We must first understand D'Arcy Thompson's own limits upon his theory. He did not deny "a principle of heredity" (p. 1023); he did not attempt to ascribe differences between rhinoceroses and watermelons to the action of physical forces. "My sole purpose," he wrote (p. 14 – my italics), "is to correlate with mathematical statement and physical law *certain* of the *simpler outward* phenomena of organic growth and structure or form." Some of these correlations are surely correct,[47] but others, including the fundamental comparison of protozoans with Plateau's figures of surface tension,[48] are almost as surely wrong.

This substantial criticism would seem to discredit *Growth and Form*, relegating it to the domain of antiquarians rather than to historians (not to mention practicing scientists). In fact, it provides the impetus for a reinterpretation that explains the book's continuing influence among a scientific community that almost matches, in its regard for past work, the journalist's maxim: "yesterday's paper wraps today's garbage."[49]

D'Arcy Thompson identified hundreds of correspondences between physical laws and organic forms. If his own theory will not explain them, then another must be sought. And that other, ironically enough, is natural selection. D'Arcy Thompson claimed a direct influence for physical forces; in fact, these forces operate indirectly by specifying the forms that provide optimal adaptation for animals subject to their influence. "Equilibrium figures are common in organic nature," Hutchinson writes, "because any organism not exhibiting them would have to be elaborately protected in other ways against deformation by the stresses and strains imposed by its environment."[50] Thus, D'Arcy Thompson's second theme (p. 10) becomes an aspect of his first – that animals are well designed.

Aristotle's great student, in short, mixed up his causes. D'Arcy Thompson's error can be epitomized this way: He viewed physical forces as the efficient cause of form; they are, in fact, formal causes or blueprints of optimum shapes that determine the direction which natural selection (the true efficient cause) must take to produce adaptation. If physical forces are not the kind of cause D'Arcy Thompson thought, they are causes nonetheless and no explanation of form is complete without reference to them. He was right to correlate physical forces with organic forms and to claim that the correspondence was no mere analogy; but he was right for the wrong reason. D'Arcy Thompson thought he had a theory for the efficient cause of good design; he gave us instead the basis for a science of form – an analytic approach to adaptation.

IV. D'ARCY THOMPSON AND THE SCIENCE OF FORM

The vindication of D'Arcy Thompson's method. Among biologists, *Growth and Form* wears the albatross of Berlioz's *Les Troyens*; for it has been regarded as an unusable masterpiece doomed by excessive length and difficulty of application. Since books are subject to no such constraint as the economics of staging, the charge of excessive length is more a lament of busy men than an argument for necessary neglect. The charge of difficulty in application, however, has been quite justified until recently. D'Arcy Thompson's mathematical analysis of form requires that all the parts of an organism be considered simultaneously – that change of shape be grasped as a whole through the transformation of coordinates imposed over entire bodies. D'Arcy Thompson drew his coordinate diagrams as pictures; he did not, because he could not in any useful way, abstract the transformations as mathematical expressions. Thus, Medawar, who attempted the only pre-computer quantification of transformed coordinates,[51] quite properly termed the technique 'analytically unwieldy'.[52]

The mathematical study of growth and form has been dominated by bivariate analysis: the plotting of one organ against another organ (or total body size) during growth. That this is unsatisfactory in theory has never been denied, for an animal grows as a whole, not as an abstracted series of pairs. Why then was the obviously preferable technique of considering all parts simultaneously – multivariate analysis – not used? The theorems of multivariate analysis are not new; many had been developed before *Growth and Form* received its second edition. But they were not widely applied because the sheer labor of calculation precluded any practical value. With the advent of electronic computers, the situation has changed completely and multivariate analysis has taken a place among the most exciting of new approaches in biology. Unfortunately, D'Arcy

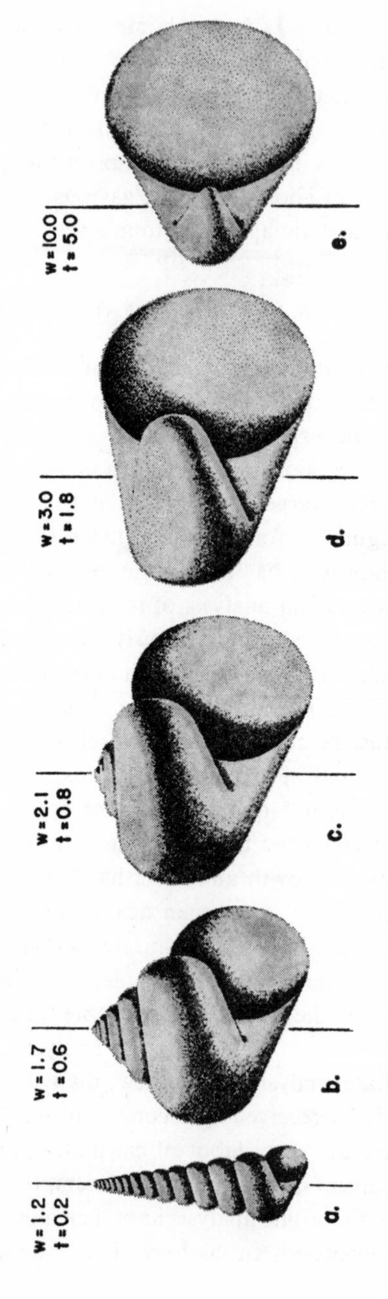

Fig. 7. Computer drawn (and shaded) 'snails'. The sequence from a to d is generated by increasing the rate of whorl expansion and decreasing the rate of translation down the axis. From D. M. Raup, 'Computer as an Aid in Describing Form in Gastropod Shells', *Science* **cxxxviii** (1962), 150–52.

Thompson was about one half century too early. His approach to form was multivariate in conception; hence it suffered the misfortune of much prophesy – it could not be used in its own time.

D'Arcy Thompson's vindication has come only now, and it has just begun. I shall cite but two among several excellent studies in multivariate analysis that were inspired directly by *Growth and Form*.[53] In his chapter on the logarithmic spiral, D'Arcy Thompson pointed out (p. 782) that the diverse and seemingly-complex forms of molluscan shells can be generated in common by just a few basic variables – the shape of the generating curve (the aperture of a snail, for example), its rate of increase in size, and its change of position relative to the axis of coiling. By varying these few measures, practically the whole range of molluscan form can be simulated. But D'Arcy Thompson could not do this, for the number of combinations among the basic measures must run into the thousands for any meaningful analysis. A computer can generate these forms in seconds, and a tame machine can even be trained to draw (and shade!) each simulated shell (Figure 7). The result is no mere exercise in mechanized wizardry, but an important contribution to our understanding of form – for it provides a matrix within which some major questions can be asked for the first time. When we generate a complete spectrum of shapes, for example, we quickly find that actual shells occupy but a small range of shapes possible in theory. Why are certain realms of shape unoccupied? David Raup has provided some fascinating explanations based upon the principles of mechanics and functional morphology.[54]

P. H. A. Sneath has recently published the first successful quantification of D'Arcy Thompson's transformed coordinates.[55] He has done this with modifications of a technique known best among geologists – trend surface analysis. In trend surface analysis, the distribution of a feature on a map (rainfall in mm per year for example) is abstracted by mathematical surfaces of increasing complexity. Sneath compares chimpanzee and human skulls. As his feature, he uses the geometric distance between each pair of corresponding points (tip of the chin for example); the undistorted coordinate net itself serves as his map. The surfaces fitted to these distances are expressions of the total difference in shape between two skulls. The differences are rendered as numbers, and these can be analyzed for patterns that express the simplest way to transform one complex object into another.

D'Arcy Thompson would have welcomed the computer more as a perpetrator of his basic attitudes than of his particular techniques; for he realized that men's theories are subtly molded by the machinery they choose (or are constrained) to use. In particular, he felt that the descriptive method of conventional taxonomy, though lauded by its practitioners as objective and atheoretic, implies the view that complexity is irreducible.

For the morphologist, when comparing one organism with another, describes the differences between them point by point and 'character' by 'character'. If he is from time to time constrained to admit the existence of 'correlation' between characters ... yet all the while he recognizes this fact of correlation somewhat vaguely, as a phenomenon due to causes which, except in rare instances, he can hardly hope to trace; and he falls readily into the habit of thinking and talking of evolution as though it had proceeded on the lines of his own descriptions, point by point and character by character (p. 1036).

In D'Arcy Thompson's approach, we do not seek a different explanation for each contrast between two organisms; by noting correlations, we try to reduce differences to the fewest factors needed to generate them. We do this not for the intellectual satisfaction provided by simplification, but because the abstracted system of factors can be linked more easily to cause. In likening the human stomach to a bubble restricted by a trammel, D'Arcy Thompson invokes the glassblower to explain its form (p. 1050):

The glass-blower starts his operations with a tube, which he first closes at one end so as to form a hollow vesicle, within which his blast of air exercises a uniform pressure on all sides; but the spherical conformation which this uniform expansive force would naturally tend to produce is modified into all kinds of forms by the trammels of resistances set up as the workman lets one part or another of his bubble be unequally heated or cooled. It was Oliver Wendell Holmes who first showed this curious parallel between the operations of the glass-blower and those of Nature, when she starts, as she so often does, with a simple tube.... Such a form as that of the human stomach is easily explained when it is regarded from this point of view; it is simply an ill-blown bubble, a bubble that has been rendered lopsided by a trammel or restraint along one side, such a trammel as is produced if the glass-blower lets one side of his bubble get cold, and such as is actually present in the stomach itself in the form of a muscular band.

A science of form. Form and diversity are the two great subjects of natural history. The study of speciation and systematics has given us a science of diversity within evolutionary theory, but we have lacked a science of form.[56] I believe that a science of form is now arising and that D'Arcy Thompson will be regarded as its godfather because he anticipated and developed the two principles upon which it will be based:

(1) Adult form, in all its complexity, shall not serve as a primary datum. It must be reduced to a smaller set of factors that can generate it during growth. Causes shall be sought among these factors, not in their results. D'Arcy Thompson rendered these factors either theoretically as the physical forces molding form or pictorially as the simple patterns of transformed coordinate nets. With the techniques of computer simulation, we can realize D'Arcy Thompson's unattained goal of quantitative expression for these factors.

(2) When the 'how' of form is explained in (1), we must achieve an equally rigorous solution for its 'why'. Yet the study of adaptation has been plagued by vague, trivial and untestable proposals. We need a criterion of relative efficiency – a way to determine which of two structures is better designed to perform the same function. D'Arcy Thompson stressed the mechanical properties of form. As we have seen (pp. 86–87), he confused his causes; but when we view physical forces not as the architects of form but as the blueprints that specify optimum shapes, we have our test for relative efficiency: the comparison of an actual structure with its optimum expressed in an engineer's terms.

With these two analytical tools, a science of form can provide insights into the major events of life's history. As one example, the attainment of strikingly similar external shapes by animals of very different ancestry[57] – the phenomenon of convergence – is recorded with great frequency in the fossil record. In fact, Sylvester-Bradley has called its recognition "the most distinctive contribution that paleontology has made to modern evolutionary synthesis."[58] Older naturalists were prone to see in convergence either an example of God's plan or an organic striving for ideal form. Today, no one would seek any other efficient cause than natural selection; yet final causes often remain as intractable as ever. D'Arcy Thompson, however, provided a general explanation: the principles of mechanics specify a limited number of good designs for the solution of common problems faced by animals. These can be determined *a priori*, and used to predict the forms of undiscovered organisms performing specified functions. What seemed mysterious can be explained and even predicted.

Our argument indicates ... that evolutionary changes, occurring on a comparatively few definite lines, or plain alternatives, of physico-mathematical probability, are likely to repeat themselves: that the 'higher' protozoa, for instance, may have sprung not from or through one another, but severally from the simpler forms; or that the worm-type, to take another example, may have come into being again and again (p. 1095).

It is easy to assess D'Arcy Thompson's influence; for that is illustrated by the continuing use of his work in the technical research of distinguished scientists, and by the location of his main ideas at the core of an emerging science of form. It is much more difficult, however, to measure his greatness; for he stated nothing truly new, and novelty in discovery is the criterion used by most scientists in assigning status. *Growth and Form* is an ingenious compendium of classical wisdom tempered with insights from the later, but scarcely modern, age of Galilean mechanics. To see it as a great work, we must abandon the usual conception of novelty and admit that the union of previously unconnected truths can be an intellectual discovery as brilliant and as important as the disclosure of something formerly unknown. And D'Arcy Thompson did not merely unite a series of truths generally known in their isolated state; he combined truths long forgotten by his colleagues.

There are some notions so abiding in man's view of nature that we can scarcely deny their status as aspects of reality or as ways we must perceive the world. I would place here, for example, the idea that animals are often exquisitely designed to do what they do. Such a notion receives different explanations at various times (God, chance, natural selection); it may even be cast aside for a time when a new scientific fashion rashly dismisses it with a properly discarded theory for its explanation.[59] Yet it returns, for it must. And, when it returns, scientists rediscover what their forgotten predecessors knew perhaps better than they. D'Arcy Thompson was great because he had not forgotten.

Museum of Comparative Zoology,
Harvard University

NOTES

[1] The outstanding example of a man of liberal education – on the occasion of his admission to the honorary degree of Civil Law. Quoted in Ruth D'Arcy Thompson, *D'Arcy Wentworth Thompson: The Scholar-Naturalist* (London, 1958), p. 238.

[2] Of D'Arcy Thompson's life and personality I shall say little in the limited space vaailable here. See the biography written by his daugher (cited in note 1) and the best three of many articles written to celebrate *Growth and Form*: J. T. Bonner, editor's introduction to abridgment of D'Arcy Thompson, *On Growth and Form* (Cambridge,

1966), pp. vii–xiv; G. Evelyn Hutchinson, 'In Memoriam, D'Arcy Wentworth Thompson', *American Scientist* **xxxvi** (1948), 577–606; P. B. Medawar, 'D'Arcy Thompson and *Growth and Form*', postscript to Ruth D'Arcy Thompson's biography of her father, pp. 219–33, and reprinted in *The Art of the Soluble* (London, 1967), pp. 21–35. As a further source of information and tribute to the synthetic attraction of D'Arcy Thompsons's ideas on form, see the books that arose from two recent art exhibitions on organic form: L. L. Whyte, ed., *Aspects of Form* (London, 1951), for the 1951 Exhibition at the Institute of Contemporary Art, London, designed as a tribute to D'Arcy Thompson; and P. C. Ritterbush, *The Art of Organic Forms* (Washington, 1968), for the 1968 Exhibition at the Museum of Natural History in Washington, D.C.

[3] A small group of marine arthropods that would, I am sure, be considered obscure by all but the very few who love (and study) them.

[4] D'Arcy Thompson's complete bibliography to 1945 will be found in: G. H. Bushnell, 'A List of the Published Writings of D'Arcy Wentworth Thompson', in W. E. Le Gros Clark and P. B. Medawar (eds.), *Essays on Growth and Form Presented to D'Arcy Wentworth Thompson* (Oxford, 1945), pp. 386–400. The list contains 279 items. This volume is a series of essays presented to D'Arcy Thompson on the occasion of his 60th year as an active professor.

[5] Medawar (see note 2), p. 232.

[6] Hutchinson (see note 2), p. 579.

[7] John Locke, 'Some Farther Considerations Concerning our Simple Ideas of Sensation', from Book 1 of *An Essay Concerning Human Understanding* (1690).

[8] James Joyce, *Ulysses* (New York, 1961), p. 37.

[9] See J. Needham's analysis in 'Biochemical Aspects of Form and Growth', in L. L. Whyte (see note 2), pp. 77–86.

[10] J. Needham, *Order and Life* (Cambridge, Mass., 1968), p. 23.

[11] I fear that the message of molecular biology has often been mistranslated in this way as it descended (or ascended according to one's orientation) from technical journals through *Scientific American* to the *New York Times* and into popular consciousness. Just as Einsteinian relativity does not preach that "everything is relative" (rather the opposite, in fact), neither does molecular biology replace the concept of form and spatial structure with the dynamics of chemical energetics or the abstraction of 'information models'; the greatest achievement of molecular biology, after all, was the elucidation by Watson and Crick of the *physical shape* of DNA, the extraordinary molecule that transcribes this genetic information.

[12] J. D. Watson, *The Double Helix* (New York, 1968).

[13] *Ibid.*, p. 38.

[14] C. C. Gillispie, 'Lamarck and Darwin in the History of Science', in B. Glass, O. Temkin and W. L. Strauss, Jr. (eds.), *Forerunners of Darwin: 1745–1859* (Baltimore, 1959), pp. 268–69.

[15] See *Origin of Species* (6th ed.): account of Crüger's experiments in chapter 6. Darwin also devoted an entire book to the subject: *On the Various Contrivances by which British and Foreign Orchids are Fertilized by Insects* (1862).

[16] Gillispie, 'Lamarck and Darwin', p. 291.

[17] M. J. S. Rudwick, 'The Inference of Function from Structure in Fossils', *British Journal for the Philosophy of Science* **xv** (1964), 39.

[18] Quoted in Ruth D'Arcy Thompson, p. 90.

[19] So many heated (and empty) arguments about 'proper' explanation would be avoided if scientists only heeded Aristotle's ancient lesson on the multiplicity of

causes (and realized that different disciplines identify different causes and then, in their limited perspective, wrongly claim a full explanation). In what D'Arcy Thompson called Aristotle's 'parable' of the house (next paragraph of main text), we must consider the various factors, in the absence of which a particular house could not have been built: the stones that compose it (material cause), the mason who laid them (efficient cause), the blueprint that he followed (formal cause), and the purpose for which the house was built (final cause).

[20] 'On the Shapes of Eggs, and the Causes which Determine Them', *Nature* **lxxviii** (1908), 111.

[21] All page numbers cited in the text refer to the 2nd edition of *On Growth and Form* (Cambridge, 1942).

[22] Quoted in Ruth D'Arcy Thompson, p. 196.

[23] 'On Aristotle as a Biologist', *Nature*, **xci** (1913), 204.

[24] *Ibid.*, p. 202: "Wise and learned as Aristotle was, he was neither artist nor poet. His style seldom rises ... above its level, didactic plane."

[25] *Ibid.*, p. 201: "I have my doubts as to his mathematics. In spite of certain formidable passages in the 'Ethics', in spite even of his treatise, *De Lineis Insecabilibus*, I am tempted to suspect that he sometimes passed shyly beneath the superscription over Plato's door." According to legend, the inscription over the door of the Academy read: "Let no one enter here who is ignorant of geometry."

[26] Quoted in Hutchinson (see note 2), p. 578.

[27] Bertrand Russell, 'What I have Lived For', in *The Autobiography of Bertrand Russell, Vol. I, 1872–1914* (Boston, 1967), pp. 3–4.

[28] In the *Metaphysics*, Aristotle wrote of the Pythagoreans: "They assumed that the elements of numbers were the elements of all things, and that the whole heavens were harmony and number." This view seems so arcane in today's context that we tend to see in it the most abstract of mysteries. In an Ionian setting, it makes for respectable science as well. R. M. McInerny, *A History of Western Philosophy*: *Vol. I, From the Beginnings of Philosophy to Plotinus* (Notre Dame, 1968), p. 45, has shown how the Greeks' very "manner of depicting numbers leads to speaking of types of numbers progressing in terms of dimensions." The Ionians had no written numerals; the Pythagoreans, in fact, represented numbers with sets of pebbles arranged in geometric form. Bodies were then generated from points (pebbles) representing units: the number 1 produced a point, 2 a line, 3 a triangle, and 4 a pyramid; moreover, the Pythagoreans had no notion of a continuum and viewed lines as series of discrete points (= units, = numbers). It is no distant extrapolation from this to a notion that the "elements of number are the elements of all things.... Since their way of depicting numbers produced plane and solid figures, bodies and even physical bodies, were taken by the Pythagoreans to be composed of units and, consequently, these bodies are numbers" (McInerny, p. 47). D'Arcy Thompson, of course, denied that numbers are the material cause of physical objects; he did, however, view number as a formal cause: numbers generate the symmetrical and regular shapes that organic forms assume under the influence of physical forces.

[29] D'Arcy Thompson's words in speaking of Pythagorean mathematics: 'The Hellenic Element in the Development of Science', *Nature* **cxxiii** (1929), 732.

[30] Discussing the honeycomb in *Growth and Form* (p. 529), D'Arcy Thompson speaks of "the two principles of simplicity and mathematical beauty as ... sure and sufficient guides." And later, in the epilogue (p. 1097): "the perfection of mathematical beauty

is such ... that whatsoever is most beautiful and regular is also found to be most useful and excellent."

[31] 'The School of Pythagoras', *Nature* **xcviii** (1916), 166. The very next sentence reads: "It was in the very spirit of Pythagorean mysticism that that great naturalist, Henri Fabre, wrote his great ode to numbeı." And it was with a quotation from that 'great ode' that D'Arcy Thompson chose to end *Growth and Form*, calling Fabre "that old man eloquent, that wise student and pupil of the ant and the bee ... who in his all but saecular life has tasted of the firstfruits of immortality ... in whose plainest words is a sound as of bees' industrious murmur; and who, being of the same blood and marrow with Plato and Pythagoras, saw in Number *le comment et le pourquoi des choses*, and found in it *la clef de voûte de l'Univers*" (p. 1097). I think we may include D'Acry Thompson among the 'some' to whom 'it is still as plain as ever...'.

[32] Hutchinson (see note 2).

[33] Galileo's major discussion is in the 'Dialogue of the Second Day' in *Dialogues Concerning Two New Sciences* (1638), trans. H. Crew and A. de Salvio (New York, 1914). Here he displays the mechanical consequences of size increase, demonstrates the impossibility of giants and explains why large animals must have relatively thicker legs than smaller relatives. Borelli's treatise, *De Motu Animalium*, was published in 1685.

[34] The differences between life in such a world and ours are brilliantly stated in George Gamow's classic *Mr. Tomkins in Wonderland* (Cambridge, 1940). Wonderland is a world in which quantum and relativity effects occur at our customary sizes and speeds.

[35] The downward force of gravity pulls animals from walls and through the surface-film of a pond; surface tension provides adhesion both to wall and pond surface. Small animals have such a high ratio of surface to volume that the force of surface tension outstrips that of gravity.

[36] The strength of a leg bone is a function of its cross-sectional area; it will increase as the square of length if shape remains constant with increase in size. Yet, the weight that the legs must support increases in proportion to volume, as the cube of length. A series of animals differing in size but not in shape will become progressively weaker at larger sizes. In nature, large animals overcome this potential problem by having relatively thick legs with a large cross-sectional area; in addition, legs tend to thicken disproportionately durir g growth. This type of growth – in which shape changes to meet the physical demands of increase in size – is called 'allometric.' Julian Huxley pioneered the quantitative study of allometric growth in the 1920's. See S. J. Gould, 'Allometry and Size in Ontogeny and Phylogeny', *Biological Reviews* **xli** (1966), 587–640.

[37] The ability to stay aloft depends upon wing surface area, but body weight increases as the cube of length. This is the classic problem of 'wing loading' in aeronautics; though Hollywood knows little of it, Pliny understood the issue when he wrote in the *Historia Naturalis* that "the heavier birds can fly only after taking a run, or else by commencing their flight from an elevated spot." At their size, the giant ants could not even have taken off from their nest near Alamagordo. In fact, they couldn't even have breathed; the tracheae (respiratory organs) of insects are invaginations of the external body surface; they must increase in constant proportion to the body weight that they service. For surface to grow as fast as weight, shape must change: the surface must become more elaborate (by increased folding or invagination). There is a limit to this elaboration, lest there be no room for internal organs. Large animals have avoided this dilemma by evolving separate internal organs (lungs) to harbor a vast amount of surface for respiration. Horror movies are notorious for the limited imagination that makes giant insects act as only normal ones could. The most incredible thing about the

Incredible Shrinking Man is that, at three inches, he continues to operate as a gravity-prone sixfooter.

[38] Strength is primarily a function of the cross-sectional areas of muscles. Since elephants have smaller areas relative to their weight than small animals, they cannot match the feats of insects when these are measured by the inappropriate criterion of weight lifted vs. body weight. That insects can drag an object many times their own weight is no testimony to any superior design or prodigious will, but merely a function of their small size.

[39] S. Sambursky, *The Physical World of the Greeks* (New York, 1956), p. 2.

[40] *Ibid.*, pp. 2–3.

[41] *Ibid.*, p. 3.

[42] That the smallest organisms are subject to forces (here electricity and the 'radiant pressure' of light) that do not affect us.

[43] For an account of the positive role of aesthetic convictions in major scientific discoveries see: E. A. Burtt, *The Metaphysical Foundations of Modern Science* (New York, 1932).

[44] I speak here only of the objections that arose from his personal theory of form. He also shared many of the common doubts of his contemporaries – especially their reluctance to grant natural selection a creative role: for if selection were only the headsman for the unfit, what created the fit; and if this creation be by 'random' mutation, how can exquisite adaptation arise from 'chance'. The achievements of population genetics in the 1930's resolved these dilemmas by showing that very small selective pressures could be effective in superintending the gradual fixation of favorable small mutations in natural populations.

[45] 'Macromutationism', in various forms, was the major evolutionary challenge to Darwinism from the 'Mutationstheorie' of de Vries (circa 1900) well into the 1940's and 1950's. The inviability of major mutations, the difficulty of spreading them through entire populations after they arise in individuals, and the demonstration that small mutations provide enough genetic variability for evolution have led to the virtual demise of this concept. D'Acry Thompson's reasons for accepting it were unique.

[46] And though we reject macromutationism today, we would not dispute this specific claim as it relates to transformations of complex adult structures. Most evolutionists would try to link these major groups through transformations of their simpler, larval forms. Though no one (to my knowledge) has ever tried to transform an adult sea urchin to a man, the larvae of echinoderms and primitive chordates are very similar. The most popular theory of vertebrate origins would seek to link the two groups at this point. Alister Hardy has referred to larval evolution as an 'escape from specialization'.

[47] Certain extinct corals (cited on p. 513), for example. When the individuals of a colony are crowded together, each assumes the hexagonal form that laws of closest packing require. When uncrowded, the individuals remain circular in outline. Since there are no genetic differences (circular and hexagonal can occur within the same colony, always correlated to the extent of crowding), direct shaping by the pressures of contact must be the efficient cause.

[48] Surface-tensions of cells have been measured directly and they are too low to shape the cell. See J. T. Bonner (note 2), p. 49.

[49] The engineering library at Harvard, for example, has exiled all its pre-1950 journals to a virtually unlit (and completely unheated) attic.

[50] Hutchinson, p. 581.

[51] P. B. Medawar, 'Size, Shape and Age', in W. E. Le Gros Clark and P. B. Medawar (see note 4), pp. 157–87.
[52] P. B. Medawar, in Ruth D'Arcy Thompson (see note 2), p. 231.
[53] A compendium of other examples will be found in my article 'Evolutionary Paleontology and the Science of Form', *Earth-Science Reviews* **vi** (1970), 77–119.
[54] D. M. Raup and A. Michelson, 'Theoretical Morphology of the Coiled Shell', *Science* **cxlvii** (1965), 1294–95. D. M. Raup, 'Geometric Analysis of Shell Coiling: General Problems', *Journal of Paleontology* **xl** (1966), 1178–90.
[55] P. H. A. Sneath, 'Trend-Surface Analysis of Transformation Grids', *Journal of Zoology* **cli** (1967), 65–122.
[56] I have mentioned several reasons for this in various parts of this work: previous lack of a technology to handle the calculations of multivariate analysis, the eclipse of functional anatomy by evolutionary theory and its use of form only for the tracing of lineages; the attitude towards form implicit in the part-by-part descriptions of conventional taxonomy. I present this general argument more fully in the article cited in note 53.
[57] Standard examples include the attainment of cup-coral shapes by a Permian brachiopod and a Jurassic clam and numerous correspondences between Australian marsupials and placentals of the major continents (mole and marsupial 'mole' for example).
[58] P. C. Sylvester-Bradley, 'Iterative Evolution in Fossil Oysters', *Proceedings, International Zoological Congress* **I** (1959), 193.
[59] It is still unfashionable, in biological circles, to use such words as 'design', 'purpose', or 'teleology'. Since final cause is so indispensable a concept in the elucidation of adaptation, and since natural selection can produce a well-designed structure without any conscious intervention of God's super-human wisdom or the sub-human intelligence of the animal in question, one would think that these terms would again be admitted into orthodoxy. Evidently, however, in our choice of words, we are still fighting the battle with theologians that we won in deeds almost a century ago.

PART II

REDUCIBILITY

KENNETH F. SCHAFFNER

THE WATSON-CRICK MODEL AND REDUCTIONISM*

I. INTRODUCTION

There are a number of interrelated problems that cluster around the issue of reduction in the sciences. The logical analysis of theory, the meaning of theoretical terms, the nature of scientific explanation, and various theses concerning the nature of scientific progress, are all closely connected with the problem of intertheoretic reduction. The recent influential and important works of T. S. Kuhn (1962) and P. Feyerabend (1962) are a testimony to the centrality of the analysis of reduction in philosophy of science.

Philosophy of science, however, is too often only the philosophy of physics, and though many philosophers pay lip service to the importance of biological theories, the examples and applications which are found in most of the work of philosophers of science are primarily taken from physics.

The purpose of this paper is to look at the problem of reduction in biology, and in addition to clarifying some problems which are peculiar to the reduction of this science,[1] to comment on the implications which this analysis has for philosophy of science in general. The strategy of the paper is to look carefully, if schematically, at the results of biochemical genetics, and to do so from the point of view of tracing the consequences and developments in molecular biology that are associated with the Watson-Crick model of DNA and its implications. Such a strategy, in the opinion of this author, makes the most fruitful use of the possible inter-relations in this area of the history of science and the philosophy of science, resulting in the mutual illumination of both disciplines. This is not to say that this is the only way of presenting some of the history of molecular biology in a philosophically relevant way. A recent paper by Stent (1968) and a collection of reminiscences by prominent molecular biologists (Cairns, Stent, Watson, eds., 1966) represent other interesting accounts. Further sources will be cited below.

II. DNA AND THE WATSON-CRICK MODEL

In a paper written in 1950 Erwin Chargaff noted that:

> The last few years have witnessed an enormous revival in interest for the chemical and biological properties of nucleic acids.... It is not easy to say what provided the impulse for this rather sudden rebirth.

Chargaff did offer some possible explanations for this renewed interest, among them the earlier paper of Avery, MacLeod, and McCarty (1944) on the genetic transformation of pneumococcal types, but his own justification of the interest was as follows:

> It is impossible to write the history of the cell without considering its geography; and we cannot do this without attention to what may be called the chronology of the cell, i.e. the sequence in which the cellular constituents are laid down and in which they develop each other. If this is done nucleic acids will be found pretty much at the beginning. An attempt to say more leads directly into empty speculations in which almost no field abounds more than the chemistry of the cell.

The situation is quite different today, in part due to Chargaff's analysis of the relative amounts of the different nucleotide types in organismal DNA. Two years later, Hershey and Chase (1952) offered persuasive evidence that the genetic information in the T2 phage of *E. Coli* was carried in its DNA, and not in a protein. In 1953 in two short papers in *Nature*, J. D. Watson and F. H. C. Crick (1953a, b) proposed a structure for DNA and considered some implications of that structure.

It is difficult to overestimate the significance of the accomplishment of Watson and Crick. Their model of the two intertwining right handed helices with sugar phosphate backbones and a varying sequence of purine and pyramidine bases on the inside not only has survived, in its essentials, the test of many and diverse experiments; it has also been instrumental in stimulating the general development toward a complete chemical explanation of biological organisms and processes.

Watson (1968) has recently published an extensive account of his and Crick's tortuous journey towards the DNA model, and the reader can easily locate literature which discusses the model, its revisions, experimental support, and various unsolved problems, in far more detail than I can do here. The purpose of this paper is, rather, to argue that the development of the implications of the Watson-Crick model have given us persuasive scientific reasons for believing that biology is nothing more than chemistry – but chemistry, nonetheless, in which the chemical systematisation of the chemical elements plays a most important role.

Watson and Crick noted in their early papers that their structure for DNA immediately suggested "how it might carry out the essential operation required of a genetic material, that of exact self duplication..." (1953b). They proposed that the double helix uncoils and forces the synthesis of a complementary chain along side of each separated helix. Some of the evidence for this account of self duplication – later termed semi-conservative replication – will be surveyed below.

In their second paper, Watson and Crick (1953b) also noted another most important consequence of their model:

> [Though] the sugar phosphate backbone of our model is completely regular, ... any sequence of the pairs of bases can fit into the structure. It follows that in a long molecule many different permutations are possible, and it therefore seems likely that the precise sequence of bases is the code which carries the genetical information.

The same paper also contained the suggestion that "spontaneous mutation may be due to a base occasionally occurring in one of its less likely tautomeric forms." This suggestion, though not yet conclusively confirmed, did focus the attention of molecular biologists on using DNA structure to explain mutation.[2]

Later I shall trace the logically relevant features of the development of fine structure genetics, the genetic code, protein synthesis, and the colinearity hypothesis to show how these are related to the Watson-Crick model, and how the findings of contemporary molecular biology support a general principle of biological reduction. Before this can be done, however, it is necessary to obtain a clear idea of what reduction is.

III. THE LOGIC AND METHODOLOGY OF REDUCTION

Various accounts of reduction have appeared in the literature of the philosophy of science.[3] Here I shall simply recount one example of reduction in the *physical* sciences, and discuss its logic and methodology. This will require a rather long digression from biology, but it is necessary in order to prepare us for an analysis of the similarities and differences involved in the more interesting and controversial example of biological reduction. The digression will also permit me to comment on some of the misconceptions which are involved in some contemporary accounts of reduction.

The example which I shall outline is the reduction of physical optics to Maxwell's electromagnetic theory.[4] The reduction was worked out in the latter part of the nineteenth century, and it is now known to be incomplete. Maxwell's theory is only adequate in the classical limit, and quantum electrodynamics is required for understanding most processes involving the interaction of light and matter. Nevertheless the example is a useful one, for it exhibits the logical structure of an actual reduction in the sciences: one that was accepted by most scientists and one which is used extensively, as an approximation, even today.[5]

In the early 1860's James Clerk Maxwell constructed a mathematical theory of the interaction of electric and magnetic fields that was based on the experimental work of Faraday (and others) and some theoretical work of Kelvin. By introducing a radically new notion of 'displacement current', Maxwell was able to show that under certain conditions an electromagnetic disturbance would be propagated through the aether. In empty space – that is, space without 'ponderable matter' in it – the velocity of the disturbance was calculated to be equal to the velocity of light. Moreover, the equations for the electric and magnetic energies of the aether bore a strong resemblance to the equations for the kinetic and potential parts of the light energy of the optical aether. Maxwell did not hesitate to identify the optical and electromagnetic aethers and to assert the identity of light with electromagnetic waves.

Maxwell and his followers were able to work out many of the consequences of these claims. Electromagnetic explanations of reflection, refraction, interference, dispersion, and diffraction were developed as the theory was extended and modified. Hertz, in a brilliant series of experiments, demonstrated that Maxwell's electromagnetic waves could be produced and detected, and that their properties agreed with the properties of light.

The logical (and methodological) features of this episode of reduction or explanation of physical optics by electromagnetic theory seem to be as follows:

(1) The basic vocabulary of physical optics is connected with the basic vocabulary of Maxwellian electrodynamics in such a way that for any descriptive expression in physical optics one can construct a correlatable

descriptive expression in electromagnetic theory. I have discussed, rather schematically, the logical form of the expressions which perform this 'correlation' in my (1967a), but it would be wise to repeat the discussion in a slightly different terminology here.

Essentially we need to introduce what I term 'reduction functions' – sometimes these are called 'bridge laws' or 'connectability assumptions' – which will identify the entities in the two theories, and which will associate the predicates. More specifically we can construct reduction functions for entity terms and predicate terms as follows:

(a) We construct reduction functions for *entity* terms such that two universal names are connected by an identity sign, this to be interpreted as 'synthetic identity.' An example in the optics-electromagnetic theory example is **Z**=**E**, where **Z** is the light vector of physical optics, and **E** is the electric force vector of Maxwell's theory.[6] The identity so constructed is in many ways analogous to the well known synthetic identity:

morning star = evening star.

All entity terms in the reduced theory must be identified with entity terms, or combinations of entity terms, of the reducing theory, but of course not the converse. (This reciprocal identification *might* occur, but it is unlikely in view of the greater generality of the reducing theory.)

The senses of the two terms which flank the identity sign are different, but are related by the reduction function. Accordingly each term has a primary sense, fixed by its own theory, and a secondary sense, ascribed to it through its coupling in a reduction function to another term with a different but not contradictory sense. Clearly the analytic-synthetic distinction is bent, if not broken in this analysis, but this is also the case in the analysis of scientific laws.[7]

This augmenting of the primary sense is one of the ways in which this analysis differs from the analyses of reduction that have been developed by Nagel (1961), Feyerabend (1962), and Sellars (1963). Nagel contends that:

> Expressions distinctive of a given science (such as the word 'temperature' as employed in the science of heat) are intelligible in terms of the rules or habits of usage of that branch of inquiry; and when those expressions are used in that branch of study, they must be understood in the senses associated with them in that branch, whether or not the science has been reduced to some other discipline.

Sellars on the other hand claims that what I have termed the primary and secondary senses can both be ascribed, without distinction, to both terms. In connection with a discussion of the reduction of chemistry to physics using identity statements Sellars wrote:

In the case of the reduction of one theory to another we can *bring it about* that the identity statements are unproblematic, both with respect to objects and with respect to properties, by making one unified vocabulary do the work of two. We give the distinctive vocabulary of chemistry a new use in which the physical and the chemical expressions which occurred in substantive correspondence rules now have the same sense.

I shall comment more extensively on Sellars, and on Feyerabend, below.

(b) We construct reduction functions for *predicate* terms – i.e. for *n*-ary predicates which could include functors – in a similar way. As above, such connections as can be established must be *referentially* specified because the senses of the predicates to be associated are different. This can be done with the help of the extensional interpretation of *n*-ary predicates in formal logic, in which ordered *n*-tuples are the extension of the predicates. Assume that entity reduction functions have been established, such that the ontology of the reducing theory T_1 ranged over by individual variables $x_1 \ldots x_n$, is identified with the ontology of T_2 represented by $y_1 \ldots y_n$. For example suppose $y_3 = \langle x_1, x_4 \rangle$. Then a reduction function for any *m*-ary predicate, G_2^m of T_2, is synthetically identified in our sense with an F_1^n of T_1, if and only if the substitution of the *x* terms for the *y* terms in the extension of G_2^m yields the extension of F_1^n. It is clear that predicates which are synthetically identified should refer to the same 'object' – in this case the ordered aggregates of entity objects.[8, 9] It should perhaps be emphasised at this point that such reduction functions for predicates do not constitute *definitions* of the reduced theory's predicates in terms of the reducing theory's predicates, though they can *become* definitions. I shall have more to say on this point later.

An example of such predicate/functor synthetic identification in the optics-electromagnetic theory case is the relation of *n*, the index of refraction, to the square root of the product of ε, the refractory media's dielectric constant, and μ, the medium's coefficient of magnetic permeability:

$$n = \sqrt{\varepsilon\mu}.$$

Unfortunately this only holds approximately because of dispersion phenomena, but the relation was important in the testing of the Maxwell theory.

It should be mentioned that such connections as are cited in (a) and (b) are not all constructed at the time of the first attempts at reduction. Some are introduced at this time, others are introduced later as the reduction is further developed. This non-simultaneous establishment of reduction functions is further complicated by the fact that not all of the problems that fall within the province of the secondary science may have been solved by the secondary theory. Such a distinction between 'science' and 'theory' is one of the reasons why I sometimes use one term, sometimes another, as the context demands.[10] For example adequate accounts for the optics of moving bodies and for dispersion phenomena were lacking in the 'domain' of physical optics when Maxwell's theory was proposed, but were not forthcoming until the development of Lorentz' and Einstein's theories.[11]

Similar complications are to be found in the reduction of biology, even of biochemical genetics, to physics and chemistry, as we shall see below. Nevertheless, if one is willing to call such a relationship as was constructed between Maxwell's theory and physical optics, a *reduction*, then such reductions are being developed between biology and physics and chemistry.

Attention must also be drawn at this point to the fact that theories change as science progresses, as the postulates of the theory and the way the theory relates to experiments shift. In an important sense, there is a strong family resemblance between Maxwell's theory in Maxwell's early papers (1861), in his later papers (1865 and 1868), and in the Hertz-Heaviside formulation (1890), so that one is justified in claiming it is the 'same' theory. In reality there are strong analogies. When reductions between theories occur, terms in the reducing and reduced theories often change their meaning: they now have new connotations and the referents are characterised in a more general way. How and why this occurs is complicated: in part it is, as was suggested above, because of the establishment of the synthetic identities, but this in turn is partly associated with a transfer of 'correspondence rules' between the two theories. I shall have more to say about this below. I now wish to stress that not only may the terms shift their meaning, but the *form* of the laws of the reduced, and even the reducing, theory can and do alter. Thus one might be led to a statistical phenomenological thermodynamics, as Szilard was,[12] or to a variant of the form of the intensity laws of reflection and refraction in physical optics, as were physicists in the late nineteenth century (Schaffner, 1967a).

To put this in the formalism of theory reduction, the reduced theory may change to T_2^* as a result of revisions in it necessitated by desiring to make it conform to evidence highlighted by the reducing theory T_1. It may happen that T_1 itself, must be altered to get a reduction of T_2 or T_2^*, as Maxwell's theory was altered and supplemented by Lorentz' electron theory to moıe adequately explain dispersion phenomena and the results of experiments done in the field of the optics of moving bodies.

(2) The synthetic identities or reduction functions discussed above require empirical support. This is achieved in a number of ways in experimental research, and it is difficult to generalise to a limited set of types of experiments. One needs experiments that show that the entities independently characterisable in terms of the reduced and reducing theories, and which are to be identified, have the same or relatable relevant properties. For example, in the Maxwell theory case, Hertz showed that the electromagnetic waves predicted by Maxwell's theory could be produced and detected, and that they had the same properties as light waves – they could be reflected, refracted, and polarised. O. Weiner and others later did experiments that indicated that it was the electric vector which was responsible for optical effects (Schaffner, 1967c). In certain more sophisticated cases in which predicates or properties are different in the two theories, as in certain areas of genetics and chemistry, what counts as empirical support becomes more difficult to characterise. Usually simultaneous and contiguous appearance of properties is taken as evidence of the referential identity of properties, and certain analogies, especially formal ones, are also relevant. The attribution of the property of valence to hydrogen atoms, and the property of unrestricted spins of the hydrogen atom, and their 'identification' is such a case in point (Schaffner, 1967c). What the reducing theory says about the causal mechanism responsible for the properties in such cases is also relevant, as we shall see in the discussion of biochemical genetics below.

(3) When these synthetic identities (reduction functions) are conjoined to the reducing theory, in our case Maxwell's theory, it is possible to derive the laws of optics and to explain the results of various optical experiments. This is where Nagel's (1961) condition of deducibility enters,

and I think it can be supported against Feyerabend's (1962) criticisms, if we allow theory modification as suggested above under (1). How the new theory explains old experiments in the reduced theory's subject area, or 'domain', without necessarily going through the old theory's postulates, and this does sometimes happen, I shall discuss in the next paragraph.

(4) In the standard analysis of scientific theories, such as one finds in the writings of Nagel and Hempel, certain theoretical terms are connected with experimental ideas, or observational terms, by means of linkages

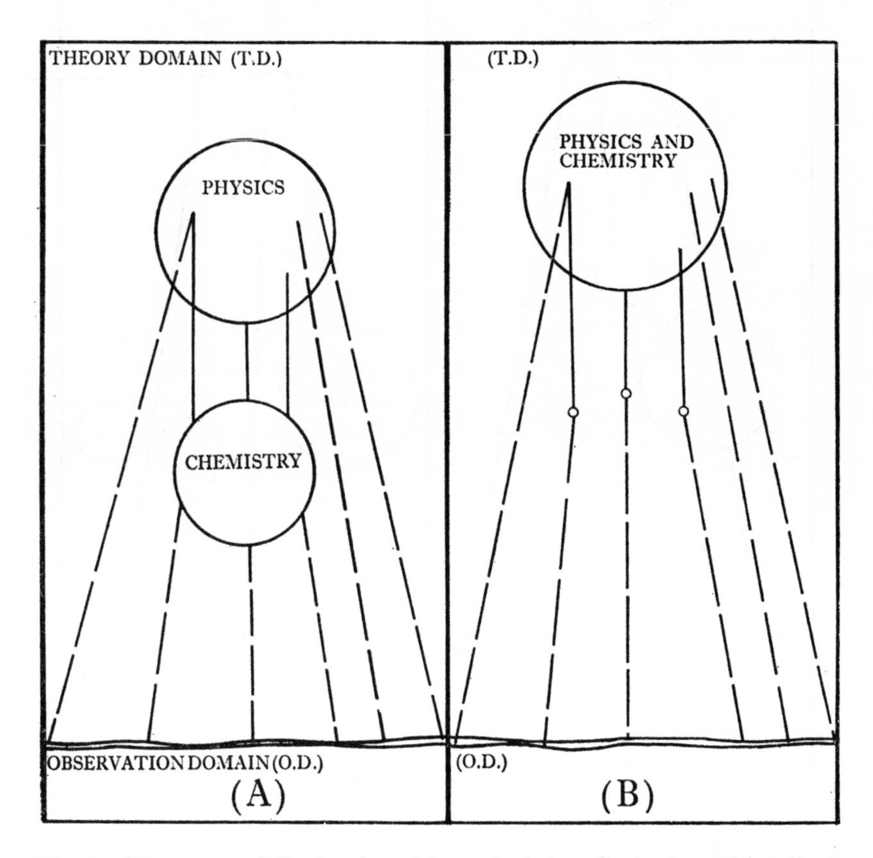

Fig. 1. (A) represents Sellars' notion of the received view of reduction; (B) is Sellars' own view of reduction. The solid vertical lines are theoretical-theoretical correspondence rules, the dashed lines are theoretical-empirical correspondence rules.

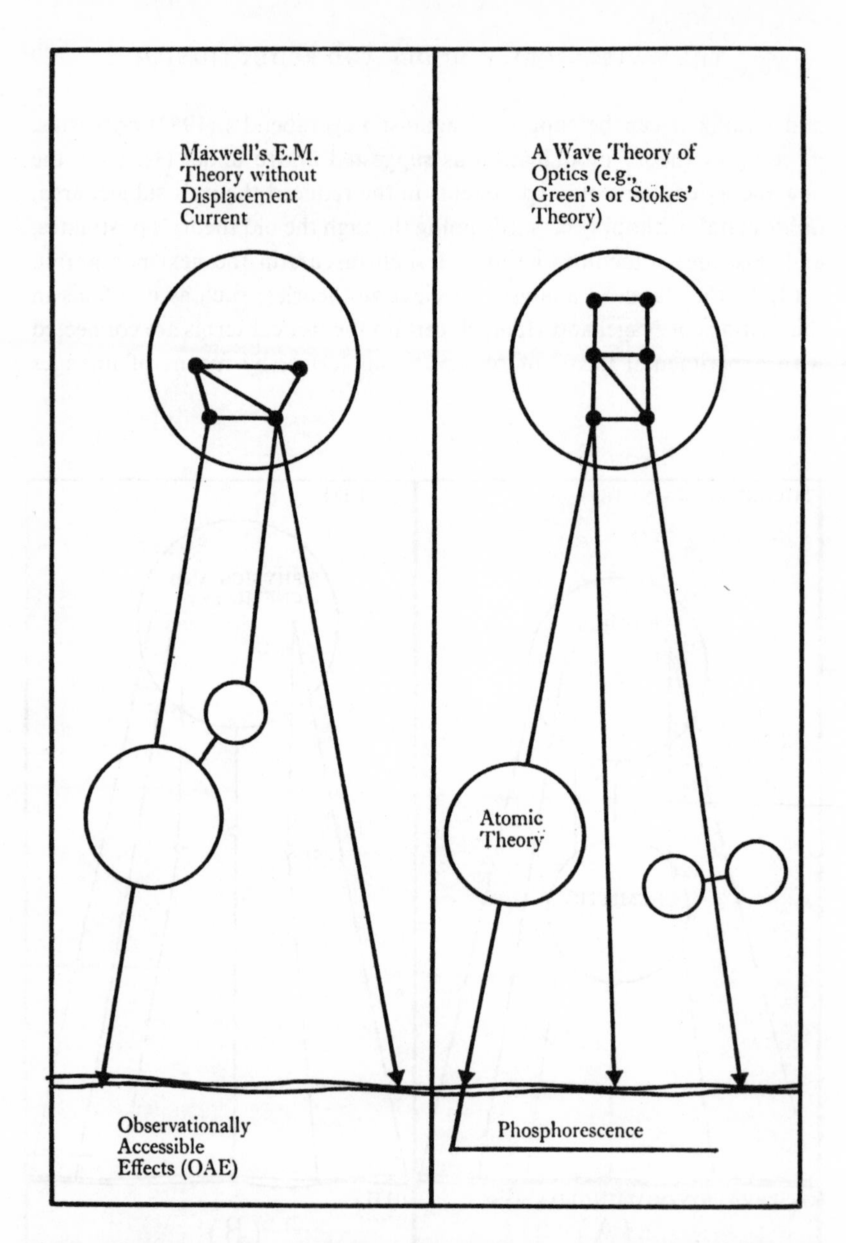

Fig. 2. The structure of two theories before a reduction. The small dark dots represent theoretical concepts and the short lines connecting them are logical connectives. The small circles represent aspects of borrowed theories; the 'vertical' arrows are the causal effects of theoretical processes which in certain cases require two or more theories to account for them.

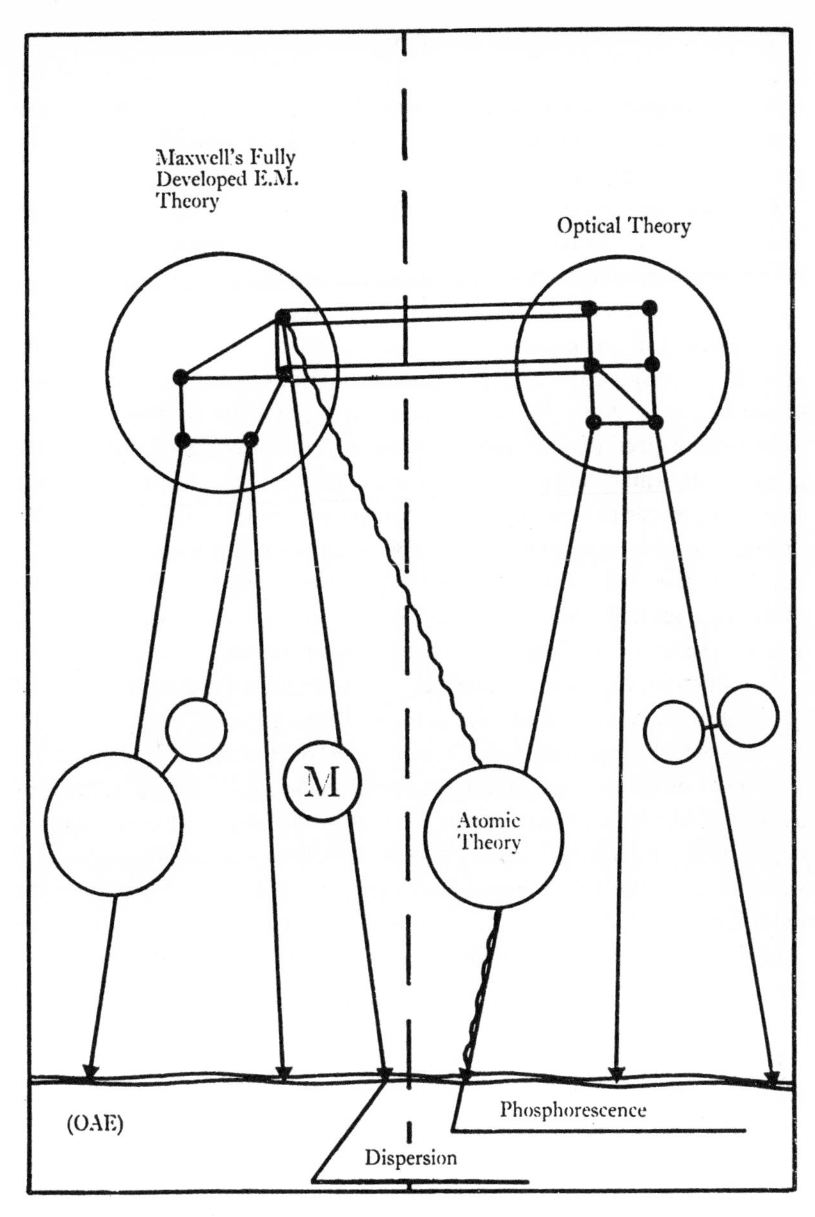

Fig. 3. After a reduction has been effected the two domains and the theories are related as shown above. The straight horizontal double lines represent two of the synthetic identities which connect theoretical concepts in the reduced and reducing theories. (In actuality, all primitive concepts of the reduced theory would have to be connected with concepts or combinations of concepts of the reducing theory.) The wavy line represents the 'analogous causal sequence' discussed in the text, and 'M' represents the science of mechanics which must be appealed to in order to account for the motion of dispersive particles in a medium.

called 'correspondence rules'. I have tried to analyse this notion of correspondence rule in another paper,[13] and here will simply refer to those aspects of that analysis that are relevant to the issues of this paper.

Correspondence rules in many instances turn out to be on closer analysis collapsed or telescoped causal sequences which connect the action of a theoretical entity or a theoretical process with the observationally accessible arena. Theoretical terms possess 'antecedent theoretical meaning', which is in an important sense, independent of the meaning ascribable by a conjunction of correspondence rules and the terms' interconnections as given in an uninterpreted postulate system. Another important aspect of this new analysis of correspondence rules is the crucial role which 'borrowed' laws and theories play in warranting the causal sequences connecting the theoretical processes of the theory under primary consideration with 'observable' phenomena. A modified Duhemian thesis falls out of such an analysis in a very natural way – the modifications essentially directed at the claims of the ability to maintain a specific theory in the face of any conflicting evidence.[14]

In reduction, the causal sequences which are associated with the action of a light vector, for example, are paralleled by analogous causal sequences which are constructed for the electric vector, with these latter sequences often utilising aspects of the physical optics causal sequences, as well as the instruments which have been devised for the testing and application of physical optics. These 'analogous causal sequences' are warranted by the two theories and the connections established by the reduction functions. There is no vicious circularity here, because such shifts in correspondence rules only occur after the synthetic identities have been supported by such considerations as were discussed under (2) above.

Sellars has drawn attention to the role of correspondence rules in theory reduction and has proposed an interesting analogy which can be generalised and extended with the help of some of the ideas proposed in the above discussion. Sellars wrote:

To reduce chemical theory to physical theory is by no means to reduce chemistry to physics. The replacement of the substantive correspondence rules which related the original theories by explicit definitions leaves relatively untouched the correspondence rules, both substantive and methodological, which tied the two theories to their empirical domains. To use a picture, a situation in which two balloons (theories), one above the other, each tied to the ground at a different set of places (theoretical-empirical correspondence rules) and each vertically tied to the other by various ropes (theoretical-theoretical correspondence rules), has been replaced by a situation in which there is one taller balloon which, however, is tied to the ground by two sets of ropes one of which connects the lower part of the balloon to the places where the lower balloon was tied, while the other connects the higher part of the balloon to the places where the upper balloon was tied (1963), p. 76.

Sellars does not give a picture of this analogy, but I think that the situation would look like what I have drawn in Figure 1.

According to the account of reduction functions and correspondence rules presented in this paper, a picture of the relations of physical optics and the electromagnetic theory before and after Maxwell's reduction is proposed in Figures 2 and 3.[15]

There are two comments that can be made concerning the revisions of the Sellarsian diagram. One is that small globes representing aspects of borrowed theories which warrant the connections of the theory under consideration with observationally accessible effects are required. In the diagram in Figure 2, these are represented by the smallest circles.

Also required is a graphical means of representing the reduction functions, the shift of the correspondence rules, and the distinction between the primary and secondary senses of meaning. Figure 3 shows a modified version of Figure 2, in which the parallel horizontal lines represent synthetic identities. Referential identity is not represented directly, but could be introduced by appropriate projection techniques that would map both senses onto a reference diagram that would be very similar to Fig. 1 (B), with some revision required to introduce the appropriate smaller circles.

Sellars' diagram is correct, according to the position taken in this paper, if it is understood to depict either (1) the referential identity as cited in the above paragraph, or (2) the situation which eventually develops when the reduction functions *become* definitional. They are not definitional when reductions are in the process of early development.

This long analysis of the logic of reduction in physics was undertaken to assist our understanding of what happens when genetics, and other biological sciences, are reduced to physics and chemistry. The utility of the reduction functions and the correspondence rule analysis will be examined in the biological context below.

IV. THE WATSON-CRICK MODEL AND GENETICS

In the 1950's further research on DNA confirmed both the essentials of the Watson-Crick structure for DNA and the mechanism of replication which they had proposed.[16] X-ray diffraction studies by Wilkins and his associates in 1955 and 1956, using a lithium salt of DNA, confirmed the double helix model, if a slight rearrangement of the bases with respect to the axis was made (Wilkins, 1956). Work by Taylor, Woods, and Hughes (1957) on the semi-conservative replication of chromosomes in the root tips of *Vicia faba* supported the hypothesis of self duplication which Watson and Crick had advanced. Meselson and Stahl (1958) used DNA from *E. coli* that was tagged with a heavy isotope of nitrogen to examine DNA reduplication. Their conclusion was that "the results of the… experiment are in exact accord with the expectations of the Watson-Crick model for DNA duplication." Experiments by Kornberg and his associates in the late 1950's on *in vitro* synthesis of DNA indicated from a nearest neighbour analysis that the DNA was replicating in accordance with the Watson-Crick predictions (Kornberg, 1960). In the early 1960's Cairns (1963), using T2 phage and *E. coli* was able to obtain auto-radiographs of DNA in the act of replicating. The visible fork structure and the analysis of the density of grains in the emulsion supported the Watson-Crick theory of replication.

These experiments, taken together with Avery, MacLeod, and McCarty's and the Hershey-Chase experiments, support the thesis that DNA is the genetic material. Numerous other experiments, for example in the area of chemically induced mutagenesis, also support this identification. Such experiments play the role of providing the empirical support for reduction functions involved in the reduction of biology, such as:

$$\text{gene}_1 = \text{DNA sequence}_1 .$$

Now, if the gene is DNA, it should be possible to construct a relation – such as was discussed earlier for light waves and electromagnetic waves – that would appropriately identify specific types of genes with specific types of DNA sequences. One of the questions that biology faced in the late 1950's was whether genetics – a purely biological science using biological methods of crossings and countings – could be so related to the chemical theory of DNA. The development of a 'fine structure genetics' in the 1950's opened up the possibility of establishing identifications between biology and chemistry.

Pontecorvo's, Demerec's and Benzer's work on the gene indicated that the precision or resolving power of genetics' methods could be so increased as to make possible the determination of the internal structure of the gene.[17] Benzer's work on the T4 phage of *E. coli* led him to suggest that the classical definition of the gene was inadequate but could be revised. Heretofore the gene had been alternatively characterised as the smallest unit of mutation, the smallest unit of recombination, and as a unit of function – in the sense of being functionally responsible for the production of a unit character. Benzer's (1957) studies on phage indicated that the unit of function (christened a 'cistron' by Benzer) was analysable into a number of sub-units of mutation ('mutons') and recombination ('recons'). Fine structure mappings continued to show linearity, as earlier genetic maps had, and the one-dimensional linearity of both the Watson-Crick model and the fine structure genetic maps suggested an identity.

In the light of the Watson-Crick model, Benzer considered the possibility of translating his 'biological' genetics into chemical terms – at least with respect to the number of base pairs of DNA in the various genetic units. Though there were certain difficulties of relating distances on genetic maps to DNA structure, Benzer was able to conclude that his evidence indicated that the muton was no longer than 5 nucleotide pairs, the recon about two, and the cistron about several hundred such pairs. In a later popular presentation of his findings, Benzer (1962) wrote that "everything that we have learned about the genetic fine structure of T_4 phage is compatible with the Watson-Crick model of the DNA molecule."

V. PROTEIN SYNTHESIS AND THE GENETIC CODE

Compatibility of the Watson-Crick model and genetics, and the *possibility* of an identification of genes with DNA sequences do not in themselves suffice to warrant a reduction of genetics to chemistry. In accordance with the account of the logic and methodology of reduction as outlined earlier, a reduction of genetics to chemistry would require an experimentally based set of identifications relating the basic vocabularies and principles of genetics and chemistry. This, when its meaning is analysed in the context of genetics and chemistry, would require (1) that specific genes be identified with specific purine and pyramidine sequences, (2) that the phenotype of the organism be identified with some chemical structure(s), and (3) that the chemical processes relating the DNA sequences to the chemically characterised phenotype be explicated. (3) is required in addition to (1) and (2) in order to satisfy a generalisation of Nagel's 'derivability condition' (Nagel, 1961, and Schaffner, 1967a). Even if such were possible, and in my analysis it isn't, simple 'definability' would not suffice here because an intricate mechanism of genetic translation of DNA to amino acid sequences is needed.

Many of these requirements were in fact met (in outline form) in the late 1950's and early 1960's by biologists and chemists working under the general rubrics of protein synthesis, the genetic code, and the colinearity hypothesis. It will be impossible in these pages to do more than give the barest outline of these developments as they relate to the problem of the reduction of biology to chemistry.

(i) *Protein Synthesis*. Experiments by many researchers have established what has been referred to as the 'central dogma' of protein synthesis. Less a dogma than a fairly well confirmed hypothesis, it states with respect to protein synthesis, that the DNA transfers coding information to RNA (ribonucleic acid) which in turn translates it into amino acid chain or a protein *via* ribosomal action. The transfer is one-way and non-reciprocal. By the middle of the 1950's, it was experimentally known that RNA was involved in protein synthesis, but it was Crick who in 1957 first predicted on theoretical grounds that at least two different types of RNA must be involved in protein synthesis. Shortly thereafter, Hoagland, Zameenik, and Stephenson (1957) discovered the second type of RNA which has come to be called transfer RNA or t-RNA, or in some literature, soluble or sRNA. Without going into the specific contributions of the various researchers, I propose to give the outlines of the current account of protein synthesis which does seem to support the 'central dogma'.

One chain of the double helix directs the synthesis of a 'complementary' messenger RNA, or m-RNA chain alongside itself. This m-RNA carries the instructions for protein synthesis to the ribosomes which are located outside the cell's nucleus in the cytoplasm. Here the shorter t-RNA molecules bring specific amino acids – the building blocks of proteins – to the ribosomes where a merger of the two types of RNA, with the assistance of still another type of RNA – (ribosomal RNA) – and a number of specialised enzymes, results in specific protein synthesis directed by m-RNA. This process is considerably more complex than I have indicated here, but it would carry us beyond the scope of the paper to discuss it in more detail.

(ii) *The Genetic Code*. That there should be a specific sequence of nucleotides that would be the codon for a specific amino acid grew out of the Watson-Crick DNA model, though the motivation for finding such a code was influenced by Pauling's (1949) work on the haemoglobin molecule which was later extended by Ingram (1956). Many types of codes were proposed in the 1950's and 1960's, with the ultimate acceptance of a degenerate or many-one codon-amino acid relation, in which reading is initiated and terminated by punctuation (other codons) along the sequence of nucleotides. Though there was a good deal of pessimism about the cracking of the genetic code in the late 1950's, when it was thought that much arduous work on the effects of mutagenic agents on simple organisms such as T4 phage would be the key to the code, a breakthrough by Nirenberg and Matthaei (1961) has resulted in a complete determination of the codon-amino acid relations.[18] Specific sequences of DNA are now known to be related to specific amino acid sequences, or proteins.

(iii) *The Colinearity Hypothesis*. The colinearity of the genetic map, the DNA sequence, and the amino acid polypeptide sequence that the gene (=DNA) produces has been established by Kaiser (1962) and Yanofsky and his associates (1964). As a result of experiments on phage chromosomes Kaiser concluded that the "genetic map sequence and physical sequence of cistrons in the DNA molecule are equivalent...." In a later paper Yanofsky and his co-workers demonstrated that gene structure and protein structure were co-linear, that is, that the gene order read off a genetic map is associated in a one-to-one correspondence with a protein order constituted out of a polypeptide chain of amino acids. Yanofsky *et al.*'s conclusion was that "genetic recombination values are representative of distances between amino acid residues in corresponding proteins." Such studies as Kaiser's and Yanofsky *et al.*'s provide evidence for specific connections between genetics and chemistry, and dovetail very nicely with the accomplishments cited under (1) and (2) above.

As regards the relationship of the phenotypes of genetics and the language of chemistry, I refer to Crick's (1958) statement that "the amino acid sequences of the proteins of an organism... are the most delicate expression possible of the phenotype of an organism...." Protein analyses have become useful tools in recent years for taxonomical classification, though there are serious questions whether such analyses are very useful in the evolutionary studies which are related to taxonomical considerations, as there are no fossil records of polypeptide chain sequence.

Molecular geneticists are now working on the problems of cell differentiation and the development of multicellular systems. Jacob and Monod's (1961) classic article suggested a 'model' by which collections of genes could be activated and repressed, and with certain minor changes and complications, their model has given biologists good reasons to believe that developmental and cytodifferentiation problems will be solved at the molecular level.[19]

The conditions for a reduction of genetics to physics and chemistry are in the process of being fulfilled. There are, to be sure, many unsolved problems – but the context of such problems and the way they are thus stated indicates that they are soluble at the molecular level. We see in the movement towards such a reduction, moreover, evidence for the analysis of reduction that was presented in connection with Maxwell's theory above. The use of synthetic identities – on both the entity and predicate levels – which move toward definitional status are being developed. In addition to characterising 'genes' in chemical terms, predicates such as 'dominant' are explicated in chemical terms. The utilisation of the causal sequence analysis of correspondence rules seems implicit in molecular biologists' use of the biochemical and biophysical techniques of autoradiography, paper chromatography, electrophoresis, and X-ray diffraction analysis, in effecting their reduction. A realistic analysis of theoretical entities, e.g. gene, which seems to be an important presupposition of the analysis of reduction given in this paper, is clearly in evidence. Finally, though there is a continual shift and alteration in both reduced and reducing theories, reduction is not ruled out by such shifts, but rather seems aided.

I think that the success of such a reduction as it has thus far proceeded in connection with one of the most central and well developed theories of biology, i.e. genetics, indicates that we have good grounds for suspecting that the remainder of biology will also admit of a chemical reduction.

VI. REDUCTION IN BIOLOGY

Though the reduction of biology to chemistry argues that biology is nothing more than chemistry, to make this assertion without further clarification might well lead us to overlook a most important feature of those chemical systems which are biological organisms – namely their organisation or structure. It is the systematisation of the chemical complexes which constitute living things which makes them so different from non-living entities. Attention to this systematisation leads to a distinction between two types of reduction.

This distinction is based on the difference between two types of aggregation of the parts which constitute an entity to be reduced. In the physical optics-electrodynamics example, a relatively simple and straightforward identification can be asserted to hold: there are not parts of wholes which themselves are independently characterisable. One entity, a light wave, is identified with *one* other entity, an electromagnetic wave – the electric and magnetic fields being interdetermined in Maxwell's theory.

As soon as there are independently characterisable parts, however, as there certainly are in biochemistry, permutations increase and often some type of order of the parts is specified in advance. For example even in the explanation of the chemical notion of the covalent bond by physics, Heitler and London considered two *atoms*, not four elementary particles. Physics, however, is interested in any arrangement of these four particles, and can explain the behaviour of the particles under any arrangement of the parts, as well as the specific reasons why such arrangements have occurred.

Biologists, however, like chemists, are interested only in certain arrangements of the parts. Loss of structure, change of arrangement, disruption of a system, often means the death of an organism. In biology, the structure of the chemical parts of the organism is extremely complex and of paramount importance for the biological properties. In molecular genetics, moreover, the reasons for the arrangement of the chemical sequence of the DNA molecules cannot be given without referring to another living organism which has passed on its – and its ancestor's – organisation.

Recall that in their second paper Watson and Crick drew attention to the chemical irregularity that was permitted in the DNA base sequence: the laws of chemistry did not require any *particular* sequence of nucleotides. In a sense it is the chemically permissible irregularity that permits the biologically relevant coding. The coding, however, is nothing *other than chemical*, and there is absolutely no reason to suppose that the coding sequence was not determined by physical and chemical conditions. The theory of evolution, numerous experiments on mutagenesis, and the experiments on spontaneous synthesis of the amino acids under primitive earth conditions, all suggest that the structure of living molecules is the result of the interaction of many impinging physical and chemical systems. Though it is clearly the task of evolutionary molecular biology to determine how self-replicating, mutable molecules arose on earth, it is almost certain that the *specific* historical facts will never be known – they are lost to time.[20]

That structure or arrangement of the parts is important is not unique to biology. Given a collection of electronic components – say a capacitor, a resistor, a coil, and an e.m.f., many combinations of these components are possible – several of which will be quite different in electrical behaviour from other systems constructed out of the same components arranged in a different way. In order to explain such electronic systems, the arrangement of the parts must be *taken as given*. Similarly, in the case of a biological system's DNA, we must take the nucleotide sequence as given. In the case of protein synthesis theory, we must take a whole chemical factory as given and perhaps nothing less than the cell will really do, though to be sure protein synthesis can be accomplished in cell-free systems.

Though the chemically specifiable co-operative interaction of enzymes, ribosomes, histones, etc., are essential, it is the nucleotide DNA sequence which is of paramount importance. The primary structure of the DNA codes the primary structure of the proteins. Perutz (1962) has analysed the cases of the three-dimensional secondary and tertiary structures of proteins and has concluded that these are determined by the primary structure which we know is determined by the DNA base sequence. It seems quite likely that as yet unanalysed organisation will be accounted for in terms of control genes and other chemically characterisable mechanisms such as histone-DNA interaction.

Biology studies highly organised historically evolved chemical systems. In contrast with the physical sciences it must take the structure of its organisms as given, though it hopes, by means of plausibility arguments connected with chemical evolutionary studies to give an account of how such highly organised systems came to be.[21,22]

Nowhere is there any evidence that there is something unique to these systems – something that would make biology an autonomous science that was irreducible to physics and chemistry. Nevertheless, because of the fact that biology is principally concerned with those arrangements of chemical parts that are living, and because this arrangement must be cited in addition to any listing of the chemical parts of a biological organism, reducing of biology to physics and chemistry possesses distinctive features not normally found in other reductions in the natural sciences.

Two kinds of reduction can then be distinguished: (1) the *simple aggregative* in which the relationship of the parts (if there are parts) plays no role, and (2) what I will term the *interactive*, in which the relationship among the parts (the parts being fully characterisable in terms of the reducing science) is not rigidly dictated by considerations of the reducing science, but which we have good reason to believe is the physically and chemically explicable result of interactions over a long period of time with other physical and chemical conditions. If I were to try and wrap the point up in a maxim, I would say that it is not that the whole is greater than the sum of its parts – it is only that the type and conditions of summation have often been disregarded.[23]

VII. CONCLUSION

The outcome of this account of the development of molecular genetics – which I have characterised as being both stimulated and unified by the Watson-Crick model of DNA – is to warrant as a working hypothesis a *biological principle of reduction.* This principle, it seems, holds not only for genetics, but also for other biological theories. The principle can be stated as follows: given an organism composed out of chemical constituents, the present behaviour of that organism is a function of the constituents as they are characterisable in isolation plus the topological causal inter-structure of the chemical constituents. (The environment must of course, in certain conditions, be specified.) The term 'topological causal interstructure' is a phrase introduced to designate both the spatial arrangements – as chemically relevant as in enzymes' action, or physically relevant, as in the lever effect of bones – as well as the causal signal influences by which one molecule or group of molecules can influence another as in a hormonal enzyme system. The phrase is left deliberately vague to allow for further types of physico-chemical connections to be specified. In the case of the reduction of genetics to physics and chemistry, the *primary* structure of the DNA sequence and the amino acid sequences have been for good reason heavily emphasised. But in many cases of reduction it will not be necessary to go back to such a 'linear' sequence, and in these cases the higher order structuring or systematisation may be taken as given, to function as initial conditions, describable in physical and chemical languages in the physico-chemical explanation of a biological system.

Now such a general principle of reduction, though it asserts that, *ultimately*, biological organisms are nothing but chemical systems, does not entail the consequence that living organisms can *only* be fruitfully studied *as* chemical systems. Just as thermodynamics can macroscopically study systems which are microscopically characterisable by statistical mechanics, so can anatomy, neurology, genetics, and so on, study the living organism in each discipline's own terms (Schaffner, 1969). Should such macroscopic study entail results that seriously conflict with molecular theories, a reassessment on both levels may be required. The lesson drawn from fine structure genetics and the DNA model, however, should lead us to suspect that one will find agreement rather than conflict in such confrontations.

University of Pittsburgh

NOTES

* An earlier version of this paper was read to the 155th National Meeting of the American Chemical Society, San Francisco, April 1968.

[1] I shall often shift between talking about the reduction of a theory and the reduction of a science. This shifting is deliberate and is context dependent. Some of the rationale for talking in this way will be discussed in the text. The shifting is necessitated by the analysis of theory, correspondence rule, and the unsolved problems in a scientific 'domain' (see p. 115, n. 11) used in this paper.

[2] For example, see Zubay's (1968), pp. 151ff., discussion on 'The Molecular Basis of Mutagenesis' for an account of mutation theory and its relation to the DNA sequence.

[3] See my (1967a) for references. I shall refer to some of these accounts in more detail below.

[4] For the history of this reduction see Whittaker (1960), Chapters 8 and 10. For an account of the logic and methodology of the reduction see my (1967c), Chapter 2.

[5] For example, see Born and Wolf (1959).

[6] The sense of 'entity' used here is that of an 'oscillating field magnitude of quantitatively determinable intensity and direction'. Toward the end of the last century a dispute developed as to whether fields were 'substances' in their own right or only modifications of a very subtle 'material' substance – the electromagnetic aether. By 1900 fields were accepted by many physicists as realtities in their own right. Einstein (1931) wrote concerning this issue:

At the turn of the century the conception of the electromagnetic field as an irreducible entity was already generally accepted and the more serious theorists had ceased to believe in the necessity of ... providing [Maxwell's equations] with a mechanical basis.

Though for Maxwell, writing in 1862, the synthetic identity utilized in his reduction identified the electromagnetic and optical *aethers* rather than the field intensity vectors, the latter interpretation seems more plausible now because of special relativity theory and the abandonment of the aether. (My forthcoming *Nineteenth Century Aether Theories*, Oxford, Pergamon Press, considers these problems in some detail.)

[7] See both Nagel's (1961) discussion, pp. 64–7, and also Putnam's (1961) essay.

[8] See N. Goodman's (1966), Chapter 1, and his account of 'extensional isomorphism' for a related analysis.

[9] A question could be raised about the adequacy of using extensional identity to relate entities and entity orderings (predicates). Both A. Fine (1967) and I. Scheffler (1967) have suggested in a somewhat different context that similar theoretical terms in different scientific theories, e.g. 'mass' in Newtonian and Einsteinian mechanics, can be considered extensionally related even though the 'meanings' of the terms are somewhat different. M. B. Hesse (1968a, b) has criticised both Fine and Scheffler and argued that extentional identity is neither necessary nor sufficient to obtain the requisite connection between scientific theories that would insure substitutivity of terms and logical comparability of theories. The problem with which these authors are concerned, however, is somewhat different from the problem in my account of theory relation which necessitates extensional relation. The difficulty for Fine and Scheffler centres about terms which are employed in sentences which are contradictory, e.g. 'Mass (Newtonian) is independent of relative velocity' and 'It is not the case that Mass (Einsteinian) is independent of relative velocity'. Since the terms are different and yet not *that* different, Fine and Scheffler wish to find some way of relating the terms. In my account of reduction, such terms cannot be logically related because the relation is one of similarity. But there are other terms, such as 'light vector' and 'oscillating electric field intensity between frequencies r and v' which can be related in non-contradictory ways. It must be recalled that the 'light vector' term may well have to change meaning so that it is not involved in such contradictory statements. Such a relation however requires extensional characterisation since the meanings are not the same: They differ in the way that the meanings of 'Tully' and 'Cicero' differ, but not in the way that the meanings of 'Newtonian mass' and 'Einsteinian mass' differ. In my formalism, the shift from T_2 to T_2^* permits requisite change in the reduced theory. Since the identities are asserted as empirical hypotheses, there is no guarantee that they are correct and they may well be refuted as science progresses.

[10] See p. 114, n. 1.

[11] I am indebted for a discussion of the notion of a 'domain' and its relation to certain adequacy criteria to Dudley Shapere's 'Discovery, Rationality, and Progress in Science:

A Perspective in the Philosophy of Science', PSA 1972, *Boston Studies in the Philosophy of Science*, Vol. XX, 1974.

[12] See Feyerabend (1965) for discussion, with references, of the relevance of Szilard's work.

[13] This is in my paper 'Correspondence Rules', *Philosophy of Science* **36** (1969) 280–90.

[14] It would be impossible in this paper to consider this modified Duhemian thesis in any detail. Suffice it to say that though the position taken here does admit the difficulty – and perhaps impossibility – of establishing relations between the actions of theoretical entities and the observable consequences without using other theories in tracing out the consequences, it does not agree with the position sometimes ascribed to Duhem (1962) and also defended by Quine (1961) that a specific hypothesis can be held, come what may, by suitably altering these other theories.

[15] The diagrams should not be taken as anything more than rough analogies. The complexity of actual science would undoubtedly require a multi-dimensional diagram to depict relations between all theories and observations with any real accuracy. I have left some of the small circles representing theories blank because I did not wish to give any suggestion of completeness in this scheme. I do believe, though, that the diagrams represent the process of theory reduction in a simplified but not misleading manner.

[16] Readers interested in more complete accounts of the historical topics discussed in this paper may refer to Carlson's (1966), Whitehouse's (1965) and Ravin's (1965) works. A series of memoirs edited by Cairns, Stent, and Watson (1966) is important, as is Stent's (1968) article. Taylor's (1965) and Zubay's (1968) anthologies are excellent collections of original papers in the fields discussed.

[17] See E. A. Carlson (1966), Chapter 22, for a historical discussion of fine structure genetics.

[18] The translation of the DNA code involved the work of a number of investigators. Especially see Whitehouse's account (1965) of the contributions of Ochoa's laboratory.

[19] I do not wish to mislead anyone on this issue of control genetics by implying that the *problems* are all that clear, much less the possible solutions. For recent discussions of cell differentiation and the role of histones in more elaborate control mechanisms than Jacob and Monod's (1961) account, see Bonner's (1965) book and also Zubay's introduction, pp. 463–7, in his (1968).

[20] See my 'Chemical Systems and Chemical Evolution: The Philosophy of Molecular Biology', *American Scientist* **57** (1969) 410–20, for detailed comments on abiogenesis, pre-biological systems, and chemically characterisable natural selection. My general thesis concerning organisation and evolution was first proposed in my (1967b, c) papers.

[21] *Ibid.*

[22] See Oparin (1962) and Jukes (1966) for rather different approaches to comparative evolutionary biochemistry.

[23] I think these paragraphs express the crux of a divergence of opinion between other authors and myself over the role of organisation and 'initial conditions' in biological reduction. See my (1967b) and also the exchange of letters in *Science* **158**, 857–62 (1967). Michael Polanyi has noted the role of what he terms 'boundary conditions' and 'structure' in the reduction of biology in his (1967) and (1968). My paper referred to in n. 19 above constitutes an indirect answer to some of his contentions, but it might be well to note here that it seems to me that Polanyi commits two basic errors in his analyses: (1) he assumes that the 'initial conditions', to use my phraseology, describing the organisation require an explanation if the explanation in which they figure is to be

legitimate, and (2) he seems to think that the employment of such initial conditions to catch the 'order' present in either a machine or a biological organism is evidence of some sort of emergence. The latter thesis only follows if there are good reasons to believe that these initial conditions are the outcome of conditions that are not themselves characterisable in the language of physics and chemistry and explainable by physics and chemistry. Not only is this something which he has not shown, but contemporary biological research is attempting to show in a rough way that such order as is present can be accounted for *via* chemically characterisable evolution. See n. 20 above for references to this research.

BIBLIOGRAPHY

Avery, O. T., MacLeod, C. M., and McCarty, M.: 1944, 'Studies on the Chemical Nature of the Substance Inducing Transformation of Pneumococcal Types', *J. Exp. Med.* **79**, 137.

Benzer, S.: 1957, 'The Elementary Units of Heredity', *The Chemical Basis of Heredity* (eds. W. D. McElroy and B. Glass), Baltimore: Johns Hopkins.

Benzer, S.: 1962, 'The Fine Structure of the Gene', *Scientific American* **206**, 70.

Bonner, J.: 1965, *The Molecular Biology of Development*, Oxford: Clarendon Press.

Born, M. and Wolf, E.: 1959, *Principles of Optics*, New York: Pergamon.

Cairns, J.: 1963, 'An Estimate of the Length of the DNA Molecule of T2 Bacteriophage by Autoradiography', *J. Mol. Biol.* **6**, 208.

Cairns, J., Stent, G. S., and Watson, J. D. (eds.): 1966, *Phage and the Origins of Molecular Biology*, Cold Spring Harbor, New York: Cold Spring Harbor Laboratory of Quantitative Biology.

Carlson, E. A.: 1966, *The Gene: A Critical History*, Philadelphia: W. B. Saunders.

Chargaff, E.: 1950, 'Chemical Specificity of Nucleic Acids and Mechanism of Their Enzymatic Degradation', *Experentia* **6**, 201.

Crick, F. H. C.: 1957, 'The Structure of Nucleic Acids and Their Role in Protein Synthesis', *Biochem. Soc. Symp.* **14**, 25.

Crick, F. H. C.: 1958, 'On Protein Synthesis', *Symp. Soc. Exp. Biol.* **12**, 138.

Duhem, P.: 1962, *The Aim and Structure of Physical Theory.* Trans. by P. P. Wiener from 1914 edition. New York: Atheneum.

Einstein, A.: 1931, 'Maxwell's Influence on the Development of the Conception of Physical Reality', *James Clerk Maxwell: A Commemoration Volume 1831–1931.* Cambridge: Cambridge University Press.

Feyerabend, P. K.: 1962, 'Explanation, Reduction, and Empiricism', *Minnesota Studies in the Philosophy of Science. Vol. 3* (eds. H. Feigl and G. Maxwell), Minneapolis: University of Minnesota Press.

Feyerabend, P. K.: 1965, 'Reply to Criticism', *Boston Studies in the Philosophy of Science. Vol. 2* (eds. R. S. Cohen and M. W. Wartofsky), New York: Humanities Press.

Fine, A.: 1967, 'Consistency, derivability, and Scientific Change', *J. Phil.* **64**, 231.

Goodman, N.: 1966, *The Structure of Appearance*, 2nd edn. Indianapolis: Bobbs-Merrill.

Hershey, A. D. and Chase, M.: 1952, 'Independent Functions of Vital Protein and Nucleic Acid in Growth of Bacteriophage', *J. Gen. Physiol.* **36, 39.**

Hertz, H.: 1890, 'On the Fundamental Equations of Electromagnetics for Bodies at

Rest' Trans. by D. E. Jones from *Göttinger Nachr.* (1890) in *Electric Waves* (1893/1962). New York: Dover.

Hesse, M. B.: 1968a, 'Fine's Criteria for Meaning Change', *J. Phil.*, **65**, 46.

Hesse, M. B.: 1968b, 'Review of I. Scheffler's', *Science and Subjectivity, B.J.P.S.* **19**, 176.

Hoagland, M. B., Zamecnik, P. C., and Stephenson, M. L.: 1957, 'Intermediate Reactions in Protein Biosynthesis', *Biochem. Biophys. Acta* **24**, 215.

Ingram, V. M.: 1956, 'A Specific Chemical Difference Between the Globins of Normal Human and Sickle-Cell Anaemia Haemoglobin', *Nature* **178**, 792.

Jacob, F. and Monod, J.: 1961, 'Genetic Regulatory Mechanisms in the Synthesis of Proteins', *J. Mol. Biol.* **3**, 318.

Jukes, T.: 1966, *Molecules and Evolution*, New York: Columbia University Press.

Kaiser, A. D.: 1962, 'The Production of Phage Chromosome Fragments and Their Capacity for Genitic Transfer', *J. Mol. Biol.* **4**, 275.

Kornberg, A.: 1960, 'Biologic Synthesis of Deoxyribonucleic Acid', *Science* **131**, 1503.

Kuhn, T. S.: 1962, *The Structure of Scientific Revolutions*, Chicago: University of Chicago Press.

Maxwell, J. C.: 1861, 'On Physical Lines of Force', *Phil. Mag.* **21**, 161, 281, 338. Also continued in (1862) *Phil. Mag.* **23**, 12, 85.

Maxwell, J. C.: 1865, 'A Dynamical Theory of the Electromagnetic Field', *Royal Soc. Phil. Trans.* **65**, 459.

Maxwell, J. C.: 1968, 'On a Method of Making a Direct Comparison of Electrostatic with Electromagnetic Force; with a Note on the Electromagnetic Theory of Light', *Royal Soc. Phil. Trans.* **158**, 643.

Meselson, M. and Stahl, F. W.: 1958, 'The Replication of DNA in *Escherida coli.*', *Proc. Natl. Acad. Sci. U.S.* **44**, 678.

Nagel, E.: 1961, *The Structure of Science*, New York: Harcourt, Brace, and World.

Nirenberg, M. and Matthaei, J. H.: 'The Dependence of Cell-Free Protein Synthesis in *E. coli* upon Naturally Occurring or Synthetic Polyribonucleotides', *Proc. Natl. Acad. Sci. U.S.* **47**, 1588.

Oparin, A. I.: 1962, *Life: Its Nature, Origin, and Development*, Trans. by A. Synge. New York: Academic Press.

Pauling, L., Itano, H. A., Singer, G. J., and Wells, I. C.: 1949, 'Sickle Cell Anemia, a Molecular Disease', *Science* **110**, 543.

Perutz, M. F.: 1962, *Proteins and Nucleic Acids: Structure and Function*, Amsterdam: Elsevier.

Polanyi, M.: 1967, 'Life Transcending Physics and Chemistry', *Chem. Eng. News* **45**, 54.

Polanyi, M.: 1968, 'Life's Irreducible Structure', *Science* **160**, 1308.

Putnam, H.: 1962, 'The Analytic and the Synthetic', *Minnesola Studies in the Philosophy of Science. Vol. 3* (Eds. H. Feigel and G. Maxwell), Minneapolis: University of Minnesota Press.

Quine, W. V.: 1961, 'Two Dogmas of Empiricism', *From a Logical Point of View*, 2nd edn., Cambridge: Harvard University Press.

Ravin, A. W.: 1965, *The Evolution of Genetics*, New York: Academic Press.

Schaffner, K. F.: 1967a, 'Approaches to Reduction', *Phil. Sci.* **34**, 137.

Schaffner, K. F.: 1967b, 'Antireductionism and Molecular Biology', *Science* **157**, 644.

Schaffner, K. F.: 1967c, 'The Logic and Methodology of Reduction in the Physical and Biological Sciences', unpublished Ph.D. dissertation, Columbia University.

Schaffner, K. F.: 1969, 'Theories and Explanations in Biology', *J. Hist. Biol.* **2**, 19.

Scheffler, I.: 1967, *Science and Subjectivity*, Indianapolis: Bobbs-Merrill.

Sellars, W.: 1963, 'Theoretical Explanation', *Philosophy of Science: The Delaware Seminar* (Ed. B. Baumrin), New York: Interscience.

Stent, G. S.: 1968, 'That Was the Molecular Biology that Was', *Science* **160**, 380.

Taylor, J. H., Woods, P. S., and Hughs, W. L.: 1957, 'The Organization and Duplication of Chromosomes as Revealed by Autoradiographic Studies Using Tritium-Labeled Thymine', *Proc. Natl. Acad. Sci. U.S.* **43**, 122.

Taylor, J. H.: 1965, *Selected Papers on Molecular Genetics*, New York: Academic Press.

Watson, J. D.: 1968, *The Double Helix*, New York: Atheneum.

Watson, J. D. and Crick, F. H. C.: 1953a, 'Molecular Structure of Nucleic Acids: a Structure for Deoxyribose Nucleic Acid', *Nature* **171**, 737.

Watson, J. D. and Crick, F. H. C.: 1953b, 'Genetical Implications of the Structure of Deoxyribose Nucleic Acid', *Nature* **171**, 964.

Whitehouse, H. L. K.: 1965, *Towards an Understanding of the Mechanism of Heredity*, New York: St. Martin's Press.

Whittaker, E. T.: 1960, *A History of the Theories of Aether and Electricity*, Vol. I, New York: Harper and Brothers.

Wilkins, M. H. F.: 1956, 'Physical Studies of the Molecular Structure of Deoxyribose Nucleic Acid and Nucleoprotein', *Cold Spring Harbor Symp. Quant. Biol.* **21**, 75.

Yanofsky, C., Carlton, B. C., Guest, J. R., Helsinki, D. R., and Henning, U.: 1964, 'On the Colinearity of Gene Structure and Protein Structure', *Proc. Natl. Acad. Sci. U.S.* **51**, 266.

Zubay, G. L.: 1968, *Papers in Biochemical Genetics*, New York: Holt, Rinehart, and Winston.

LIFE'S IRREDUCIBLE STRUCTURE

If all men were exterminated, this would not affect the laws of inanimate nature. But the production of machines would stop, and not until men arose again could machines be formed once more. Some animals can produce tools, but only men can construct machines; machines are human artifacts, made of inanimate material.

The *Oxford Dictionary* describes a machine as "an apparatus for applying mechanical power, consisting of a number of interrelated parts, each having a definite function." It might be, for example, a machine for sewing or printing. Let us assume that the power driving the machine is built in, and disregard the fact that it has to be renewed from time to time. We can say, then, that the manufacture of a machine consists in cutting suitably shaped parts and fitting them together so that their joint mechanical action should serve a possible human purpose.

The structure of machines and the working of their structure are thus shaped by man, even while their material and the forces that operate them obey the laws of inanimate nature. In constructing a machine and supplying it with power, we harness the laws of nature at work in its material and in its driving force and make them serve our purpose.

This harness is not unbreakable; the structure of the machine, and thus its working, can break down. But this will not affect the forces of inanimate nature on which the operation of the machine relied; it merely releases them from the restriction the machine imposed on them before it broke down.

So the machine as a whole works under the control of two distinct principles. The higher one is the principle of the machine's design, and this harnesses the lower one, which consists in the physical-chemical processes on which the machine relies. We commonly form such a two-leveled structure in conducting an experiment; but there is a difference between constructing a machine and rigging up an experiment. The experimenter imposes restrictions on nature in order to observe its behavior under these restrictions, while the constructor of a machine restricts nature in order to harness its workings. But we may borrow a term from physics and describe both these useful restrictions of nature as the imposing of *boundary conditions* on the laws of physics and chemistry.

Let me enlarge on this. I have exemplified two types of boundaries. In the machine our principal interest lay in the effects of the boundary conditions, while in an experimental setting we are interested in the natural processes controlled by the boundaries. There are many common examples of both types of boundaries. When a saucepan bounds a soup that we are cooking, we are interested in the soup; and, likewise, when we observe a reaction in a test tube, we are studying the reaction, not the test tube. The reverse is true for a game of chess. The strategy of the player imposes boundaries on the several moves, which follow the laws of chess, but our interest lies in the boundaries – that is, in the strategy, not in the several moves as exemplifications of the laws. And similarly, when a sculptor shapes a stone or a painter composes a painting, our interest lies in the boundaries imposed on a material, and not in the material itself.

We can distinguish these two types of boundaries by saying that the first represents a test-tube type of boundary whereas the second is of the machine type. By shifting our attention, we may sometimes change a boundary from one type to another.

All communications form a machine type of boundary, and these boundaries form a whole hierarchy of consecutive levels of action. A vocabulary sets boundary conditions on the utterance of the spoken voice; a grammar harnesses words to form sentences, and the sentences are shaped into a text which conveys a communication. At all these stages we are interested in the boundaries imposed by a comprehensive restrictive power, rather than in the principles harnessed by them.

I. LIVING MECHANISMS ARE CLASSED WITH MACHINES

From machines we pass to living beings, by remembering that animals move about mechanically and that they have internal organs which perform functions as parts of a machine do – functions which sustain the life of the organism, much as the proper functioning of parts of a machine keeps the machine going. For centuries past, the workings of life have been likened to the working of machines and physiology has been seeking to interpret the organism as a complex network of mechanisms. Organs are, accordingly, defined by their life-preserving functions.

Any coherent part of the organism is indeed puzzling to physiology – and also meaningless to pathology – until the way it benefits the organism is discovered. And I may add that any description of such a system in terms of its physical-chemical topography is meaningless, except for the fact that the description covertly may recall the system's physiological interpretation – much as the topography of a machine is meaningless until we guess how the device works, and for what purpose.

In this light the organism is shown to be, like a machine, a system which works according to two different principles: its structure serves as a boundary condition harnessing the physical-chemical processes by which its organs perform their functions. Thus, this system may be called a system under dual control. Morphogenesis, the process by which the structure of living beings develops, can then be likened to the shaping of a machine which will act as a boundary for the laws of inanimate nature. For just as these laws serve the machine, so they serve also the developed organism.

A boundary condition is always extraneous to the process which it delimits. In Galileo's experiments on balls rolling down a slope, the angle of the slope was not derived from the laws of mechanics, but was chosen by Galileo. And as this choice of slopes was extraneous to the laws of mechanics, so is the shape and manufacture of test tubes extraneous to the laws of chemistry.

The same thing holds for machine-like boundaries; their structure cannot be defined in terms of the laws which they harness. Nor can a vocabulary determine the content of a text, and so on. Therefore, if the structure of living things is a set of boundary conditions, this structure is extraneous to the laws of physics and chemistry which the organism is harnessing. Thus the morphology of living things transcends the laws of physics and chemistry.

II. DNA INFORMATION GENERATES MECHANISMS

But the analogy between machine components and live functioning organs is weakened by the fact that the organs are not shaped artificially as the parts of a machine are. It is an advantage, therefore, to find that the morphogenetic process is explained in principle by the transmission of information stored in DNA, interpreted in this sense by Watson and Crick.

A DNA molecule is said to represent a code – that is, a linear sequence of items, the arrangement of which is the information conveyed by the code. In the case of DNA, each item of the series consists of one out of four alternative organic bases.[1] Such a code will convey the maximum amount of information if the four organic bases have equal probability of forming any particular item of the series. Any difference in the binding of the four alternative bases, whether at the same point of the series or between two points of the series, will cause the information conveyed by the series to fall below the ideal maximum. The information content of DNA is in fact known to be reduced to some extent by redundancy, but I accept here the assumption of Watson and Crick that this redundancy does not prevent DNA from effectively functioning as a code. I accordingly disregard, for the sake of brevity, the redundancy in the DNA code and talk of it as if it were functioning optimally, with all of its alternative basic bindings having the same probability of occurrence.

Let us be clear what would happen in the opposite case. Suppose that the actual structure of a DNA molecule were due to the fact that the bindings of its bases were much stronger than the bindings would be for any other distribution of bases, then such a DNA molecule would have no information content. Its codelike character would be effaced by an overwhelming redundancy.

We may note that such is actually the case for an ordinary chemical molecule. Since its orderly structure is due to a maximum of stability, corresponding to a minimum of potential energy, its orderliness lacks the capacity to function as a code. The pattern of atoms forming a crystal is another instance of complex order without appreciable information content.

There is a kind of stability which often opposes the stabilizing force of a potential energy. When a liquid evaporates, this can be understood as the increase of entropy accompanying the dispersion of its particles. One takes this dispersive tendency into account by adding its powers to those of potential energy, but the correction is negligible for cases of deep drops in potential energy or for low temperatures, or for both. We can disregard it, to simplify matters, and say that chemical structures established by the stabilizing powers of chemical bonding have no appreciable information content.

In the light of the current theory of evolution, the code-like structure of DNA must be assumed to have come about by a sequence of chance variations established by natural selection. But this evolutionary aspect is irrelevant here; whatever may be the origin of a DNA configuration, it can function as a code only if its order is not due to the forces of potential energy. It must be as physically indeterminate as the sequence of words is on a printed page. As the arrangement of a printed page is extraneous to the chemistry of the printed page, so is the base sequence in a DNA molecule extraneous to the chemical forces at work in the DNA molecule. It is this physical indeterminacy of the sequence that produces the improbability of occurrence of any particular sequence and thereby enables it to have a meaning – a meaning that has a mathematically determinate information content equal to the numerical improbability of the arrangement.

III. DNA ACTS AS A BLUEPRINT

But there remains a fundamental point to be considered. A printed page may be a mere jumble of words, and it has then no information content. So the improbability count gives the *possible*, rather than the *actual*, information content of a page. And this applies also to the information content attributed to a DNA molecule; the sequence of the bases is deemed meaningful only because we assume with Watson and Crick that this arrangement generates the structure of the offspring by endowing it with its own information content.

This brings us at last to the point that I aimed at when I undertook to analyze the information content of DNA: Can the control of morphogenesis by DNA be likened to the designing and shaping of a machine by the engineer? We have seen that physiology interprets the organism as a complex network of mechanisms, and that an organism is – like a machine – a system under dual control. Its structure is that of a boundary condition harnessing the physical-chemical substances within the organism in the service of physiological functions. Thus, in generating an organism, DNA initiates and controls the growth of a mechanism that will work as a boundary condition within a system under dual control.

And I may add that DNA itself is such a system, since every system conveying information is under dual control, for every such system restricts and orders, in the service of conveying its information, extensive resources of particulars that would otherwise be left at random, and thereby acts as a boundary condition. In the case of DNA this boundary condition is a blueprint of the growing organism.[2]

We can conclude that in each embryonic cell there is present the duplicate of a DNA molecule having a linear arrangement of its bases – an arrangement which, being independent of the chemical forces within the DNA molecules, conveys a rich amount of meaningful information. And we see that when this information is shaping the growing embryo, it produces in it boundary conditions which, themselves being independent of the physical-chemical forces in which they are rooted, control the mechanism of life in the developed organism.

To elucidate this transmission is a major task of biologists today, to which I shall return.

IV. SOME ACCESSORY PROBLEMS ARISE HERE

We have seen boundary conditions introducing principles not capable of formulation in terms of physics or chemistry into inanimate artifacts and living things; we have seen them as necessary to an information content in a printed page or in DNA, and as introducing mechanical principles into machines as well as into the mechanisms of life.

Let me add now that boundary conditions of inanimate systems established by the history of the universe are found in the domains of geology, geography, and astronomy, but that these do not form systems of dual control. They resemble in this respect the test-tube type of boundaries of which I spoke above. Hence the existence of dual control in machines and living mechanisms represents a discontinuity between machines and living things on the one hand and inanimate nature on the other hand, so that both machines and living mechanisms are irreducible to the laws of physics and chemistry.

Irreducibility must not be identified with the mere fact that the joining of parts may produce features which are not observed in the separate parts. The sun is a sphere, and its parts are not spheres, nor does the law of gravitation speak of spheres; but mutual gravitational interaction causes the parts of the sun to form a sphere. Such cases of holism are common in physics and chemistry. They are often said to represent a transition to living things, but this is not the case, for they are reducible to the laws of inanimate matter, while living things are not.

But there does exist a rather different continuity between life and inanimate nature. For the beginnings of life do not sharply differ from their purely physical-chemical antecedents. One can reconcile this continuity with the irreducibility of living things by recalling the analogous case of inanimate artifacts. Take the irreducibility of machines; no animal can produce a machine, but some animals can make primitive tools, and their use of these tools may be hardly distinguishable from the mere use of the animal's limbs. Or take a set of sounds conveying information; the set of sounds can be so obscured by noise that its presence is no longer clearly identifiable. We can say, then, that the control exercised by the boundary conditions of a system can be reduced gradually to a vanishing point. The fact that the effect of a higher principle over a system under dual control can have any value down to zero may allow us also to conceive of the continuous emergence of irreducible principles within the origin of life.

V. WE CAN NOW RECOGNIZE ADDITIONAL IRREDUCIBLE PRINCIPLES

The irreducibility of machines and printed communications teaches us, also, that the control of a system by irreducible boundary conditions does not *interfere* with the laws of physics and chemistry. A system under dual control relies, in fact, for the operations of its higher principle, on the working of principles of a lower level, such as the laws of physics and chemistry. Irreducible higher principles are *additional* to the laws of physics and chemistry. The principles of mechanical engineering and of communication of information, and the equivalent biological principles, are all additional to the laws of physics and chemistry.

But to assign the rise of such additional controlling principles to a selective process of evolution leaves serious difficulties. The production of boundary conditions in the growing fetus by transmitting to it the information contained in DNA presents a problem. Growth of a blueprint into the complex machinery that it describes seems to require a system of causes not specifiable in terms of physics and chemistry, such causes being additional both to the boundary conditions of DNA and to the morphological structure brought about by DNA.

This missing principle which builds a bodily structure on the lines of an instruction given by DNA may be exemplified by the far-reaching regenerative powers of the embryonic sea urchin, discovered by Driesch, and by Paul Weiss's discovery that completely dispersed embryonic cells will grow, when lumped together, into a fragment of the organ from which they were isolated.[3] We see an integrative power at work here, characterized by Spemann and by Paul Weiss as a 'field'[4], which guides the growth of embryonic fragments to form the morphological features to which they embryologically belong. These guides of morphogenesis are given a formal expression in Waddington's 'epigenetic landscapes.'[5] They say graphically that the growth of the embryo is controlled by the gradient of potential shapes, much as the motion of a heavy body is controlled by the gradient of potential energy.

Remember how Driesch and his supporters fought for recognition that life transcends physics and chemistry, by arguing that the powers of regeneration in the sea urchin embryo were not explicable by a machine-like structure, and how the controversy has continued along similar lines, between those who insisted that regulative ('equipotential' or 'organismic') integration was irreducible to any machine-like mechanism and was therefore irreducible also to the laws of inanimate nature. Now if, as I claim, machines and mechanical processes in living beings are themselves irreducible to physics and chemistry, the situation is changed. If mechanistic and organismic explanations are both equally irreducible to physics and chemistry, the recognition of organismic processes no longer bears the burden of standing alone as evidence for the irreducibility of living things. Once the 'field'-like powers guiding regeneration and morphogenesis can be recognized without involving this major issue, I think the evidence for them will be found to be convincing.

There is evidence of irreducible principles, additional to those of morphological mechanisms, in the sentience that we ourselves experience and that we observe indirectly in higher animals. Most biologists set aside these matters as unprofitable considerations. But again, once it is recognized, on other grounds, that life transcends physics and chemistry, there is no reason for suspending recognition of the obvious fact that consciousness is a principle that fundamentally transcends not only physics and chemistry but also the mechanistic principles of living beings.

VI. BIOLOGICAL HIERARCHIES CONSIST OF A SERIES OF BOUNDARY CONDITIONS

The theory of boundary conditions recognizes the higher levels of life as forming a hierarchy, each level of which relies for its workings on the principles of the levels below it, even while it itself is irreducible to these lower principles. I shall illustrate the structure of such a hierarchy by showing the way five levels make up a spoken literary composition.

The lowest level is the production of a voice; the second, the utterance of words; the third, the joining of words to make sentences; the fourth, the working of sentences into a style; the fifth, and highest, the composition of the text.

The principles of each level operate under the control of the next-higher level. The voice you produce is shaped into words by a vocabulary; a given vocabulary is shaped into sentences in accordance with a grammar; and the sentences are fitted into a style, which in turn is made to convey the ideas of the composition. Thus each level is subject to dual control: (i) control in accordance with the laws that apply to its elements in themselves, and (ii) control in accordance with the laws of the powers that control the comprehensive entity formed by these elements.

Such multiple control is made possible by the fact that the principles governing the isolated particulars of a lower level leave indeterminate conditions to be controlled by a higher principle. Voice production leaves largely open the combination of sounds into words, which is controlled by a vocabulary. Next, a vocabulary leaves largely open the combination of words to form sentences, which is controlled by grammar, and so on. Consequently, the operations of a higher level cannot be accounted for by the laws governing its particulars on the next-lower level. You cannot derive a vocabulary from phonetics; you cannot derive grammar from a vocabulary; a correct use of grammar does not account for good style; and a good style does not supply the content of a piece of prose.

Living beings comprise a whole sequence of levels forming such a hierarchy. Processes at the lowest level are caused by the forces of inanimate nature, and the higher levels control, throughout, the boundary conditions left open by the laws of inanimate nature. The lowest functions of life are those called vegetative. These vegetative functions, sustaining life at its lowest level, leave open – both in plants and in animals – the higher functions of growth and in animals also leave open the operations of muscular actions. Next, in turn, the principles governing muscular actions in animals leave open the integration of such actions to innate patterns of behavior; and, again, such patterns are open in their turn to be shaped by intelligence, while intelligence itself can be made to serve in man the still higher principles of a responsible choice.

Each level relies for its operations on all the levels below it. Each reduces the scope of the one immediately below it by imposing on it a boundary that harnesses it to the service of the next higher level, and this control is transmitted stage by stage, down to the basic inanimate level.

The principles additional to the domain of inanimate nature are the product of an evolution the most primitive stages of which show only vegetative functions. This evolutionary progression is usually described as an increasing complexity and increasing capacity for keeping the state of the body independent of its surroundings. But if we accept, as I do, the view that living beings form a hierarchy in which each higher level represents a distinctive principle that harnesses the level below it (while being itself irreducible to its lower principles), then the evolutionary sequence gains a new and deeper significance. We can recognize then a strictly defined progression, rising from the inanimate level to ever higher additional principles of life.

This is not to say that the higher levels of life are altogether absent in earlier stages of evolution. They may be present in traces long before they become prominent. Evolution may be seen, then, as a progressive intensification of the higher principles of life. This is what we witness in the development of the embryo and of the growing child – processes akin to evolution.

But this hierarchy of principles raises once more a serious difficulty. It seems impossible to imagine that the sequence of higher principles, transcending further at each stage the laws of inanimate nature, is incipiently present in DNA and ready to be transmitted by it to the offspring. The conception of a blueprint fails to account for the transmission of faculties, like consciousness, which no mechanical device can possess. It is as if the faculty of vision were to be made intelligible to a person born blind by a chapter of sense physiology. It appears, then, that DNA *evokes* the ontogenesis of higher levels, rather than *determining* these levels. And it would follow that the emergence of the kind of hierarchy I have defined here can be only evoked, and not determined, by atomic or molecular accidents. However, this question cannot be argued here.

VII. UNDERSTANDING A HIERARCHY NEEDS 'FROM-AT' CONCEPTIONS

I said above that the transcendence of atomism by mechanism is reflected in the fact that the presence of a mechanism is not revealed by its physical-chemical topography. We can say the same thing of all higher levels: their description in terms of any lower level does not tell us of their presence. We can generally descend to the components of a lower level by analyzing a higher level, but the opposite process involves an integration of the principles of the lower level, and this integration may be beyond our powers.

In practice this difficulty may be avoided. To take a common example, suppose that we have repeated a particular word, closely attending to the sound we are making, until these sounds have lost their meaning for us; we can recover this meaning promptly by evoking the context in which the word is commonly used. Consecutive acts of analyzing and integrating are in fact generally used for deepening our understanding of complex entities comprising two or more levels.

Yet the strictly logical difference between two consecutive levels remains. You can look at a text in a language you do not understand and see the letters that form it without being aware of their meaning, but you cannot read a text without seeing the letters that convey its meaning. This shows us two different and mutually exclusive ways of being aware of the text. When we look at words without understanding them we are focusing our attention on them, whereas, when we read the words, our attention is directed to their meaning as part of a language. We are aware then of the words only subsidiarily, as we attend to their meaning. So in the first case we are looking at the words, while in the second we are looking *from* them *at their meaning*: the reader of a text has a *from-at* knowledge of the words' meaning, while he has only a *from* awareness of the words he is reading. Should he be able to shift his attention fully toward the words, these would lose their linguistic meaning for him.

Thus a boundary condition which harnesses the principles of a lower level in the service of a new, higher level establishes a semantic relation between the two levels. The higher comprehends the workings of the lower and thus forms the meaning of the lower. And as we ascend a hierarchy of boundaries, we reach to ever higher levels of meaning. Our understanding of the whole hierarchic edifice keeps deepening as we move upward from stage to stage.

VIII. THE SEQUENCE OF BOUNDARIES BEARS ON OUR SCIENTIFIC OUTLOOK

The recognition of a whole sequence of irreducible principles transforms the logical steps for understanding the universe of living beings. The idea, which comes to us from Galileo and Gassendi, that all manner of things must ultimately be understood in terms of matter in motion is refuted. The spectacle of physical matter forming the basic tangible ground of the universe is found to be almost empty of meaning. The universal topography of atomic particles (with their velocities and forces) which, according to Laplace, offers us a universal knowledge of all things is seen to contain hardly any knowledge that is of interest. The claims made, following the discovery of DNA, to the effect that all study of life could be reduced eventually to molecular biology, have shown once more that the Laplacean idea of universal knowledge is still the theoretical ideal of the natural sciences; current opposition to these declarations has often seemed to confirm this ideal, by defending the study of the whole organism as being only a temporary approach. But now the analysis of the hierarchy of living things shows that to reduce this hierarchy to ultimate particulars is to wipe out our very sight of it. Such analysis proves this ideal to be both false and destructive.

Each separate level of existence is of course interesting in itself and can be studied in itself. Phenomenology has taught this, by showing how to save higher, less tangible levels of experience by not trying to interpret them in terms of the more tangible things in which their existence is rooted. This method was intended to prevent the reduction of man's mental existence to mechanical structures. The results of the method were abundant and are still flowing, but phenomenology left the ideal of exact science untouched and thus failed to secure the exclusion of its claims. Thus, phenomenological studies remained suspended over an abyss of reductionism. Moreover, the relation of the higher principles to the workings of the lowest levels in which they are rooted was lost from sight altogether.

I have mentioned how a hierarchy controlled by a series of boundary principles should be studied. When examining any higher level, we must remain subsidiarily aware of its grounds in lower levels and, turning our attention to the latter, we must continue to see them as bearing on the levels above them. Such alternation of detailing and integrating admittedly leaves open many dangers. Detailing may lead to pedantic excesses, while too-broad integrations may present us with a meandering impressionism. But the principle of stratified relations does offer at least a rational framework for an inquiry into living things and the products of human thought.

I have said that the analytic descent from higher levels to their subsidiaries is usually feasible to some degree, while the integration of items of a lower level so as to predict their possible meaning in a higher context may be beyond the range of our integrative powers. I may add now that the same things may be seen to have a joint meaning when viewed from one point, but to lack this connection when seen from another point. From an airplane we can see the traces of prehistoric sites which, over the centuries, have been unnoticed by people walking over them; indeed, once he has landed, the pilot himself may no longer see these traces.

The relation of mind to body has a similar structure. The mind-body problem arises from the disparity between the experience of a person observing an external object – for example, a cat – and a neurophysiologist observing the bodily mechanism by means of which the person sees the cat. The difference arises from the fact that the person observing the cat has a *from*-knowledge of the bodily responses evoked by the light in his sensory organs, and this *from*-knowledge integrates the joint meaning of these responses to form the sight of the cat, whereas the neurophysiologist, looking at these responses from outside, has only an *at*-knowledge of them, which, as such, is not integrated to form the sight of the cat. This is the same duality that exists between the airman and the pedestrian in interpreting the same traces, and the same that exists between a person who, when reading a written sentence, sees its meaning and another person who, being ignorant of the language, sees only the writing.

Awareness of mind and body confront us, therefore, with two different things. The mind harnesses neurophysiological mechanisms and is not determined by them. Owing to the existence of two kinds of awareness – the focal and the subsidiary – we can now distinguish sharply between the mind as a 'from-at' experience and the subsidiaries of this experience, seen focally as a bodily mechanism. We can see then that, though rooted in the body, the mind is free in its actions – exactly as our common sense knows it to be free.

The mind itself includes an ascending sequence of principles. Its appetitive and intellectual workings are transcended by principles of responsibility. Thus the growth of man to his highest levels is seen to take place along a sequence of rising principles. And we see this evolutionary hierarchy built as a sequence of boundaries, each opening the way to higher achievements by harnessing the strata below them, to which they themselves are not reducible. These boundaries control a rising series of relations which we can understand only by being aware of their constituent parts subsidiarily, as bearing on the upper level which they serve.

The recognition of certain basic impossibilities has laid the foundations of some major principles of physics and chemistry; similarly, recognition of the impossibility of understanding living things in terms of physics and chemistry, far from setting limits to our understanding of life, will guide it in the right direction. And even if the demonstration of this impossibility should prove of no great advantage in the pursuit of discovery, such a demonstration would help to draw a truer image of life and man than that given us by the present basic concepts of biology.

IX. SUMMARY

Mechanisms, whether man-made or morphological, are boundary conditions harnessing the laws of inanimate nature, being themselves irreducible to those laws. The pattern of organic bases in DNA which functions as a genetic code is a boundary condition irreducible to physics and chemistry. Further controlling principles of life may be represented as a hierarchy of boundary conditions extending, in the case of man, to consciousness and responsibility.

NOTES

[1] More precisely, each item consists of one out of four alternatives consisting in two positions of two different compound organic bases.

[2] The blueprint carried by the DNA molecule of a particular zygote also prescribes individual features of this organism, which contribute to the sources of selective evolution, but I shall set these features aside here.

[3] See P. Weiss, *Proc. Nat. Acad. Sci. U.S.* **42**, 819 (1956).
[4] The 'field' concept was first used by Spemann (1921) in describing the organizer; Paul Weiss (1923) introduced it for the study of regeneration and extended it (1926) to include ontogeny. See P. Weiss, *Principles of Development* (Holt, New York, 1939), p. 290.
[5] See, for example, C. H. Waddington, *The Strategy of the Genes* (Allen & Unwin, London, 1957), particularly the graphic explanation of 'genetic assimilation' on page 167.
[6] See, for example, M. Polanyi, *Amer. Psychologist* **23** (Jan. 1968) or ——, *The Tacit Dimension* (Doubleday, New York, 1967).

PART III

PROBLEMS OF EXPLANATION IN BIOLOGY

A. LEVELS OF ORGANIZATION

CLIFFORD GROBSTEIN

ORGANIZATIONAL LEVELS AND EXPLANATION

The problem of levels is especially critical for certain areas of biological investigation, particularly that of development. The question is posed whether such biological phenomena require special forms of explanation because the phenomena themselves are hierarchically organized. By explanation I mean explication of the properties of a phenomenon in logically interrelated and substantively relevant statements. By 'hierarchically organized' I refer to the fact that biological phenomena can be dissected conceptually and physically into subsets which are again dissectable into subsets (and this several times repeated), with each subset having a reasonably integral set of properties. The question of whether this requires explanatory principles that are fundamentally peculiar to biology is, I believe, to be answered negatively. However, the question of whether there are special difficulties to be taken into account and special conceptual problems to be solved, I believe must be answered positively. The special difficulty is the relationship between properties at successive subset levels.

The special difficulty is heightened by the fact that biological subsets are not the same in isolation and in combination, and the difficulty is compounded when new sets and subsets are appearing during the process under investigation with emergence of new properties within a set. The problem was classically presented in the phenomenon of development, where the seeming miracle of emerging order has confounded biological thought since Aristotle. We are no longer confounded, though the miracle is not yet thoroughly illuminated. What we now see is a process involving conversion of properties to successively more inclusive sets. More substantively, we see information contained in the linear sequence of certain macromolecules transformed into spatial, temporal, macromolecular patterns, and these in turn into cell behaviors which give rise to tissue organization, organ formation, and ultimate organismic and supraorganismic order. Developmentally emerging order involves continuous translation of the properties at lower levels to higher, with subsets interacting with each other even as their properties are interlocked in some sense with properties of sets above and below.

What are the rules which interlock properties of subsets, sets, and supersets? Is a new form of explanation required to comprehend this? Is there a miracle of indeterminacy at any step in the process? Is the notion of emergence of new properties a figment of vitalism, or is there a valid phenomenon to be subjected to causal analysis? This is the way I see the substantive problem which we are forced to deal with in development. It is essential to understand the translation rules between successive levels, because such translation is a mechanism of development. Let me illustrate the problems with a particular case.

The development of the kidney is almost diagrammatic in illustrating the situation. Let me state certain facts about the phenomenon and then return to the more formal aspects. The adult kidney of the mouse is an intricate structure; it has a very complex form and function that are essential to the life of the whole organism of which the kidney is a part. We may say that the organ is a subset of the whole organism, which in turn can be dissected into subsets called glomeruli, secretory tubules, collecting tubules, and so on. These parts are normally in a characteristic array and each can be further dissected into subsets (the cells) which again are in characteristic array. The cells too can be dissected into organelles, and these into macromolecular arrays. The important first point to make is that this impressive complexity has been reproduced over successive and countless generations of mice in essentially identical form. The elaborate duplicative performance is a regular and predictable phenomenon, which appears to occur *de novo* in each generation. However, kidneys do not produce kidneys. Rather, the mouse produces an egg, the egg an embryo, and within the embryo the kidney appears from nonkidney. Our problem is the genesis of this intricate order. Confessing immediately that we cannot fully explain this genesis, let me state some facts that give hope that ultimately we will explain it, and without qualitatively new forms of explanation.

I have noted that successive generations of mice produce essentially identical kidneys. There are, however, significant exceptions. There is a strain of mice which on suitable breeding yields offspring which are abnormal in a number of respects. Included among the abnormalities are defects of kidneys. The breeding behavior indicates conclusively that there is a relatively simple genetic factor involved in the kidney genesis. This is no surprise, in view of our general knowledge of the role and mechanism of action of hereditary factors in development. We summarize this by asserting that there is a replicative program transferred from generation to generation, and that within this program there is information whose proper interpretation leads to kidney formation. Localized alterations or defects in the program lead to developmental defects of the kidney; hence there must be particular portions of the program which have special relevance to the kidney.

Whatever the mechanism of registry of the information – and I think that very few biologists today will doubt that it resides initially in important part in base sequence in DNA – all of the information is contained in the kidney rudiment of an 11-day embryo. I mean that one can remove such a rudiment from the embryo and this isolated subset under suitable conditions will continue essentially its normal development: the rudiment contains all of the necessary information for completion of the complex, intricate genesis. The rudiment itself is dissectable into two components – the socalled ureteric bud and the surrounding mesenchyme, and each of these is a cellular system. A whole rudiment placed in culture will continue reasonably normal development. Separated components, however, do not, although they do if recombined. Clearly some interaction of the components is required. At this stage the genetic information, initially present in the egg in the form of linear sequences in DNA, has been processed sufficiently so that neither component without the other has an effective program to continue development. Something exchanged between them (and it can be shown to be separable from cells and to be macromolecular) is essential to continued expression of the genetic program. The exchanged material is known not to be DNA itself, although we do not know exactly what it is. Since it contains essential information for continued development it must be a translation product of DNA, at least in the sense that there is some systematic correspondence between its properties and those of DNA. It might, for example, be a protein whose amino acid sequence is determined directly or indirectly by the base sequence of DNA. This would conform to what is now regarded as the fundamental translation step in genetic read-out. Let me add that I refer to it as protein for simplicity. It could equally well be RNA, and the same argument would apply.

To relate the material to the emerging order of kidney development we must describe some further biological facts. If we treat the kidney rudiment properly we can dissociate it without significant damage to a suspension of individual cells; that is, we can progress in our dissection from rudiment to subset components and then to subset cells. These interspersed cells will recohere into components and these in turn will interact to give continued kidney development. The separated cells contain information that enables them to reform the next higher level, which involves components having properties that can interact to continue the development. Analysis shows that the cells have surface properties which cause them to sort as to kind through differential adhesion; the surface materials involved are macromolecular and at least in some instances separable from the cells.

The macromolecular materials promote aggregation and contain the recognition specificity. They appear to be similar in general chemical type to those referred to earlier as exchanging between the components. Moreover, if one examines the development of the kidney mesenchyme into tubule rudiments one notices that cellular aggregation is one of the first consequences of the action of one component on the other. Cells within the affected mesenchyme cohere more tightly to form packages, or aggregates, and these gradually transform into tubules. The first step in the formation of the new subset is cohesion to form a package, and this step is promoted by exposing dissociated mesenchyme to macromolecular extracts of tubule-inducing sources. Hence, assembly of supercellular systems appears to be dependent on macromolecular materials associated with cell surfaces. Tubules emerge in combinations of suitable cells and macromolecules.

It is well known that complex macromolecular patterns can self-assemble from dissociated molecules; that is, what we have been talking about at the cellular level also occurs at the molecular one. Native collagen, for example, can be solubilized in dilute acid so as to completely lose its fibrous character. On return to neutrality, the still intact collagen monomers reassemble into fibers and these can be shown to have a characteristic ultrastructural banding pattern whose detail can be altered by the conditions of the reassembly. Depending upon relatively simple conditions the reassembly can duplicate or alter the original macromolecular assemblage.

It happens that the detailed structure of the reassembled material is important in development, because, if it is normal, calcium will deposit on the collagen to form normal bone. If the substructure of collagen is not normal, deposition of calcium is not normal and bone formation will be defective.

Thus large molecules specified in their linear sequence by the linear sequence of DNA can self-assemble in arrays and these in turn can influence cellular assembly. The emergence of tubules in the developing kidney is an epigenetic conversion, in the proper context, of properties at the molecular level. I am not anxious to press the point too far as to detail because there are still large gaps in our knowledge; rather, what I have been sketching is a somewhat impressionistic example of what most of us now see as the essential course of development. It is a series of translations of information stored in chemical terms, but read and registered by scanners of increasing complexity as the developmental course progresses. The process is frightening in its intricacy but it does not appear to contain elements requiring new intellectual feats to comprehend. The basic translations may be broadly indicated as:

micromolecule → macromolecule → polymer →
ultrastructural array → cell → tissue → organ → organism.

I have tried to make it clear that we see ways in which cellular behavior is altered by the specificity of molecules, in turn laid down by genetically transmitted chemical information. Cellular behavior, like specific aggregation and specific synthesis, leads to specific structure and activity in cellular systems, and interaction among these leads to specific behavior at the level of the organism. It is clear that we move upward in levels with conversion of information from one level to the next. The rules, however, we only partly understand. Moreover, I have emphasized only the outflow of genetic information. There are a whole series of feedback operations as well which I have not touched on.

The issue posed, however, seems to me to be worthy of joint consideration by biologists and philosophers. Are there general rules by which properties at a given level are summed, integrated, and transformed, so as to emerge into properties at the level above? To illustrate what I am asking we can write the relationship as follows (S=set; [ ]=combining relations):

$$S_1 + S_2 \cdots + S_n \rightarrow [S_1 + S_2 + \cdots + S_n] = S_S$$

The issue is the relationship between the properties of S_S and the properties of any one of its subsets (P_{Sn}). There are a number of cases of increasing complexity which can serve as models in trying to make some sense of the rules of conversion. The simplest case may be referred to as 'collections'. Professor Scriven made reference to sand dunes, and I will also use these as an example. In this case the properties of subsets are identical whether they are within or outside of the set. In other words, it makes virtually no difference in the properties of a sand grain whether it is by itself or whether it is within the sand dune. One can further define a collection as the case in which association is fully reversible; you can pile up the sand or you can scatter it, and the grains are otherwise unaffected. What is the relation in this simplest case of properties at the level of the individual sand grain (subset) and the level of the sand dune? One can make certain obvious statements; for example, that the mass properties of the sand grains are summed in the properties of the dune. Further elementary interactions do go on between grains, and these lead to characteristic packing of the sand and hence to dune properties. There are more complex effects as well. If one examined the temperature gradients in the dunes in the early morning or late at night, I suppose that the gradients from the exterior to the interior would be different depending on the size and shape of the dune. This is a mass property of the whole dune which is not readily predictable from properties of grains. Beyond this, there is the effect of content and vantage point – the sand dune has different significant properties depending on where we are and where it is in relation to other objects. The dune does not change at all, but its properties are defined very differently from the inside, from immediately above, or from 2000 feet above. This is not an inconsequential matter because there is an often unappreciated vantage-point effect in looking at more complex systems like biological ones.

Even this simplest case of collections shows a striking new dimension when heterogeneity becomes a factor. As soon as more than one kind of subunit is involved in the collection, a whole new range is opened in terms of higher-level properties. Differential distribution of kinds, as happens when wave patterns move over the beach and different colors of sand grains are sorted out, produce patterns – properties which are predominantly relational and relatively independent of unit properties.

One can say a good deal in this relatively simplest case about the relationship of the properties at the upper and lower levels, and the kinds of rules which relate them. This becomes much more difficult when the case gets even a little more complex; for example, when there is some kind of coherence among the subsets. Assume that the only differences in the properties of a subset are those immediately resulting from coherence with a neighbor subset. Assume further that coherence is not fully reversible and that the coherence is limited to a particular point on the subset. Packing of subsets is immediately strongly influenced, and patterns of coherence appear (crystallization). Now assume a second kind of complication, that coherence involves affinity, for example, complementarity between antigen and antibody or between enzyme site and substrate. Relationships depend now on properties stemming from confoımation; that is, pattern itself becomes a source of binding force.

Real complexities obviously occur, however, when elements moving into a set are strongly altered in their properties by the relation itself. A whole new level of transformation is introduced. The kidney behavior I referred to is a case in point. The two components are altered very importantly in their properties when brought into interaction. Such cases may be reversible: the properties are altered within the set but return to initial state on disassociation of the set. The slime mold, for example, moves sequentially from a higher ordered aggregative state to a lower ordered disaggregative state. It is only in the higher ordered aggregative state that one finds significant biological differences of properties between the cells.

Such a system may be called facultative in that interacting elements change their properties in interaction, but can very readily be returned to initial state. The case that presents most of the mystery is the obligatory system, the one in which interaction has reached the point where subsets cannot, without very great difficulty, be induced to show their characteristic properties because these are 'bound' to the set. The requirements for the characteristic properties of the subset are intricate and subtle, and very difficult to reproduce outside of the set. Cells of multicellular organisms belong here; they have only recently and incompletely yielded to the blandishments of controlled environmental conditions *in vitro*. In such obligatory systems one nearly always finds an internal environment which is controlled and separated from the external environment so as, in a sense, to acquire a life of its own. The internal milieu begins to develop properties which are partly self-controlled, not independently of the cells but not fully under their control. This kind of phenomenon occurs not only in complex organisms with respect to cells, but in ecologic systems like kelp communities where a controlled environment is set up with the kelp in which only certain kinds of organisms will be found. The combining conditions at this point acquire their own dynamic properties and their own stability. This means new properties in the total complex resulting from this controlled environment.

The objective of this discussion has been to pose a question. Is there a useful formal approach to the problem of the conversion of properties from one level to those of another? The question is no longer idle for there is a wealth of new data in the last few decades stemming specifically from the success of the reductionist approach. The reductionist approach has multiplied problems for the constructionist approach: how do we put things back together again conceptually, what are the rules operating within the whole system that are abrogated in the process of reduction? Can we look at vertical transformations of properties in simple systems and generalize rules that will allow us to 'see through' more complex systems? If there are general rules, do they constitute a new kind of explanation? I think not, but in part this may be a matter of definition. Perhaps, again, we deal with levels. Perhaps it is a new level of explanation which emerges from the interaction of explicatory process with the complexities of hierarchical order. It is not new in kind, but it has a new power. Perhaps it will prove to be the explanatory counterpart of the enormous biological power of hierarchical organization.

University of California, San Diego

H. H. PATTEE

PHYSICAL THEORIES OF BIOLOGICAL CO-ORDINATION

The more constraints one imposes, the more one frees one's self of the chains that shackle the spirit ... and the arbitrariness of the constraint serves only to obtain precision of execution. Igor Stravinsky, *Poetics of Music*.

I. BACKGROUND

Biological organization is manifestly different from the order of the non-living world, and the study of biology is largely a search for the nature of this difference. The perspective and style with which we see this difference has changed in many respects as our knowledge of living systems has grown. Today, following the molecular biological revolution, we commonly find the opinion that there is no real problem left. This attitude was recently expressed by Delbrück (1970) in his Nobel Lecture: "Molecular genetics, our latest wonder, has taught us to spell out the connectivity of the tree of life in such palpable detail that we may say in plain words, 'This riddle of life has been solved'." Many prominent molecular biologists have expressed similar views (Watson, 1965; Crick, 1966; Kendrew, 1967; Stent, 1968).

However, in spite of our knowledge of the 'palpable detail' which is said to be normal chemistry for all known cellular reactions, the origin and nature of the *co-ordination* of these reactions remains an obscure and evasive question. It is an old question for biologists. Sir Charles Sherrington (1953) over thirty years ago gave this picturesque description of the cell's activities: "We seem to watch battalions of catalysts like Maxwell's 'demons' lined up, each waiting, stopwatch in hand, for its moment to play the part assigned to it, yet each step is understandable chemistry.... In this great company, along with the stopwatches run dials telling how confrères and their substrates are getting on, so that at zero time each takes its turn." Of course Sherrington did not know the details, but he appreciated the necessity of distinguishing the ordinary chemistry of each step with the extraordinary spatial and temporal co-ordination of these steps.

Co-ordination in biological organisms takes the form of hierarchical control levels which at each level provide greater and greater freedom or adaptability for the whole organism by selectively adding more and more constraints to its component parts. Therefore the physical nature of an individual constraint does not make sense if it is studied out of the context of the hierarchical control level in which it occurs. Since hierarchical control levels begin at least with the genetic code and pass throughout the entire range of biological organization, even to the language structures of man, there is no hope for a comprehensive review of the co-ordination problem.

In this paper I shall only try to clarify some of the problems of the most primitive biological co-ordination in terms of the language of physics. In particular I shall explore possible physical principles which would explain or predict the origin of co-ordinated constraints from initially incoherent processes. Many biologists will at once express doubt that physical explanations are relevant for what appears to be the results of an evolutionary principle, namely that natural selection does not favor unco-ordinated reactions. But this evades both the problem of the origin of life, by which I mean the minimal level of co-ordination which will support natural selection in the first place, as well as the problem of the origins of evolutionary innovations: that is, of entirely new hierarchical levels of co-ordinated functions.

II. BIOLOGICAL EXAMPLES OF THE PROBLEM

1. *The Genetic Code*

The most universal, though probably not the most primitive example of a co-ordinated system of molecular constraints, is the genetic code. I would guess that the origin of the genetic code is a problem of the future which in some ways will replace in importance the problem of the structure of the gene during the past fifty years. However, it is clearly not the same type of problem. It shares with other biological co-ordination and control problems the apparent paradox that the more we learn of the molecular details, the less easily we can formulate reasonable hypotheses or mechanisms for their origin. Furthermore, as the effectiveness of the co-ordinated function increases, there also appears to be an increase in the arbitrariness of the constraints which execute these functions. This arbitrariness is implicit in frozen accident theories of the code (Crick,

1968) which attempt to explain the universality of the code by the observation that any change in what was apparently an arbitrary codon amino acid assignment would have been lethal. However, this says nothing about why any code at all should arise. Several very thoughtful discussions of the origin of the code have appeared recently (Woese, 1967; Jukes, 1966; Crick, 1968; Orgel, 1968) as well as a review (Caskey, 1970). These discussions offer some reasonable evolutionary mechanisms for selected aspects of the code's structure, but they all recognize the severe unsolved difficulty of the origin of the threshold level of co-ordinated constraints necessary for any set of RNAs or transfer enzymes to function as an effective code.

2. *Developmental Controls*

If we think of the genetic code as a co-ordinated set of constraints that creates a basic language structure in which genetic messages can take on constructive meaning, then developmental controls represent a new hierarchical level of constraints imposed on the order of the expression of these genetic messages. In return for these additional constraints, single cells become free to evolve additional functions of greater specificity and effectiveness. The central point is that lacking the co-ordination of the developmental control level, the additional functions become senseless in the individual cell and pathological or even lethal for the multicellular organism.

While the effect of loss of co-ordination is particularly obvious in developmental growth in highly evolved organisms, we must look more generally at what this means for a theory of evolution. It is difficult to imagine any individual structure or catalyst, such as a primitive membrane or protoenzyme, whose sudden appearance in a complex chemical network would favour an increase in co-ordinated events. Indeed, Haldane (1965) was not attracted to theories of the origin of life that require millions of years of chemical evolution for this very reason: that any unco-ordinated set of protoenzymes would appear to be worse than no enzymes at all as a favourable environment for the first life. So it is at all hierarchical levels, that loss of control or loss of co-ordination of functions is often worse than the absence of any individual function.

3. *Integrated Cognitive Systems*

With the imposition of co-ordinated developmental constraints and the subsequent freedom to differentiate tissues and specialize functions, evolution again added a new hierarchical level of control which we recognize as nervous activity. In addition to their role in development the constraints of the nervous system effectively free some of the responses of the organism from the diffusion-limited time scale of the genetic and developmental messenger molecules and at the same time, through the perfection of the senses and muscles, allow freedom to respond to a far greater variety of environmental challenges.

At no other level of biological organization does the concept of co-ordination take on such profound meaning or display such elegant examples as in the reflex or sensorimotor activities which integrate the acquisition of data through the sense organs, the classification of this data by the nervous system, and the decisive muscular responses of the organism based on this classification. The subtlety of such coherent systems has made it difficult to simulate even the most elementary sensorimotor tasks by co-ordinating artificial visual detectors, large computers and mechanical output devices (Michie, 1970). Consequently, this example may not seem appropriate for our discussion of physical principles of primitive co-ordination. On the other hand, more is known of the stages in the evolution of the components of integrated cognitive systems than is known for the genetic code or the developmental control system. Furthermore, it is a level of co-ordinated activity which is in fact being actively studied through the use of artificial devices, and consequently it is an area where experiments can usefully be done to test theories of the origin of co-ordination. As yet, the experiments on simulated cognitive behavior are not aimed at theories of origin but at understanding or duplicating existing biological behavior. Considering the difficulties which have been encountered in these simulations, it could be that what is lacking is a theory of how hierarchical co-ordination evolves rather than how it presently operates.

4. *Language Structures*

As a final example of biological co-ordination I choose the sets of constraints which form a language structure. In the narrow sense the meaning of the word language is limited to human communication and as such it represents the highest and latest set of constraints that evolution has imposed on living organisms. Even in this narrow sense, the origin and nature of language constraints is a formidable unsolved problem. However, I want to use the concept in the widest possible sense. Biologists already speak of the genetic code as the language of the cell, and although this might be considered a metaphorical usage, I would prefer in this discussion to broaden the concept of language structures to include literally all co-ordinated sets of constraints which allow what are physically quite ordinary molecules to function as messages. For example, nucleic acid molecules do not inherently possess symbolic properties. No amount of chemical or physical analysis of these molecules would reveal any message unless one presupposes the co-ordinated set of constraints which reads nucleic acids. The same is true for message molecules at all levels of organization. At the developmental level we may speak of the regulation of specific genes by repressor and derepressor molecules as if only chemical properties were significant; but again, without presupposing a co-ordinated set of such molecular constraints 'regulation' would have no more meaning in the developmental sense than 'translation' of a gene would have meaning outside the context of a complete genetic code.

In these examples of biological co-ordination I have emphasized how the imposition of new constraints results in some corresponding freedom in the behavior of the organism. I do not use this language to sound poetic or philosophical about the intricacies of living matter. My purpose is simply to guard against an evasion of the fundamental problem of co-ordination which I regard as the physical basis of life. The behavior of non-living matter is usually described in the language of physics as deterministic, spontaneous, ordered, disordered, or even coherent or correlated, but never as free in its interactions or responses. The problem is to say more clearly what the physical nature of this freedom of living matter is, and in particular to find its origin in the growth of physical constraints. In the next section, then, we turn to the problem of what a constraint means in physical language.

III. THE PHYSICAL NATURE OF CONSTRAINTS

1. *Constraints as Alternative Descriptions*

In common language the concept of constraint is a kind of forceful confinement which limits our freedom. The same general concept may also hold in physical systems where the constraint is a fixed boundary, like the box which confines the molecules of a gas. However, the physical nature of constraints leaves room for much more subtle forms of influence on the motions of a system. Constraints cannot be treated in physics as fundamental properties of matter. This is because the laws of motion are assumed to specify the detailed behavior of matter as completely as possible. In fact the laws of motion do not leave any freedom at all for alternative behavior at the microscopic level of description. So what is the meaning of imposing more 'constraints' when the laws of motion already impose what amounts to total constraint on the detailed behavior of the system?

The answer is that the physical concept of constraint is not expressible in the same language as the microscopic description of matter. A constraint is an *alternative description* which generally ignores selected microscopic degrees of freedom in order to achieve a simplification in predicting or explaining the motion. Perhaps the most common and useful type of alternative description is the replacement of the microscopic description of the particles at the surface of a solid by the equation of a geometrical surface. This kind of constraint may be called a boundary condition, but the essential point is that this is an alternative description of what may also be recognized as a very large collection of molecules following their detailed laws of motion. The constraint description is the only useful one in such cases, since there is no possibility of following so many molecules dynamically.

This type of equation of constraint may also be interpreted as a limiting case of the average positions of the particles over a long period of time, although no explicit averaging process is employed to derive the equation. This is an important point, since one of the problems of the origin of the constraints of co-ordinated systems is to account for an objective embodiment of an alternative description. In other words, constraints are most easily explained as the invention of the physicist who sees a new way of looking at a problem which is much simpler or more useful than taking into account all degrees of freedom with equal detail. How this can happen spontaneously is more difficult to imagine.

A third interpretation of constraint which is a generalization of our first two ideas of alternative description and averaging process, is the idea of classification of degrees of freedom. A microscopic physical system is largely defined by the choice of degrees of freedom. Once they are chosen, they enter into the equations of motion according to prescribed rules which in no way serve to interpret the relative significance of the variables. Any form of classification of the degrees of freedom must arise from additional rules over and above the laws of motion. These rules may be considered as constraints. For example, some variables may be particularly sensitive to perturbations while others are relatively stable. However, both sensitivity and perturbation must be defined by an additional set of rules since they are not a part of the detailed dynamical description. Some form of classification of variables is the most general requirement for any process of pattern recognition, feature extraction or functional description.

One final point. Since we assume that the microscopic description is complete, or as complete as possible, the alternative description of the constraint language must be less complete in some sense, otherwise it would simply be equivalent or redundant. In other words, the concept of constraint must represent a selective loss of detail or a predetermined rule of what is to be ignored. Speaking in this way emphasizes what appears to be a strong subjective element, since the implication is that someone or something must choose what to ignore about the system. Very much the same problem arises when we speak of measurement or making records of physical events, since again there must be a clear separation between the event and the observer or measuring device which creates the record of the event. In fact, a measuring or recording device may itself be considered as a special constraint which is designed to ignore all degrees of freedom except those to which the measured variable is sensitive.

The basic physical explanation of the origin of constraints must therefore account for the spontaneous separation of a physical system into a part in which we recognize collective behavior that we can describe without using all details, that nevertheless exhibits in the forms of its collective behavior a strong sensitivity to selected details of the other part of the system. Furthermore, in order not to beg the question we must begin with initial conditions that are chaotic, or that at least exhibit no inherent sensitivities. However, before outlining some mechanisms of the origin of constraints, I would like to give some more explicit examples of the types of constraint to which the concept of co-ordination can be applied in living systems.

2. *Examples of Constraints*

In analytical mechanics constraints are often called auxiliary conditions because they cannot be derived from the basic axioms or laws of mechanics. They are generally classified by the mathematical form in which they are expressed. Thus there are scleronomic and rheonomic constraints that are time-independent and time-dependent respectively. Constraints are also holonomic or non-holonomic depending on whether the auxiliary conditions are relations between the co-ordinates of the system or whether they are non-integrable differentials of the co-ordinates respectively. These classifications are important for the mathematical solution of mechanics problems since general methods of incorporating these auxiliary conditions are known (Whittaker, 1944; Lanczos, 1949). However, as is usual in physics, only the simplest types of constraints can be treated generally, while almost all of the artifacts and machines which man has designed to harness nature can only be described functionally. Unfortunately, the same is true of the simplest natural biological constraints. Thus we find, for example, that all forms of switches and ratchets as well as allosteric enzymes can be described only as rheonomous, non-holonomic constraints for which no general analytical treatment is possible.

From our point of view it is therefore more useful to classify constraints in terms of structure and function, and in terms of how they are actually constructed. A basic elementary constraint in physics is the chemical bond, and it is probably no accident that biological co-ordination begins at this level. It is important to understand in what sense a chemical bond is a constraint, that is, an alternative description. The microscopic description is, of course, the quantum mechanics of the elementary particle system. In this description chemical molecules are simply disjoint stationary states of the system which are quite incidental to the dynamics. These states also may have an exceedingly complex structure in this description. The ordinary chemical concept of a molecule, on the other hand, not only ignores all the dynamics but also most of the detailed structure. More significant is the physically arbitrary emphasis which the alternative chemical description puts on a particular stationary state which is the molecule under discussion (Golden, 1969). From the quantum mechanical, microscopic point of view the original system of elementary particles does not include this emphasis. In other words, there is no recognition or classification of the alternative stationary states formed by this system. In a later section we shall use the selective catalyst or enzyme molecule as a primitive example of what we mean by a co-ordinated set of constraints, since such molecules are described as recognizing alternative structures as well as inherently classifying degrees of freedom into those that are sensitive with respect to the catalytic dynamics and those that are not.

In physical terms we might normally think of the next level of constraint above the chemical bond as the interphase boundary – that is, a collection of molecules which on the average forms a two-dimensional structure which separates two states or phases of matter. The condensation of separate phases as well as the surface activity of polyelectrolytic molecules at the boundary can be described in various levels of detail. For non-living systems the thermodynamic or free energy level of description may serve adequately. For living systems there are numerous alternative descriptions of membranes which ignore different details of the molecular structure and dynamics. However, in some cases, such as the mitochondrial membrane or the ribosome, the essential function may be described only in terms of a co-ordinated set of individual molecules since the cost of ignoring molecular detail is a failure to achieve a quantitative or even qualitative model of what is going on.

The essential concept I wish to make clear in the above discussion of the physical nature of constraints is that contrary to common usage a constraint is not a definite, fixed, objective structure, but an alternative simplified description of an underlying, complex dynamical process. The use of such simplified descriptions is absolutely essential for all but the most elementary physical problems, not only because the detailed underlying dynamics becomes intractable, but because the crucial functional aspects of the system are hidden by the mass of microscopic detail.

IV. THE SPONTANEOUS GENERATION OF CONSTRAINTS

I have used the word 'constraint' in this discussion because it is a part of the language of classical physics, and can be used to characterize whatever auxiliary conditions must be appended to the fundamental equations of motion in order to predict more easily how a system will behave. In effect, however, the constraint is simply some additional regularity or order which is not explicitly found in the initial conditions. How a collection of matter with chaotic initial conditions spontaneously generates new regularities and patterns of order that we can then clearly represent as constraints is the first problem we must discuss before going on to the next level of order which we have called the co-ordination of constraints.

1. *The Problem of Subjectivity*

One common difficulty with this concept of a constraint as an alternative description as well as any distinction between order and disorder is the subjective element of the observer who apparently must choose when to use an alternative description or who must choose what details to ignore. How do we distinguish inherent regularity from regularity which is only in the mind of the observer? One may of course take the position of naive materialism that there is only one objective, microscopic, physical reality and that all the rest is our attempt to describe the complexity of matter at different levels. Thus constraints and consequently the behavior of living matter become only a useful alternative description of the complex objective motions of non-living matter. Surprisingly enough, this is still a popular view among biologists. I do not want to pursue this metaphysical question here, except to point out that the opposite view is equally consistent, and in fact is more common among modern physicists than biologists.

For example, it is possible to interpret the microscopic laws of quantum mechanics as only an alternative description of the results of primary observations or measurements which are interpreted as the fundamental reality. Thus Born (1964) argues that observation is primary and that in spite of the apparent causality of the laws of motion chance is a more fundamental conception than causality: "for whether in a concrete case a cause-effect relation holds or not can only be judged by applying the laws of statistics to the observations." Wigner (1967) has entertained an even more active form of solipsism in which the description of quantum events is altered essentially upon entering the observer's consciousness. He concludes, "Solipsism may be logically consistent with present quantum mechanics, monism in the sense of materialism is not."

The interpretation of events and the observation of these events has played a crucial role in the development of quantum mechanics and remains under active and controversial discussion (Ballantine, 1970, and references therein). I would not like to leave the impression that the nature of measurement and description (i.e. the symbolic behavior of matter) is anything less than the most fundamental problem of both physics and biology (Pattee, 1967, 1970, 1971a) but for this discussion I shall simply assume that events and descriptions of events are primitive concepts without assigning more or less reality to either. Nevertheless the origin problem remains. Living matter, I assume, obeys precisely the same laws of motion as non-living matter, and since these laws are assumed to be complete the very possibility of records and measurements requires alternative descriptions. As outside, intelligent observers we can invent such descriptions, but that introduces too strong a subjective element for a theory of origins. The only useful objective criterion that I have found for distinguishing living from non-living matter is to say that *the constraints of living matter must contain their own descriptions.* This is the closest I have come to a physical basis for the evolutionary necessity of a separation of genotype and phenotype – the genotype being the cell's description of its own constraints. But as we shall see, to give any meaning to this concept of description we must presuppose the existence of a co-ordinated set of constraints.

2. *Topological Dynamics as a Source of Constraints*

We return now to the question of the spontaneous generation of constraints. How do physical systems with chaotic or homogeneous initial conditions develop definite structures? Or to be more explicit, how does a physical system which is assumed to follow dynamical equations that completely determine the unique trajectories of all possible systems generate entirely new types of regularities or structures?

Every level of description in nature has its particular state variables and dynamical equations which are used to predict from the present state of the system what will happen in the future. In fact one may argue that the definition of state variables at any level of description is that convenient summary of the past information about the system which is sufficient for predicting its future behavior within chosen limits of observational significance and precision. The dynamical equations then become the rules for transforming the state variables from the past to the future. In terms of the origin of new constraints, this apparently poses a problem since within this strict protocol there is no specified procedure for selectively ignoring degrees of freedom or altering the dynamical equations (which are assumed to be invariant with respect to initial conditions).

The topological approach simply ignores all distinctions between the initial conditions as well as the time, and considers the system as the family of all possible trajectories defined by the dynamical equations. In other words, the dynamical equations define a vector field in an appropriate space, and it is the structure of this vector field which is interpreted as describing the global behavior of the physical system. In this sense, the structure of the vector field is a constraint or alternative description of the motion of individual particles. Constraints appear because we take a broader temporal and spatial perspective of the microscopic dynamics. Instead of asking what happens if we start with these initial conditions at time t_0, we ask what will happen no matter what the initial conditions or time happen to be.

These methods of general dynamics have played vital roles in many branches of physics, beginning with stability theory in celestial mechanics (Poincaré, 1957) and in machines (Maxwell, 1868), and are now highly developed in a broad area often called systems theory or control theory (Bellman and Kalaba, 1964). However, the applications of general dynamics to the origin of biological structure are more limited. One of the earliest applications of stability theory to biology was Lotka's and Volterra's studies of ecological stability (Lewontin, 1969), as well as Turing's (1952) paper on the chemical basis of morphogenesis in which he demonstrates how an initially homogeneous reaction system may develop standing waves of concentration.

The most recent attempts to generate biological organization from the singularities of a dynamical topology by Thom (1968, 1970) are still in a very early stage. Although the elegant, general concepts of differential topology supply rich heuristic images for picturing the development and control processes in biological systems, the present weakness of the method lies in its failure to connect clearly with the specific molecular constraints which execute the developmental and control events. To some extent the introduction of statistical mechanical concepts, as outlined in the following section, may help bridge the gap.

3. *Dissipative Structures as a Source of Constraints*

Living systems depend for their organization on a continuous flow of energy which keeps them far from equilibrium. A large body of theory has developed as a branch of non-equilibrium statistical mechanics which illustrates possible mechanisms for generating and maintaining new structures and patterns in initially homogeneous collections. The paradigm is convectional instability which generates regular patterns of flow from an initially uniform temperature gradient. A particular example which has been studied experimentally and theoretically is the convection pattern in a horizontal cell of liquid heated uniformly on the bottom surface and cooled uniformly on the top surface (Bénard, 1901; Chandra, 1938). Under favourable conditions the fluid in such a cell will develop convection in a pattern of uniform polygonal cells.

Burgers (1963) in his paper, "On the emergence of patterns of order" invokes "the principle of the most unstable solution" for such dissipative patterns, since many patterns are possible solutions of the dynamical equations, but only those patterns which grow or propagate with the greatest rate will become predominant. The *rate of growth* of an instability then becomes a selection mechanism which operates on random initial conditions to establish a definite constraint on the motion. Clearly this principle applies to more general systems – for example, a network of catalytic reactions where initially random cycles develop into a predominant reaction path within which the maximum overall rate is greatest (cf. Levins, 1971).

General investigations of ordering processes in dissipative systems involving chemical reactions with specific application to the generation of biological organization have been carried out, notably by Prigogine and his colleagues (Prigogine *et al.*, 1969) and Morowitz (1968). The dissipative constraints of these studies make much closer contact with the molecular level than do the structures of topological dynamics described in the last section. However, so little is known of the origins of biological co-ordination that we cannot yet evaluate the relative importance of each level. For this very reason, however, the study of all levels where spontaneous generation of constraints occurs would appear to be the only strategy. As a final section on the generation of constraints, I shall therefore return to the chemical bond level. Following this summary of how constraints can be generated at different physical levels of description, I shall then be able to discuss the possible meanings of co-ordination of these constraints.

4. *Chemical Bonds as Constraints*

I used the chemical bond as the first example of a constraint which we can derive generally from the underlying dynamical equations as long as we do not ask if one stationary state is more important than another. The chemical bond is also an example of a constraint that played a very fundamental role in predicting the chemical behaviour of matter long before its dynamical basis was understood. What happens as we continue up the scale of molecular complexity? How much can we predict from the dynamics just what forms of regularity or order will arise?

Platt (1961) has written a stimulating paper on this question in which he argues that dynamics at any level is always an approximation and that if one attempts to extend the description beyond its particular scale of size or energy or number then the errors of approximation are always magnified so far out of proportion that a new alternative description becomes essential. Platt lists many properties of molecules in the 5- to 50-atom range (after Kasha) that have no counterpart in diatomic and triatomic molecules, and then goes on to give new properties for the 50- to 500-atom range and beyond. His emphasis is not on the predictable changes which result from change of scale, but on "the unexpected changes represented by major differences in important principles of behavior. How can such differences develop?" Platt (following Burton) suggests that the general answer to this question lies in the "organized or non-random complexity in which each atom or group has a specifiable role". Such 'fully complex' systems are contrasted with repetitive systems, such as crystals, and systems with 'meaningless' imperfections, such as liquids, which can be treated with some success by statistical methods.

I do not think that these ideas of complexity get to the root of the physical problem. In the first place, as we have pointed out earlier, the complexity at any given level of description is a property of the number of degrees of freedom and the dynamical equations for that level. What we recognize as a system with a 'specifiable role' depends not on how 'fully complex' the system appears but on the type of behavior of the system when it is fully simplified by constraints or alternative descriptions. For example, the microscopic description of the particle dynamics of a crystal are fully complex. It is only in the alternative description which ignores detail (or emphasizes regularities) that we find simplification.

But still we find no 'specifiable role' in the simplified description. Similarly, the dynamical description of gas or liquid particles is fully complex, and only by the alternative statistical description do we see simpler behavior. But again, we find no 'specifiable role' in this behavior.

Intuitively I believe it is clear that a crystal has too many constraints to play an interesting role in the context of biological co-ordination whereas the liquid and gas have too few. Equivalently we could say that too much detail is specified in the crystal, while in the liquid or gas too many details are ignored (by averaging). This suggests that a specifiable role or a co-ordinated function requires some optimum loss of detail or optimum degree of constraint in between these extremes (Pattee, 1971b). We see this at all levels of function. Man-made machines are an obvious example. A watch consists of many co-ordinated constraints, and there are many possible sets of such constraints which will perform the same time-keeping function, but in any case, adding more constraints or removing constraints either stops the movement or allows such irregular movement that the function is destroyed. The set of constraints which form an enzyme molecule also appear to be optimum in the same sense that either adding or removing constraints generally decreases their activity. The limits of this optimum in the degree of constraint become even sharper as the functions become discrete. For example, any set of constraints which constitutes a code or grammar must have enough constraints to cover all possible messages but no extra constraints which could conflict with the complete set.

V. THE FUNCTION OF BIOLOGICAL CO-ORDINATION

Unfortunately we are not talking about the type of optimization problem which normally occurs in physics where the constraints are given and we seek the maximum or minimum value of some well-defined mathematical function. Clearly there is no meaning to optimizing a collection of constraints unless there is also a measure of effectiveness of function, and to make matters worse the meanings of function in biology are so broad and ill-defined that any generalization would appear impossible. Nevertheless, since I see no other way to study the problem of the evolution of co-ordinated constraints, I am going to suggest a general function which I find essential for all biological activity. This I call the function of hierarchical control. It exists at all levels of biological organization, and the role of co-ordinated constraints is to establish and execute the hierarchical controls.

I have emphasized the idea of a constraint as a simplified alternative description of a more detailed, underlying process. This description classifies these details by emphasizing some and ignoring others. But what is the objective meaning of description? What is the actual physical embodiment of a device which classifies details? Or we can ask what type of physical organization must exist to give an objective meaning to the concept of description?

I consider the four examples of co-ordinated constraints I gave in section II as such examples of physical organizations in which the objective idea of description makes sense. As I implied earlier, it is only through the constraints of the genetic code that we can speak of portions of the DNA molecules as a description of a protein. And at the next level, it is only through the additional constraints of developmental controls that the DNA can be said to describe the organism. In other words, I want to say that the most fundamental function of co-ordination in biology is to establish generalized languages which allow structures at one level to be recognized and executed as descriptions from a higher level. It is in this sense that co-ordinated constraints establish and execute hierarchical controls. We also know from direct experience that without the constraints of a language we have no opportunity to describe alternative behavior, and it is precisely this opportunity which distinguishes the freedom of living matter from the order and disorder of non-living matter (Pattee, 1971c, d).

VI. THE ORIGIN PROBLEM

The association of co-ordinated constraints with language structures and hierarchical controls does not directly alleviate the problem of the origin of life, though I believe it points to the central difficulty. There is still the basic chicken-egg paradox at whatever hierarchical level we choose: we say that a constraint is an upper level alternative description of the dynamics at the lower level, yet in order to give any meaning to the concept of description we need constraints: i.e. the co-ordinated constraints of a language.

On the other hand we know of properties of languages and hierarchies which appear to be universal, and since we also have some knowledge of the evolution of the higher levels of co-ordinated constraints we may hope for some hints on how to explain specific origin problems by applying general theories which can be tested at different available levels of organization. I shall conclude by giving two examples of such universal properties.

1. *The Property of Self-Description*

All language systems have self-referent capability on at least two levels – they can generate their own grammar or code, and they can interpret statements about the language, i.e. they contain their own metalanguage (Harris, 1968). For example, the genetic code and selected parts of the gene together form a language system which not only generates the enzymes and RNAs that form the grammar, but which can also interpret control statements having to do with the expression of structural genes. This is a fundamental requirement of what Orgel (1968) has called 'natural selection with function' since self-description is necessary for self-replication and meta-statements are necessary for functional co-ordination in development. Consequently, the self-referent property appears to be a requirement for what we now generally regard as biological evolution.

One basic question which needs to be studied is how simple a set of physical constraints can be and still exhibit the self-referent property. This in turn may require a clearer understanding of the minimum logical conditions for a self-referent grammar. A second question is whether there is an even simpler set of constraints which does not have the self-referent property, but which can be shown to evolve this property spontaneously. There are certainly simpler systems which are co-ordinated. The enzyme molecule is a set of co-ordinated constraints which classifies its collisions with other molecules – the only sensitive collisions being with the substrate, but the evolutionary potential for a chaotic collection of enzymes is very low. In general we know that highly specialized co-ordination or function is the end result of evolution – not the origin of it.

2. *The Property of Arbitrariness*

Physicists have long recognized the distinction between initial conditions and laws of motion and they do not study initial conditions alone because these are incidental to the basic theories of matter. In living matter it is not easy to distinguish arbitrary, frozen accident structures from the underlying necessary constraints. This property of arbitrariness is characteristic of all symbol systems where the physical structure of the symbol vehicle has no direct relationship to what it signifies. Yet it is obvious that only definite structures can serve as symbol vehicles. It is now well established that many amino acid sequences will function to catalyze a specific reaction, and although there is no real evidence one way or the other, it does not appear unreasonable that life could exist just as well with entirely different codon-amino acid assignments. The reason or necessity for the number of bases and amino acids is equally obscure.

We expect, then, that at all hierarchical levels there are a multiplicity of physical constraints which effectively perform the same function, and that there are also a multiplicity of descriptions which can represent the same physical structure. But what is likely to be significant in our study of origins is the fact that the multiplicity of constraints which performs given functions will decrease as the specificity and precision of function increases. In the same way, the multiplicity of descriptions will decrease as the specificity and precision of its representation increases. Consequently, in the beginning stages of the evolution of co-ordinated constraints, at any level, we must expect inherent classifications with only the minimum specificity or resolution, even though as external observers we normally resolve much greater detail. Also, we must be particularly careful in our experiments with complex pre-co-ordinated systems not to impose our own clear forms of intelligent, but subjective, simplification, since this can easily override the primitive, diffuse forms of internal classifications. Experimental searches for the origin of co-ordination in evolution must therefore be of a different type than searches for existing structures and functions.

3. *Experimental Strategy*

Most experiments on the origin of life have been searches for increasingly complex molecules, characteristic of living cells, from simpler molecules. But this strategy requires the experimenter to impose artificial simplification on the primeval matrix. In these experiments it is generally assumed that by bringing together several diverse lines of synthesis there will emerge an even greater complexity which will be even more lifelike than the stage before. However, the difficulty is that if we assume that particular molecules such as polypeptides possess an inherent descriptive or functional property solely as a result of their particular chemical structure, then the 'bringing together' of a co-ordinated set of such molecules as descriptions and codes appears incredibly unlikely.

But from what I have said this may be asking the question backwards. We must also try experiments which begin with realistically complex but incoherent networks of reactions with no implication of functional behavior. The question, then, will not be one of 'bringing together' more and more complex functional parts, but of how any set of constraints separates out of the complex matrix to persist in a simpler form of behavior (Pattee, 1969, 1971b).

VII. SUMMARY

Life is distinguished from inanimate matter by the co-ordination of its constraints. The fundamental function of this co-ordination is to allow alternative descriptions to be translated into alternative actions. The basic example of this function is the co-ordinated set of macromolecules which executes the genetic coding. It is useful to think of such co-ordinated constraints as generalized language structures that classify the detailed dynamical processes at one level of organization according to their importance for function at a higher level. In this sense, co-ordinated constraints, language structures, alternative descriptions and hierarchical controls are inseparably related concepts. According to this picture, biological evolution is the product of natural selection not only within the external constraints of the environment but within the internal constraints of a generalized language structure.

The spontaneous origin of co-ordinated constraints at all hierarchical levels remains a difficult problem. I feel that we still have too vague a concept of language and too narrow a concept of natural law to come to grips with the key questions. We do not understand the physical basis of symbolic activity. Moreover, it is still not at all clear how serious a problem this may be. The history of the matter-symbol paradox and the problem of interpretation of measurement in quantum theory should give us great respect for the difficulties. Nevertheless, I do not see any way to evade the problem of co-ordination and still understand the physical basis of life.

This study has been supported by NSF grant no. GB16563.

State University of New York at Binghamton

BIBLIOGRAPHY

Ballantine, L. E.: 1970, 'The Statistical Interpretation of Quantum Mechanics', *Rev. Mod. Phys.* **42**, 358.

Bellman, R. and Kalaba, R.: 1964, *Selected Papers on Mathematical Trends in Control Theory*, New York: Dover Publications.

Bénard, H.: 1901, 'Les tourbillons cellulaires dans une nappe liquide transportant de la chaleur en régime permanent', *Ann. chim. Phys.* (7e série) **23**, 62.

Born, M.: 1964, *Natural Philosophy of Cause and Chance*, New York: Dover, p. 47.

Burgers, J. M.: 1963, 'On the Emergence of Patterns of Order', *Bull. Am. math. Soc.* **69**, 1.

Caskey, C. T.: 1970, 'The Universal RNA Genetic Code', *Q. Rev. Biophys.* **3**, 295.

Chandra, K.: 1938, 'Instability of Fluids Heated from Below', *Proc. R. Soc.* **A 164**, 231.

Crick, F. H. C.: 1958, 'The Origin of the Genetic Code', *J. Molec. Biol.* **38**, 367.

Crick, F. H. C.: 1968, *Of Molecules and Man*, Seattle: University of Washington Press.

Delbrück, M.: 1970, 'A Physicist's Renewed Look at Biology: Twenty Years Later', *Science, N.Y.* **168**, 1312.

Golden, S.: 1969, *Quantum Statistical Foundations of Chemical Kinetics*, Oxford University Press.

Haldane, J. B. S.: 1965, 'Data Needed for a Blueprint of the First Organism', *The Origins of Prebiological Systems* (ed. S. Fox), New York: Academic Press, p. 11.

Harris, Z.: 1968, *Mathematical Structures of Language*, New York: Interscience (John Wiley).

Jukes, T. H.: 1966, *Molecules and Evolution*, New York: Columbia Univ. Press.

Kendrew, J. C.: 1967, Review of *Phage and Origins of Molecular Biology* (ed. J. Cairns, G. Stent and J. Watson), Cold Springs Harbor Laboratory of Quantitative Biol. *Scient. Am.* **216**, 141.

Lanczos, C.: 1949, *The Variational Principles of Mechanics*, Toronto: University of Toronto Press.

Levins, R.: 1971, 'The Limits of Complexity', *Biological Hierarchies: Their Origin and Dynamics* (ed. H. Pattee), New York: Gordon and Breach.

Lewontin, R. C.: 1969, 'The Meaning of Stability', *Diversity and Stability in Ecological Systems*, Brookhaven Symposia on Biology, no. 22, New York.

Maxwell, J. C.: 1868, 'On Governor's, *Proc. R. Soc.* **16**, 270.

Michie, D.: 1970, 'Future for Integrated Cognitive Systems', *Nature* **228**, 717.

Morowitz, H. J.: 1968, *Energy Flow in Biology*, New York and London: Academic Press.

Orgel, L. E.: 1968, 'Evolution of the Genetic Apparatus', *J. Molec. Biol.* **38**, 381.

Pattee, H. H.: 1967, 'Quantum Mechanics, Heredity and the Origin of Life', *J. Theoret. Biol.*, **17**, 410.

Pattee, H. H.: 1969, 'Now Does a Molecule Become a Message?' *Devl Biol.*, Suppl.**3**, 1.

Pattee, H. H.: 1970, 'The Problem of Biological Hierarchy', *Towards a Theoretical Biology*, vol. 3 (ed. C. H. Waddington), Edinburgh University Press, p. 117.
Pattee, H. H.: 1971a, 'Can Life Explain Quantum Mechanics?', *Quantum Theory and Beyond* (ed. T. Bastin), Cambridge University Press, p. 307.
Pattee, H. H.: 1971b, 'The Physical Basis and Limits of Hierarchical Control', in *Biological Hierarchies: Their Origin and Dynamics* (ed. H. H. Pattee), New York: Gordon and Breach, p. 161.
Pattee, H. H.: 1971c, 'The Nature of Hierarchical Controls in Living Matter', *Textbook of Mathematical Biology* (ed. R. Rosen), New York: Academic Press.
Pattee, H. H.: 1971d, 'The Recognition of Description and Function in Chemical Reaction Networks', *Chemical Evolution and the Origin of Life* (ed. R. Buvet and C. Ponnamperuma), Amsterdam: North-Holland.
Platt, J. R.: 1961, 'Properties of Large Molecules that Go Beyond the Properties of Their Chemical Sub-Groups', *J. Theoret. Biol.* **1**, 342.
Poincaré, H.: 1957, *Méthodes nouvelles de la mécanique céleste*, New York: Dover Publications.
Prigogine, I., Lefever, R., Goldbetter, A., and Herschkowitz-Kauffman, M.: 1969, 'Symmetry Breaking Instabilities in Biological Systems', *Nature* **223**, 913.
Sherrington, C.: 1953, *Man On His Nature* (2nd ed.), New York: Doubleday Anchor Books, p. 80.
Stent, G.: 1968, 'That Was the Molecular Biology that Was', *Science* **160**, 390.
Thom, R.: 1968, 'Une théorie dynamique de la morphogenèse', in *Towards a Theoretical Biology*, vol. 1 (ed. C. H. Waddington), Edinburgh University Press, p. 152.
Thom, R.: 1970, 'Topological Models in Biology', *Towards a Theoretical Biology*, vol. 3 (ed. C. H. Waddington), Edinburgh University Press, p. 88.
Turing, A. M.: 1952, 'The Chemical Basis of Morphogenesis', *Phil. Trans. R. Soc.* **B 237**, 37.
Watson, J. D.: 1965, *The Molecular Biology of the Gene*, New York: Benjamin, p. 47.
Whittaker, E. T.: 1944, *A Treatise on the Analytical Dynamics of Particles and Rigid Bodies*, New York: Dover Publications.
Wigner, E. P.: 1967, *Symmetries and Reflections*, Bloomington and London: Indiana University Press, p. 167.
Woese, C. R.: 1967, *The Genetic Code*, New York: Harper and Row.

WILLIAM C. WIMSATT

COMPLEXITY AND ORGANIZATION*

In his now classic paper, 'The Architecture of Complexity', Herbert Simon observed that "... In the face of complexity, an in-principle reductionist may be at the same time a pragmatic holist." (Simon, 1962, p. 86.) Writers in philosophy and in the sciences then and now could agree on this statement but draw quite different lessons from it. Ten years ago pragmatic difficulties usually were things to be admitted and then shrugged off as inessential distractions from the way to the *in principle* conclusions. Now, even among those who would have agreed with the *in principle* conclusions of the last decade's reductionists, more and more people are beginning to feel that perhaps the ready assumption of ten years ago that the pragmatic issues were not interesting or important must be reinspected. This essay is intended to begin to indicate with respect to the concept of complexity how an *in principle* reductionist can come to understand his behavior as a pragmatic holist.

I. REDUCTIONISM AND THE ANALYSIS OF COMPLEX SYSTEMS

A number of features of the reductionistic orientation contribute to a point of view which is ill-suited to an adequate treatment of the concept of complexity:

(1) There is a bias towards theoretical monism. In biology and the social sciences, there is an obvious plurality of large, small, and middle-range theories and models, which overlap in unclear ways and which usually partially supplement and partially contradict one another in explaining the interaction of phenomena at a number of levels of description and organization.[1] In spite of this plurality all of the models, phenomena and theories in a given area (however that be defined)[2] tend to be treated as ultimately *derivative* from one primary theory. This means that questions concerning their relationships *to one another* tend to be ignored on the supposition that all will be made clear when their relationships to the perhaps as yet unknown reducing theory are determined.

But scientists must work with their plurality of incompletely articulated and partially contradictory, partially supplementary theories and models. The requirements of this situation have not been extensively investigated by philosophers, though this kind of theoretical pluralism has played an important role in the analyses of some biologists,[3] and I will argue that it is central to the analysis of our intuitive judgements of complexity.

(2) Given the difficulty of relating this plurality of partial theories and models to one another, they tend to be analyzed in isolation, with frequent unrealistic assumptions of system-closure (to outside 'disturbing' forces) and completeness (that the postulated set of variables are assumed to include all relevant ones.)[4] But these incomplete theories and models have, *individually*, impoverished views of their objects. Within each, the objects of the theory are just logical receptacles for the few predicates the theory can handle with manageable degrees of theoretical simplicity, accuracy, closure, and completeness. Nobody attempts to put these views together to see the 'resultant' objects. It is as if the five blind men of the legend not only perceived different aspects of the elephant, but, conscious of the tremendous difficulties of reconciling their views of the same object, decided to treat their views as if they were of *different* objects. The net result is often not to talk about objects at all, but to emphasize predicates, or the systems of predicates grouped together as theories or models.[5]

Thus, although biologists, social scientists, and others who work in areas where 'complexity' is a frequent term talk almost invariably of the complexity of *systems* (thereby meaning the objects, in the full-blooded sense, which they study), most analyses of complexity in the philosophical literature have been concerned with the simplicity or complexity of sets of predicates or of theories involving those predicates in a manner jumping off from the pioneering analyses of Nelson Goodman (1966, pp. 66–123).[6] But the Goodmanian complexity of a theory even if generally acceptable is a poor measure of the complexity of the objects of that theory unless the theory gives a relatively complete view of those objects. Short of waiting for the ultimate all encompassing reduction to an all-embracing theory, one can only talk about the internal complexity of our different theoretical perspectives or 'views' of an object. Nor could one avoid this conclusion by taking the complexity of the object as some aggregate of the complexities of the different views of the object, since part of its complexity would be located at the interfaces of these views – in those laws, correlations, and conceptual changes that would be necessary to relate them – and not in the views themselves.

(3) It thus would be profitable to see how we tend to relate our different views or theoretical perspectives of objects and in particular of complicated objects. This would be an enormous task for even two views where these views are theories or theoretical perspectives if we take that task to be equivalent to relating those theories conceptually, thus unifying them into a single theory. Fortunately, there are ways of relating the different views through their common referents or objects – if we are willing to assume, *contra* Berkeley and modern conceptual relativists (for different reasons in each case) that these different views *do* have common referents. An appreciable amount of work has been done by modern psychologists and others on the identification, reification (or, as Donald Campbell says, 'entification') delineation, and localization of objects and entities.[7] Most interesting in the present context is the emphasis on the importance of boundaries of objects. This work would have been ignored just a few years ago as irrelevant to philosophy of science and appeals to it would have been regarded as 'psychologism'. Nonetheless, it has an important bearing on the ways in which we decompose a system into subsystems and upon how we conceive the results.

II. COMPLEXITY

There are a number of factors relevant to our judgments of the complexity of a system, though I will here discuss only two, which I will call 'descriptive' and 'interactional' complexity, respectively.[8]

Kauffman (1971) advances the idea that a system can be viewed from a number of different perspectives, and that these perspectives may severally yield *different non-isomorphic* decompositions of the system into parts. A modification of his point has an application in the analysis of complexity: systems for which these different perspectives yield decompositions of the system into parts whose boundaries are not *spatially coincident* are properly regarded as more descriptively complex than systems whose decompositions under a set of perspectives are spatially coincident.[9]

Assume that it is possible to individuate the different theoretical perspectives, T_i, applicable to a system. Each of these T_i's implies or suggests criteria for the identification and individuation of parts, and thus generates a 'decomposition' of the system into parts. These decompositions, $K(T)_i$, I will call 'K-decompositions'. The different $K(T)_i$ may or may not give spatially coincident boundaries for some or for all of the parts of the system. The boundaries of two parts are spatially coincident if and only if for any two points in a part under $K(T)_j$ these points are in a single part under $K(T)_k$, and conversely. This is, of course, spatial coincidence defined relative to $K(T)_j$ and $K(T)_k$, but it can be generalized in an obvious manner. If all of a set of decompositions, $K(T)_i$, of a system produce coincident boundaries for all parts of the system, the system will be called *descriptively simple* relative to those $K(T)_i$.[10]

If two parts from different $K(T)_i$ are not coincident, but have a common point which is an interior point of at least one of them, then there are a number of different mapping relations which can hold between their boundaries, each of which contributes to its *descriptive complexity*. Specifying these mapping relations for all parts of the system under both decompositions gives a complete description of this complexity from a set theoretic point of view.[11]

Different level theories of the same system (e.g., classical *versus* statistical thermodynamics) generally exhibit many-one mappings from the microlevel to the macro level. Far more interesting, however, is the relation between different $K(T)_i$'s which apply at roughly the same spatial order of magnitude. Thus, the decomposition of a piece of granite into subregions of roughly constant chemical composition and crystalline form, $K(T)_1$; density, $K(T)_2$; tensile strength (for standard orientations relative to the crystal axes), $K(T)_3$; electrical conductivity, $K(T)_4$; and thermal conductivity, $K(T)_5$, will produce at least roughly coincident boundaries. The granite is thus descriptively simple relative to these decompositions (see Figure 1).

By contrast, decomposition of a differentiated multi-cellular organism into parts or systems along criteria of being parts of the same anatomical, physiological, biochemical, or evolutionary functional system; into cells having common developmental fates or potentialities, or into phenotypic features determined by common sets of genes will, almost part by part and decomposition by decomposition, result in mappings which are not

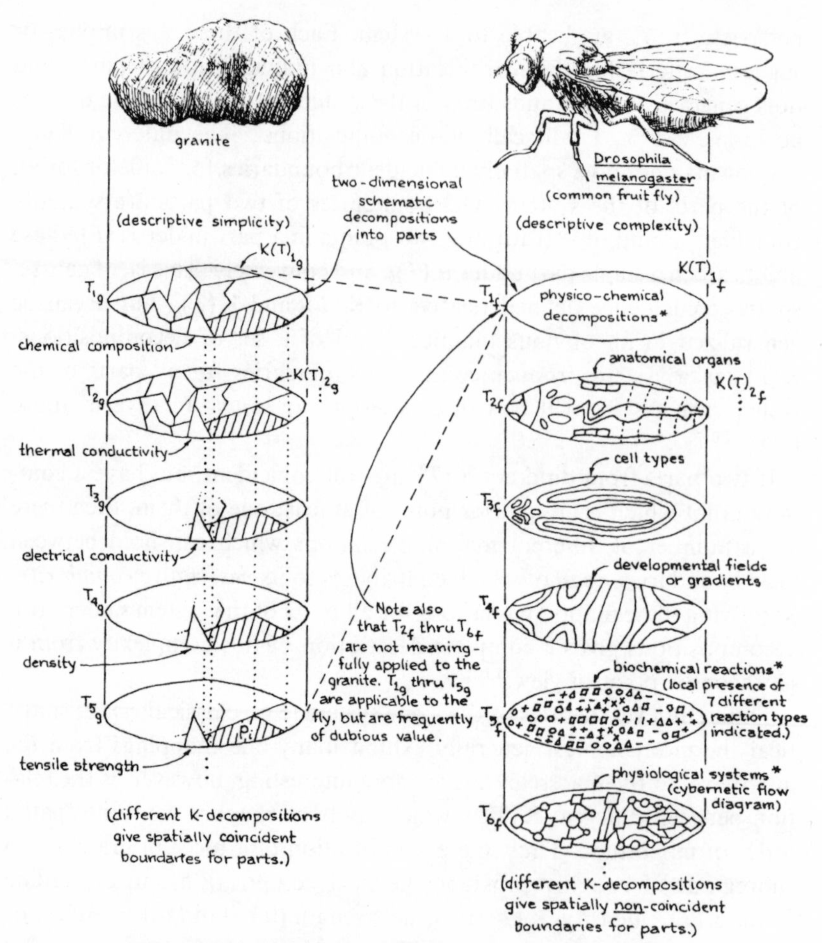

Fig. 1. Descriptive simplicity and complexity.

1-1 – which are not even isomorphic, much less coincident. This surely involves substantial 'descriptive complexity'.

In biology, at least, the picture is further complicated by another factor – that different theoretical perspectives are not nearly as well individuated as in the physical sciences. Thus, anatomical, physiological. developmental, and biochemical criteria, not to mention paleontological information and inferences of phylogenetic relations and homologies, all interact with criteria of evolutionary significance in the analysis of organisms into functional systems and subsystems. This borrowing of criteria for individuation of parts from different and diverse theoretical perspectives is one of the factors which make functional organization in general and biology in particular such a conceptual morass at times. This is further discussed in Wimsatt (1971, Chapters 6 and 7).

Descriptive complexity has a point, in large part because of the existence of what I will call 'interactional complexity'. This is a kind of measure of the complexity of the causal interactions of a system, with special attention paid to those interactions which cross boundaries between one theoretical perspective and another.

Many systems can be decomposed into subsystems for which the *intra-systemic* causal interactions are all much stronger than the *extra-systemic* ones (see Figure 2). This is the concept of 'near-complete decomposability' described by Simon and others (see, e.g., Ando *et al.*, 1963; Levins, 1970, and Simon, 1962). Such systems can be characterized in terms of a parameter, ϵ_c, which depends upon the location of the system in phase space and is a measure of the relative magnitudes of inter- and intra-systemic interactions for these subsystems.[12] This notion will be called S-ϵ_c-decomposition, and the subsystems produced according to such a decomposition will be denoted by $\{s^i_{\epsilon_c}\}$. A system is *interactionally simple* (relative to ϵ_c) if none of the subsystems in $\{s^i_{\epsilon_c}\}$ cross boundaries between the different K-decompositions of a system, and *interactionally complex* in proportion to the extent to which they do (see Figures 2b and c).

The importance of interactional complexity is as follows: The parameter ϵ_c can also be used as a measure of the accuracy of a prediction of the behavior of the system under a given decomposition if inter-systemic interactions are (perhaps counterfactually) assumed to be negligible. The larger ϵ_c is for that system under that decomposition, the less accurate the prediction. Alternatively, if a specific value of ϵ_c, say ϵ_c^*, is picked in order to achieve a certain desired accuracy of prediction of the behavior of a system, and the system turns out to be interactionally complex for that value of ϵ_c, then the investigator *must* consider the system from more than one theoretical perspective if he is to be able to make predictions with the desired level of accuracy.

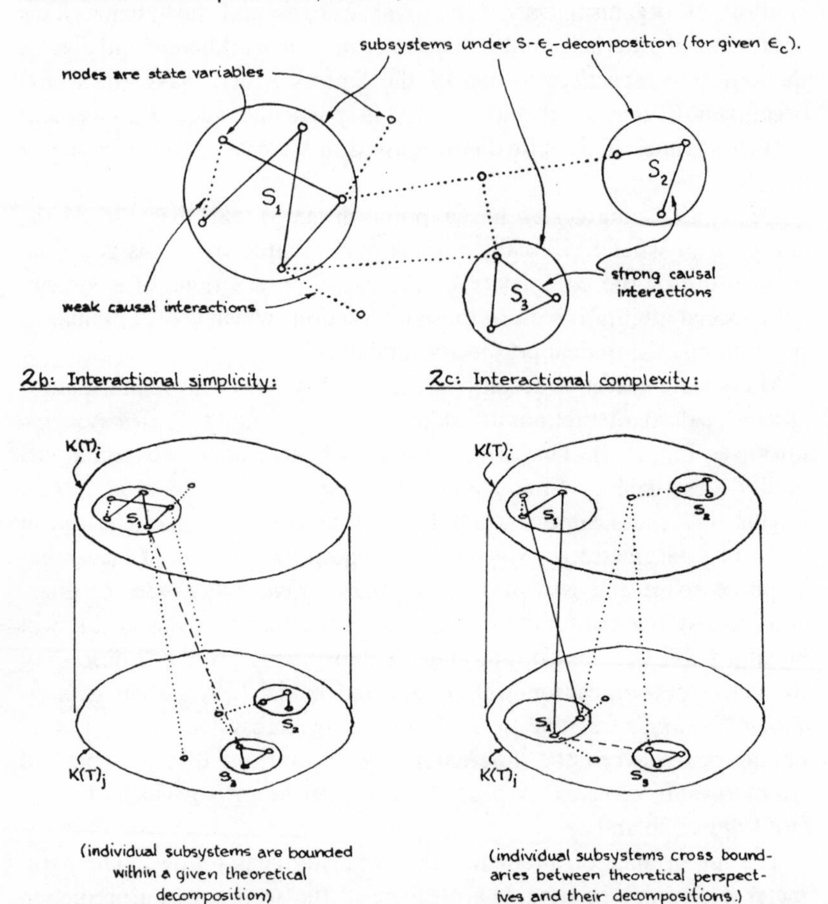

Fig. 2. Near-decomposeability and interactional complexity. Note that views in this figure represent decompositions into *sets of state variables* in different perspectives, whereas those in the preceding figure were in terms of *sets of parts*. The two are not to be confused, although they are not unrelated. Thus, strong causal correlation among state variables would amount to a generalized version of Campbell's (1958) 'common fate' criterion for the individuation of objects (or parts of objects), though the other criteria he discusses, such as spatial proximity and similarities in other properties would conflict with this, and the entities we pick out as parts would represent a compromise.

Obviously, the value of ϵ_c^* is an important factor here. If a system is interactionally simple for a given value of ϵ_c^* it will remain so for all larger values of that parameter, since larger values of ϵ_c^* denote lower standards of predictive precision. A system which is interactionally complex for one value of ϵ_c^* may be interactionally simple for larger values. Thus the interactional simplicity of a system also increases as the minimum value of ϵ_c^* for which it is interactionally simple decreases.

In any case of interactional complexity, the investigator is forced to attempt to relate the different *K*-decompositions in order to trace and analyze the causal networks in the system. This is a relatively straightforward task if the system is descriptively simple, since the spatial decomposition of the system into parts in one perspective automatically gives the spatial decompositions (but not all the properties!) for the other perspectives. But if the system is descriptively complex and is also interactionally complex for more than a very small number of interactions, the investigator is forced to analyze the relations of parts for virtually all parts in the different decompositions, and probably even to construct connections between the different perspectives at the theoretical level.

Many investigators of biological, social, and other complicated systems have claimed that no one perspective appears to do justice to their objects of study, or, somewhat more obscurely, that their systems are unanalyzeable into component parts – or at least that there is no clear way to perform this analysis. Kauffman (1971) and Levins (1970) both claim that in complex systems there are a number of different possible decompositions and often no way of choosing between them. Levins' remarks suggest something even stronger:

> ... [for] a system in which the component subsystems have *evolved* together [the subsystems] are not even obviously separable. ... It may be conceptually difficult to decide what are the really relevant component sub-systems.... This decomposition of a *complex* system into sub-systems can be done in many ways... it is no longer obvious what the proper subsystems are. (Levins, 1970, p. 76)

It seems reasonable to construe these claims in part as claims that the systems in question are interactionally and descriptively complex. Levins' claim about evolved systems raises further questions about the origins of complexity.

III. EVOLUTION, COMPLEXITY, AND FUNCTIONAL ORGANIZATION

Why should some systems be interactionally complex? If this question is examined in an *a priori* manner, it is perhaps more amazing that some systems should be interactionally simple. The systems, as analyzed to apply S-decompositions, are composed of state variables and causal relations between them. State variables are properties, and different properties, picked at random, would be expected to be found as properties of parts or systems in different theoretical perspectives. Unless causal relations or state variables were organized in a rather specific way relative to the different theoretical perspectives, one would expect systems to be interactionally complex.

But isn't this what successful theories do for us – that is, isn't it the mark of a successful theory of a range of phenomena that it unites and embraces the causally relevant parameters and state variables within a single theoretical perspective? This question suggests that if our theories are successful, then they should produce descriptions of systems according to which the systems are interactionally simple. I think that this would be to put the conceptual cart before the phenomenal horse. As the criterion (one of many) for the adequacy of a theory of a system, this statement seems correct but it is hardly sufficient. Also, one should not automatically assume that our *existing* theories are adequate theories of complex systems. The belief that they are is based largely on a still unfilled reductionist promise.[13]

It is true that our existing theories work well on simple systems – simple in part (but only in *part*) because these theories are constructed so as to render them interactionally simple. But one cannot assume that it is always possible to find a theory which will render a given system interactionally simple. As W. Ross Ashby said some years ago:

> Science stands today on something of a divide. For two centuries it has been exploring systems that are either intrinsically simple or that are capable of being analyzed into simple components. The fact that such a dogma as 'vary the factors one at a time' could be accepted for a century shows that scientists were largely concerned in investigating such systems as *allowed* this method; for this method is often fundamentally impossible in complex systems. (Ashby, 1956, p. 5)

Nonetheless, the *a priori* argument for the interactional complexity of systems given at the beginning of this section is intrinsically defective, for it ignores facts which every scientist takes for granted – namely, that systems *are* constrained and that state variables are not causally related at random. Thus, some of the arguments of 'The Architecture of Complexity' (Simon, 1962) appear to suggest that there *are* physical constraints on evolving systems which would render them interactionally simple and that they are descriptively complex only after the manner of a multi-level theory, with many-one mappings of parts from lower to higher levels. It is an open question whether this is indeed the conclusion Simon intends, but in any case, it seems to me to be mistaken.

Simon makes an elegant case for the conclusion that all evolved systems containing many causally interrelated parts will be hierarchially organized. This is *via* an argument that for two systems of roughly equal complexity, each to be built out of simple components (and each subject to perturbations tending to cause decomposition), one which arises out of the successive aggregation of individually stable subassemblies into larger subassemblies will have a much higher probability (or lower expected time) of formation than one which does not. (See his parable of the watchmakers, 'Tempus' and 'Hora' in 1962, pp. 90–95.) Thus, one would expect that at least the vast majority of complex evolved systems would be hierarchially organized.

But Simon's use of the concept of near-decomposability in the same article sometimes appears to suggest that he believes such hierarchial systems to be nearly-decomposable in a nestable manner – with smaller subassemblies (at lower levels) having successively stronger interactions and no S-decompositions crossing boundaries between levels. Indeed, if the subassemblies which go to make up a hierarchially organized system are stable, isn't it the case that these subassemblies at all levels must be the subsystems which emerge for various characteristic values of ε_c in the different level S-decompositions of a system?[14]

To accept this opinion is to fail to distinguish the decomposability or stability of the subassemblies before they aggregate from their decomposability or stability (in isolation) after they have aggregated – especially a long time afterwards, when they have had time to undergo a process of mutually coadaptive changes under the optimizing forces of natural selection. The optima and conditions of stability for a system of aggregated parts are in general different in a non-aggregative way from the optima and conditions of stability for its parts taken in isolation.[15]

Naive design procedures in engineering, in which the organization of the designed system was made to correspond to the conceptual breakdown of the design problem into different functional requirements, with a 1-1 correspondence between physical parts and functions, have given way to more sophisticated circuit minimization and optimal design techniques. These methods have led to increases in efficiency and reliability by letting several less complicated parts jointly perform a function that had required a single more complicated part, and simultaneously letting these simpler parts perform more than one function (in what might before have been distinct functional subsystems) where possible. This has the effect of making different functional subsystems more interdependent than they had been before, and of encouraging still further specialization of function, and interdependence of parts. It is reasonable to believe that the optimizing effects of selection do just this for evolving systems, and if so that hierarchially aggregating systems will tend to lose their neat S-decomposability by levels, and become interactionally complex.[16]

This argument is buttressed empirically by considering what happens when natural organized systems are artificially decomposed into subassemblies which are the closest modern equivalents of the subassemblies from which the systems presumably came. Few modern men (or better, couples, for bisexual organisms) could survive for long outside of our specialized society. The same goes for mammalian cells – at least under naturally occurring conditions, even though multi-cellular organisms are presumably descended from uni-cellular types. Even many bacteria cannot survive and reproduce outside of a reasonably sized culture of similar bacteria. The current belief of some biologists is that mitochondria and chloroplasts originated as separate organisms, and acquired their present role in animal and plant cells via parasitic or symbiotic association. According to this view these once independent organisms (or 'subassemblies') are now so totally integrated with their host that only their independent genetic systems are a clue to their origin (Lynn Margulis, 1971).

With increasing differentiation of function in systems, different subsystems become dependent, not only upon the presence of other subsystems, but also upon their arrangement. Experiments with the transplantation of imaginal discs in the larvae of holometabolous insects demonstrates that the developmental fate of these discs (which develop into organs in the pupal stage) depends not only upon the disc and its 'age' but in some cases also upon its location. Slime moulds are remarkable among organisms for their ability to function undisturbed if they are pushed through a sieve in their undifferentiated form. This is quite unusual for multicellular animals. An adult man or mouse would do considerably less well under similar conditions.

The end result, I think, is that one cannot expect hierarchial organization resulting from selection processes to be S-decomposable into different levels – or at least, not into the different levels of organization relevant at the time of aggregation. Nor could one expect that such organization would be interactionally simple when decomposed according to any other theoretical perspectives bearing no intrinsic relation to the selection forces acting upon it.

Nor, unfortunately, is there even any guarantee that functional organization in terms of the operation of selection mechanisms (Wimsatt, 1971, 1972) is the road to a descriptively and interactionally simple analysis of such systems. The use of functional criteria might lead to more simplicity perhaps, (I would argue strongly that they would!) but functional systems are still subject to physical, chemical, and biological constraints at a number of levels, and never completely lose the marks of the systems which they have evolved from – even down to the level of the basic chemical elements of which they are composed. These simultaneous constraints seem almost certain to result in interactional complexity.

IV. COMPLEXITY AND THE LOCALIZATION OF FUNCTION

In the last section, I attempted to suggest how considerations of efficiency in evolution would lead to the co-adaptation and increased interdependence of parts of a functional system, and that this would lead to increases in the descriptive and interactional complexity of that system. One aspect of this increase in complexity is a trend away from 1 to 1 mappings between functions and recognizeable physical objects. It seems plausible to suggest that one of the main temptations for vitalistic and (more recently) anti-reductionistic thinking in biology and psychology is due to this well-documented failure of functional systems to correspond to well-delineated and spatially compact physical systems. (Richard Gregory's excellent and suggestive remarks (1959, 1961, 1962) on the problem of localization of function in the brain (and his humorous illustrations with engineering examples) offer too many riches to mine them superficially here, but they are heartily recommended to the reader!)

It is only too tempting to infer from the fact that functional organization does not correspond neatly to the most readily observable physical organization – the organization of physical objects – to the howling *nonsequiter* that functional organization is not physical.[17] A tantalizing explanation for our tendencies in this direction is hinted at in Donald Campbell's (1958) suggestion that our willingness to 'entify' things as real is directly proportional to the number of coincident boundaries we find.[18]

Thus, organisms count as systems because of the coincidence of a number of boundaries at, roughly, the surface of the skin or its functional equivalent. In addition to the relatively discontinuous change of a number of physical and physiological variables at what is taken to be the organism/environment interface, the systems thus picked out are usually relatively independent agents biologically, since what we call organisms usually live, die, metabolize, mate, and move relatively independently of one another. Indeed, we tend to marvel at the problem cases – eucaryotic cells, slime moulds and social insects – where one or more of these boundaries do not coincide, and have problems deciding whether a given unit is an organism, assemblage of organisms, or part of an organism.

But what holds true of an organism (that many boundaries coincide at its skin) need not hold true of its parts. Inside the complex system, there is a hegemony of different constraints and perspectives and boundaries. If what Campbell says is correct, this hegemony leads us to be slow or dubious about objectifying the parts of such a system. What is unobjectifiable is to that extent unphysical, and so functional organization becomes a thicket for vital forces and mental entities. It is no accident that those systems for which vitalisms and mentalisms have received spirited defenses are those systems which are also paradigmatically complex.

The difficulties with the spatial localization of function in complexly organizes systems suggest a more positive approach to at least one aspect of the psycho-physical identity thesis.[19] In 1961, Jerome Shaffer took account of the frequently discussed non-spatiality of mental events and proposed that the spatial location of corresponding brain events could, as a *convention*, be taken as the location of the corresponding mental events. Norman Malcolm (1964, p. 119, note 6) argued that as this would be just a convention, talk of the location of mental events would just be taken as a shorthand way of locating the corresponding physical events.[20] Malcolm and Shaffer's later discussions (Malcolm, 1971; Shaffer, 1965) raise other issues that lead off in other directions, but neither of them seem to take seriously the implications of the interactional and descriptive complexity of functional organization.

It is not merely that functionally characterized events and systems are spatially distributed or hard to locate exactly. That much can be said for bulk terms like water, or even more, like fog and smoke. But ordinary bulk matter, like ordinary fields, can be conceived of as homogeneous. The problem is that a number of different functionally characterized systems, each with substantial and different powers to affect (or effect) behavior appear to be as interdigitated and intermingled as the infinite regress of qualities-within-qualities of Anaxagoras' seeds. Furthermore, the high degree of redundancy and plasticity of the cortex – pointed to by the work of Lashley and his doctrine of 'equipotentiality' – make it seem as if functional systems are not essentially located *anywhere*. The apparent contradiction of having a number of (functionally) distinct organized systems, each of which appears to occupy all or most of the same space, and at the same time, none of it, leads to the tendency to deny spatiality at all,[21] or in less extreme forms, to invent special kinds of quasi-physical predicates. The non-physicalistic and anti-reductionistic strains in the writings of Jerry Fodor (1965, 1968), Hilary Putnam (1967), and Margaret Boden (1970, 1972) who speak with a strange ambivalence about their 'functional roles', 'programs', and 'internal models' reflect this no less than did 19th-century vitalism.

This tendency is I think at least partially explained by the 'pathological' behavior of boundaries in complex systems and what this does to our normally workable criteria for spatial objectification. This way of putting it may be misleading for it could be argued that we conceived of the mental as non-spatial long before we had any idea (gleaned from neurophysiology) about the problems with localization of function in the cerebral cortex. But there are alternatives to suggesting that active awareness of these problems led us to conceive of the mental as non-spatial. One might suppose, for example, that spatial objectification is an *active* hypothesis that we apply to those groups of phenomena which tie up into sufficiently neat packages in the right ways. On this account, the mental realm was not denied spatiality. It just has not yet been added to the list.

The denial of spatiality as a category to mental entities has the ring of a philosopher's invention. Now that we see one reason why we might not have been able to attribute an exact location to mental events, we can wonder perhaps whether the common man need ever have been more than an agnostic about their spatiality. If he has asserted more, he has probably only succumbed to the seductions of philosophy – deifying as conceptually true a hypothesis which future writers may well decide was empirically false.[22]

Dept. of Biology, Dept. of Philosophy, and Committee on the Conceptual Foundations of Science, University of Chicago

NOTES

* Parts of this paper are based on my doctoral dissertation (Wimsatt, 1971) and on work done during the tenure of a Woodrow Wilson Dissertation Fellowship at the University of Pittsburgh and a post-doctoral fellowship with the Committee on Evolutionary Biology at the University of Chicago, supported by the Hinds Fund for Studies in Evolution. I gratefully acknowledge their support.

[1] This also seems to be true in physics in active research areas such as (but not limited to) meteorology and magnetohydrodynamics, though those arguments for complexity based upon evolutionary phenomena have no obvious application in these areas.

[2] The problem of how to delineate the domain of a scientific theory has received relatively little attention until recently. See Shapere (1971) for a close examination of these issues.

[3] It is an intriguing fact that discussions of theoretical pluralism have arisen in the context of scientific change, but implications of this pluralism (including the hotly debated problems of translation and meaning variance) have generally not been investigated for *simultaneously held* partially overlapping theories and models. Richard Levins' views on the nature and use of theories and models in biology (Levin, 1966, and 1968, Chapter 1) are a notable exception, and Stuart Kauffman's (1971) views on the plurality of ways of seeing or describing systems are also suggestive. It is tempting to dismiss this as a kind of pre-paradigm or multi-paradigm science, but this is to ignore the fact that many of the scientists must then be viewed as simultaneously (or alternatively) using several of the paradigms at any given time.

[4] See Bergmann (1957), pp. 93–96 for discussion of the assumptions of closure and completeness of a system.

[5] On this, see also Bishop Berkeley (1709), paragraphs 41–51, especially 48–50. There

should be obvious application of these remarks to the problem of interrelating information from different sensory modalities in the construction of 'external objects'. Why, and when, do we objectify?

[6] In the light of this fact, it is quite ironic that Goodman's calculus of individuals or something quite like it (*Ibid.*, pp. 46–61) appears to be admirably suited to the analysis of what I have described below as 'descriptive complexity'. I thank Leonard Linsky for this reference.

[7] Some of the most interesting contributions in this area are Campbell's (1958, 1959, 1972). Campbell's most relevant point in the present context is an attention to the boundaries of systems. He argues, in part, that we consider objects the more real and substantial the more there is a coincidence of boundaries on different criteria of individuation. He applies this, among other things to the individuation of social units (1958) and the order in which we learn different kinds of concepts (1972).

For more analyses of the importance of the boundaries of systems on different criteria, see Platt (1970) and Simon (1962, 1969). On the localization of functions and questions concerning whether functions are objects, see Gregory (1959, 1961, 1962) and Wimsatt (1971, 1972) and this article.

[8] These two factors represent in part an attempt to give a conceptual basis and motivation for some of the ideas expressed by Richard Levins (1970, 1973), and personal conversations.

[9] Kauffman is primarily concerned to argue that these perspectives may not be reducible, one to another, and so he emphasizes the possibility of non-isomorphism. Spatial coincidence of the parts' boundaries under two decompositions implies spatial isomorphism of the decompositions but not conversely, and it is coincidence, not isomorphism, which is important here.

[10] The importance of being able to unambigously identify and individuate the different theoretical perspectives appropriate to a system cannot be underestimated. *Any* system is trivially descriptively simple relative to just a single T_i and $K(T)_i$. This case thus must be ruled out as a case of descriptive simplicity, with the requirement that two or more theoretical perspectives and correlative decompositions be considered. But if criteria for individuating these are in doubt, then it might be open for someone to claim that a system purportedly classified as descriptively simple relative to a set of perspectives (like the granite case discussed below) is not, because the perspectives are not in fact distinct, but are parts of the *same* perspective.

I believe that it is possible to give criteria for identifying and individuating theoretical perspectives, and in a way that a given T_i, for a given system and set of conditions of and on that system, has a unique $K(T)_i$. Further work needs to be done on this matter, however.

[11] For two such parts considered as an ordered pair, the mapping relations 1-1, 1-many, 1-part, 1-many part, many-1, part-1, and many part-1 are exhaustive. There are also many other considerations which are important, such as whether the mappings are continuous. Any discontinuities in a mapping will greatly increase the complexity of description of a system.

[12] A number of measures are possible, and different ones would be preferable under different circumstances. For illustrative purposes, ε_c can be thought of as the average strength of intersystemic interactions divided by the average strength of intrasystemic interactions. This ranges between 0 and 1, approaching 0 as intersystemic interactions become neglegible.

[13] I wish to emphasize that I am *not* an anti-reductionist. There is not even any reason to believe that these theories are *incorrect* for what we intuitively regard as complex systems. On the other hand, there is no reason to believe that they will describe other systems as simply as the systems they were generated to explain, and adequate theories of complex

systems may require new categories if they are to be described and analyzed more simply. Statistical mechanics owes its acceptability to just such considerations.

[14] The view criticized here is Levins' reading of Simon, and this line of attack is basically an elaboration and adaptation of a point Levins has made many times in conversation. Levins' interpretation seems not to be too idosyncratic: Howard Pattee (1970) also reports Simon's analysis as showing that hierarchially organized structures are nearly-decomposa-ble (see note 10, p. 136). Nonetheless, I am not convinced that this is a fair reading of Simon. The evidence is at least equivocal: In addition to the 'stable subassemblies' argument, this point of view is also suggested by Simon's discussion of near-decomposability in physical systems (pp. 103–104) and the first part (pp. 114–115) of his discussion of genetic control of development in terms of hierarchially organized computer programs. (The latter part, pp. 116–117, appears to suggest that Simon is aware of the issues raised here.)

[15] Some of the most interesting cases of this are to be found in so-called 'cooperative phenomena' in polymers. Thus, it has been suggested that hemoglobin (a tetramer) might have evolved from its monomeric precursor because interactions among the 4 subunits facilitate binding of the hemes (which carry oxygen). See Jukes (1966; chapter 5) for further details.

[16] If this argument holds, it produces firmer grounds than the 'randomness' assumption for a belief that hierarchially aggregating systems will become interactionally complex through co-adaptation. In fact the argument here seems to be just the other side of the argument made by von Neumann and Morgenstern (e.g., (1946), pp. 11–12) that a game-theoretic interaction is not a simple maximum or minimum problem. Their point there is that with two or more individuals acting, there are two (or more, as appropriate) functions each of which some individual is trying to maximize, subject to the constraints of the actions of the others, and there is no overall function which tends to an extremal value.

In our case, we begin with independent systems with their own (relatively) independent optima. When they aggregate and begin to coevolve, their individual survival becomes dependent upon the 'collective welfare' with something more akin to a single optimum. I will assume, as von Neumann and Morgenstern do, that it is overwhelmingly improbable that the conditions for individual and collective optima coincide. The failure of this coincidence results in selection for interactional complexity.

Indeed, the situation is more complicated than indicated here, in which it is suggested that either individual optima or the group optimum dominates absolutely. In many cases (see Lewontin, 1970) selection can operate independently upon units at two or more levels of the *same* hierarchy, in concert or in opposition. In these cases, we have a hegemony of forces, and a compromise among conflicting optima is inevitable.

[17] I have attempted detailed analysis of functional organization in Chapters 6 and 7 of Wimsatt (1971). Some of the conceptual complexities inherent in this problem are discussed in Wimsatt (1972).

[18] Campbell has argued in conversation that even for those systems which I have chosen to call complex in one of the two above ways, there is still probably a coincidence of boundaries for the vast majority of properties. This seems reasonable: if there were an arbitrary thicket of overlapping boundaries we probably would not be able to pick out *any* systems. Furthermore, the use of modular construction techniques (both as new as third generation computers and as old as multicellularity) would appear to imply the coincidence of at least *many* boundaries at the boundaries of the modules.

Simon has recently made much the same point, arguing that these remarks on complexity are a second-order approximation to modify the first-order approximation of describing systems as nearly decomposable (personal communication). I strongly agree with both of these remarks, though not with the implication one might draw that '2nd order' effects are, in the relevant sense, always ignorable relative to '1st order' effects. Whether they are or not depends upon which phenomena you are interested in. Weak hydrogen bonding is, chemically speaking, energetically negligible. It happens however to be crucial to the proper functional behavior of biologically important macro-molecules.

Another interesting and powerful line is suggested by Levins (1973) when he argues that an initially arbitrarily complex system will tend towards greater simplification, and perhaps this is implicit in Campbell and Simon's arguments. Thus bounded (by coevolution) away from aggregative simplicity and (by the need to have parts of the system responding semi-autonomously to selection pressures) away from total interactional complexity, it would appear that there must be a relatively stable intermediate level of complexity. Whether this would involve an intuitively and immediately recognizable degree of modularity is an open and important question. Might it be, for example, that there would be different modes of modularity or near decomposability for systems that arise by aggregation of stable sub-assemblies (*à la* Simon) and for those which arise by specialization and differentiation of sub-parts of a single system (*à la* Levins)? This is not an 'academic' question. A recent challenge to Margulis's (1971) aggregative account of the origin of eucaryotic cells is offered by Raff and Mahler (1972), who suggest that eucaryotic cells evolved by specialization. It would be extremely useful to have criteria for adjudicating this and many other similar disputes.

[19] Jaegwon Kim (Kim, 1971, pp. 329–334) discusses views which both suggest that problems with spatial localization of function has been a problem in this context and seems himself (p. 334) to suppose that events of this type must have a precise location.

[20] Obviously, talk about the location of mental events in this sense is not intended to apply to those cases where we most frequently *do* talk about location – cases such as feeling an 'itch in the leg' or a 'pain in the tooth'. These kinds of cases are probably best explicated as Margaret Boden does (1970, pp. 207–209), as events which are occurrences in an 'internal model' representing states of the organism. As such the events can refer to occurrences at the locations in question without themselves occurring at those locations. I discuss this problem and several others not mentioned here in Wimsatt (1973). See also Globus (1972).

[21] Keith Gunderson (1970, p. 303) has suggested that one of the problems with conceptualizing the self arises from the feeling that the concept of the self requires that one entity be in *two* places at the same time (In one place as observ*er* and in another place as observ*ed*.) It seems reasonable to suggest that the belief that such spatial paradoxes bedevil the mental realm is or was influential in denying its spatiality.

[22] I think that this will be the judgment of the future even though I admit that these points about functional organization and localization go but one small and not even very important part of the way there. In Wimsatt (1973) I argue that there are three classes of paradoxical phenomena to be handled in attributing spatiality to the mental realm: those respectively associated primarily with 1st person knowledge, with 3rd person knowledge, and with interactions between 1st and 3rd person knowledge. The terms, '1st person knowledge' and '3rd person knowledge' must first be reanalyzed in terms of assymmetries among and limitations on the locational information given by the various sensory modalities. The functional localization problems discussed here are then seen to be pure 3rd person problems, whereas most of the interesting problems discussed by philosophers are seen to be one of the other two types. Interestingly, there is a sense in which 1st person knowledge of the mental realm *is* non-spatial, though this fact turns out to be of no comfort for those who would wish to use it against materialism. This argument builds substantially upon key points raised by Gunderson (1970) and Globus (1972).

BIBLIOGRAPHY

Ando, Albert, Fisher, F. M., and Simon, H. A., 1963, *Essays on the Structure of Social Science Models*, MIT Press, Cambridge.

Ashby, W. R., 1956, *An Introduction to Cybernetics*, Chapman and Hall, Ltd., reprinted in 1960 by W. J. Wiley, New York.

Berkeley, Bishop George, 1709, *An Essay Towards a New Theory of Vision*.

Bergmann, Gustav, 1957, *Philosophy of Science*, University of Wisconsin Press, Madison.

Boden, Margaret, 1970, 'Intentionality in Physical Systems', *Philosophy of Science* **37**, 200–214.

Boden, Margaret, 1972, *Purposive Explanation in Psychology*, Harvard University Press, Cambridge.

Campbell, Donald T., 1958, 'Common Fate, Similarity, and Other Indices of the Status of Aggregates of Persons as Social Entities', *Behavioral Science* **3**, 14–25.

Campbell, Donald T., 1959, 'Methodological Suggestions from a Comparative Psychology of Knowledge Processes', *Inquiry* **2**, 152–182.

Campbell, Donald T., 1972, 'Ostensive Instances and Entitativity in Language Learning', forthcoming in N. D. Rizzo (ed.), *Unity Through Diversity*, vol. III, Gordon and Breach, New York.

Fodor, Jerry, 1965, 'Functional Explanation in Psychology', in Max Black (ed.), *Philosophy in America*, Allen and Unwin, London, pp. 161–179.

Fodor, Jerry, 1968, *Psychological Explanation*, Random House, New York.

Globus, Gordon G., 1972, 'Biological Foundations of the Psychoneural Identity Hypothesis', *Philosophy of Science* **39**, 291–301.

Goodman, Nelson, 1966, *The Structure of Appearance*, second edition, Bobbs-Merrill, Indianapolis.

Gregory, R. L., 1959, 'Models and the Localization of Function in the Central Nervous System', *National Physical Laboratory Symposium No. 10*, pp. 671–681. Reprinted in C. R. Evans and A. D. J. Robertson (eds.), *Key Papers: Cybernetics*, Butterworths, London, pp. 91–102.

Gregory, R. L., 1961, 'The Brain as an Engineering Problem', in W. H. Thorpe and O. L. Zangwill (eds.), *Current Problems in Animal Behavior*, Cambridge University Press, London, pp. 307–330.

Gregory, R. L., 1962, 'The Logic of the Localization of Function in the Central Nervous System', in R. Bernard and B. Kare (eds.), *Biological Prototypes and Synthetic Systems*, vol. 1, Plenum Press, New York, pp. 51–53.

Gregory, R. L., 1969, 'On How So Little Information Controls so much Behavior', in C. H. Waddington (ed.), *Towards a Theoretical Biology*, vol. 2, University of Edinburgh Press, Edinburgh, pp. 236–247.

Gunderson, Keith, 1970, 'Asymmetries and Mind-Body Perplexities', in M. Radner and S. Winokur (eds.), *Minnesota Studies in the Philosophy of Science*, vol. 4, University of Minnesota Press, Minneapolis, pp. 273–309.

Jukes, Thomas H., 1966, *Molecules and Evolution*, Columbia University Press, New York.

Kauffman, Stuart A., 1971, 'Articulation of Parts Explanations in Biology', *Boston Studies in the Philosophy of Science* (ed. by R. C. Buck and R. S. Cohen), vol. 8, pp. 257–272.

Kim, Jaegwon, 1971, 'Materialism and the Criteria of the Mental', *Synthese* **22**, 323–345.

Levins, Richard, 1966, 'The Strategy of Model Building in Population Biology', *American Scientist* **54**, 421–431.

Levins, Richard, 1968, *Evolution in Changing Environments*, Princeton University Press, Princeton, New Jersey.

Levins, Richard, 1970, 'Complex Systems', in C. H. Waddington (ed.), *Towards a Theoretical Biology*, vol. 3, University of Edinburgh Press, Edinburgh, pp. 73–88.

Levins, Richard, 1973, 'The Limits of Complexity', in H. Pattee (ed.), 1973, pp. 111–127.

Lewontin, R. C., 1970, 'The Units of Selection', *Annual Review of Ecology and Systematics* **1**, 1–18.

Malcolm, Norman, 1964, 'Scientific Materialism and the Identity Theory', *Dialogue* **3**, 115–125.

Malcolm, Norman, 1971, *Problems of Mind*, Harper and Row, New York.

Margulis, L., 1971, 'The Origin of Plant and Animal Cells', *The American Scientist* **59**, 230–235.

Pattee, Howard, 1970, 'The Problem of Biological Hierarchy', in C. H. Waddington (ed.), *Towards a Theoretical Biology*, vol. 3, Aldine, Chicago, pp. 117–136.

Pattee, Howard (ed.), 1973, *Hierarchy Theory: The Challenge of Complex Systems*, Braziller, New York.

Platt, John R., 1967, 'Theorems on Boundaries in Hierarchial Systems', in Albert G. G. Wilson, L. L. Whyte and Donna Wilson (eds.), *Hierarchial Structures*, American Elsevier, New York, pp. 201–213.

Putnam, Hilary, 1967, 'The Mental Life of Some Machines', in H. Castañeda (ed.), *Intentionality, Minds, and Perception*, Wayne State University Press, Detroit, pp. 177–213.

Raff, R. A. and Mahler, H. R., 1972, 'The Nonsymbiotic Origin of Mitochondria', *Science* **177**, 575–582.

Shaffer, Jerome, 1961, 'Could Mental States Be Brain Processes?', *J. Phil* **58**, 813–822.

Shaffer, Jerome, 1965, 'Recent Work on the Mind Body Problem', *Amer. Phil. Quart.* **2**, 81–104.

Shapere, Dudley, 1971, 'Scientific Theories and Their Domains', forthcoming in Frederich Suppe (ed.), *The Nature of Scientific Theories*, University of Illinois Press, Urbana.

Simon, Herbert A., 1962, 'The Architecture of Complexity', reprinted in Simon (1969), pp. 84–118.

Simon, Herbert A., 1969, *The Sciences of the Artificial*, MIT Press, Cambridge.

Von Neumann, Johann and Morgenstern, Oskar, 1946, *Theory of Games and Economic Behavior*, reprinted 1964 by Wiley, New York.

Wimsatt, W. C., 1971, *The Conceptual Foundations of Functional Analysis*, Ph.D. dissertation, University of Pittsburgh, Department of Philosophy.

Wimsatt, W. C., 1972, 'Teleology and the Logical Structure of Function Statements', *Studies in History and Philosophy of Science* **3**, 1–80.

Wimsatt, W. C., 1973, 'Spatiality and the Mental Realm', in preparation (dittoed draft, December, 1972.)

B. FUNCTION AND TELEOLOGY

MORTON BECKNER

FUNCTION AND TELEOLOGY

I. INTRODUCTION

In trying to decide whether teleology in the sciences is good, bad, or indifferent, philosophers have tended to focus on three sorts of cases. They are exemplified in these paradigms:

(1) The function of the heart is to pump blood. (Call this a *functional ascription*)

(2) The goal of the rat is to reach food at the end of the maze. (*Goal-ascription*)

(3) Jones intends to retire early by working hard. (*Intention-ascription*)

These sorts of cases are sometimes confounded, but they should be clearly distinguished. In a rough-and-ready fashion, some important differences and connections can be described in the following way.

1. *Function*

Functional ascriptions describe the role played by a part or process in the activities of a larger or more inclusive system. Standard examples are the ascriptions of roles to the organs, tissues, cellular parts, biochemical processes, and so on, in the growth, regulation, maintenance, and reproduction of organisms. Functions are also ascribed to the parts of artifacts, especially such objects as machines, pieces of furniture, and so forth. In these cases we antecedently identify a system S and activity ϕ such that the whole of S can be said to do ϕ; and functions are then assigned to the parts P of S or to the activities ϕ' of P, only if ϕ' or P do contribute to the ϕing of S. In general, function is always function *in a whole* system.

2. *Goals and Intentions*

We ascribe goals to persons whenever we ascribe intentions; indeed, anything describable as an intention is also describable as a goal. But not all goals are intentions – or so people commonly argue. For example, we might acknowledge (with Sartre) that a person's goal is to become God, but he has no *intention* of becoming God. Another sort of case: we may say that the rat's goal is food, but that it has no intentions. Finally, we might say that a self-regulated system, such as a target-tracking missile, has a goal but no intentions.

These examples may seem dubious. I think that the concepts of 'goal' and 'intention', as we have them, always leave room for doubt; there are no completely convincing arguments that establish the possibility of goals without intentions. One can maintain that rats and missiles have no intentions, but that they also have no goals; or that rats and persons indeed have goals, but that they also have the corresponding intentions.

Many philosophers have bypassed this point by introducing a technical concept of 'goals' which insures their existence in such cases as the rat and missile. The application of the technical concept is defined so as to leave open the question of intention. Its introduction is guided, however, by examination of selected paradigms, namely, cases which (1) involve intention, and also (2) exhibit certain 'behavioral marks' of directedness. The technical concepts 'goal', 'goal-directed', 'goal-seeking', 'directive correlation' (Sommerhoff), and 'directive organization' (Nagel) are yielded, in a fairly obvious fashion, by considering the behavioral marks and/or some very general features of the organization of a system that would exhibit them. An instance of such a paradigm would be a man trying to reach a destination in the face of a series of obstacles. Writers differ in the details of their accounts of the behavioral marks, but they have in common some reference to persistence and to the range of variation in the disposition of obstacles under which the goal still tends to be reached. Again, there are differences in detail in the accounts of the organization of goal-directed systems, but the acceptable ones have in common (1) some reference to the power of the system to compensate for environmental changes that might impede the system's progress toward the goal, and (2) some reference to the independence of the variables that define the system and its environment. Sommerhoff, I think, first saw the necessity of the latter reference for ruling out various unwanted cases. If conditions (1) and (2) are satisfied, the system is self-regulated by means of feedback.

One aim of introducing such a concept is to provide for cases of goal-seeking that are not goal-intended. The analysis does apply to the rat in a maze, the self-guided missile, and to a large set of other biological and technological cases.

3. *Goals and Functions*

In typical biological cases, achievement of a goal (in the above sense) does have a function in the system. Suppose, for example, that the movement of the water flea, *Daphnea*, toward the surface is goal-directed; this movement also serves the function of respiration in *Daphnea.* In a general way we may say that any biological machinery capable of goal-directed activity is (or at one time was) also capable of performing some function or other. But it is easy to imagine cases of functionless goal-direction. We could, for example, build an ingenious mechanism, regulated by feedback, that pumped sea water out of and back into the sea at a constant rate. The activity of the machine would be goal-directed; but achievement of the goal serves no function. (Of course, the parts of the machine would serve the function of pumping sea water.) I think that many cases of human goal-seeking fall in this category.

Moreover, there are processes that do serve functions without being goal-directed. The blink reflex is an example. Thus, there is a clear distinction between activities that serve functions, and activities that are, in the technical sense, goal-directed. To challenge this point would, I think, be quixotic. To insist that we ought to say the function of any activity that has a function is also its goal would amount to no more than a rejection of the technical terminology.

There is a sense, hard to get at precisely, in which functions must be fulfilled, whereas goals need not be reached. Suppose that people are constantly trying to reach the moon by climbing ladders. It is no objection to saying that reaching the moon is their goal to point out that they will never make it that way. But it is a conclusive objection to the statement 'The function of the brain is to cool the blood' to point out that the brain does not cool the blood. This point is complicated, however, by the following two considerations.

(1) In the case of ostensibly non-intentional biological activity, we would not identify something as the *goal*, for example, of an organism, unless organisms of that sort sometimes achieved it.

(2) We sometimes say that something has a function, but is not performing it (or is performing it poorly). For example, if my heart stops

pumping blood it does not thereby lose its function. So it is not strictly true that if ϕ is a function of P, then P contributes to ϕ. The following, however, is true: if the members of the class of P's never ϕ, then ϕ is not a function of P.

Thus we readily apply the vocabulary of success and failure to both goal-directed activities and to functions.

4. *Functions, Goals, and Intentions*

All three concepts share the following feature: we may say of intentional, goal-directed, and functional activities that these all take place 'for the sake of' something and 'in order to' do something.

Consider the following statements: (1) The heart beats in order to pump blood (for the sake of pumping blood). (2) The rat snuffles along the maze in order to get food. (3) Jones works hard in order to retire early. I think (not everyone would agree) that all these remarks are entailed by the corresponding functional, goal-, and intention-ascriptions, and are therefore on occasion true. Not only might (1) and (2) be true; so far from presupposing a theology or discredited metaphysic, they do have important scientific uses. Clarification of these uses is one aim of this paper.

II. FUNCTIONAL ASCRIPTIONS

The task of this section is to give an account of the meaning of functional ascriptions; I shall concentrate on sentences of the form 'A function of P (or of ϕ') is ϕ', although functions may obviously be ascribed in a variety of alternative vocabularies.

I divide functional ascriptions into three classes; my working examples, one from each class, are the following:

(1) (Call it FA_1): 'A function of the heart is to pump blood'.
(2) (FA_2): 'A function of the heart is to produce heart sounds'.
(3) (FA_3): 'A function of the earth is to intercept passing meteorites'.

FA_1 is clearly true. The heart has many functions that we know of (others are distributing oxygen, removing wastes, etc.), and, no doubt, some we know nothing about; but pumping blood is certainly one of them.

FA_2 is in all probability false. The heart does produce heartsounds; the production of heart sounds is even a necessary condition of the heart's pumping blood. But this is not one of its functions.

In the case of FA_3, something has gone wrong. The earth does intercept passing meteorites, but very few people, if any, would be willing to say that this is one of its functions. Is, then, FA_3 false? Some would say so, on the grounds that meteorite interception is not a function of the earth. Others would say, not that FA_3 is false, but that it is somehow inappropriate – that it is pointless, or nonsensical, or involves a category error. I do not think much turns on this difference. If FA_3 is regarded as false, we do have to distinguish two ways in which functional ascriptions may be false, that is, between FA_2 and FA_3. I shall call the third sort of case 'inappropriate'; the reader may please himself as to whether inappropriate functional ascriptions are also false.

In the analysis of statements of the form 'A function of P is ϕ', we may distinguish two questions: (1) If we suppose that a functional ascription is appropriate, what is the relation between P and ϕ? (2) What distinguishes those which are appropriate from those which are not?

First consider question (1). One answer is that P, or an activity ϕ' of P, is a necessary (or perhaps both necessary and sufficient) condition of ϕ. For example, it might be held that 'a function of the heart-beat is to pump blood' states that 'the heart-beat is a necessary condition of blood-pumping'.

This answer can be construed as no more than a first approximation, in view of two sorts of difficulties. First, if the heart-beat is necessary for blood-pumping, then so are heart-sounds, since the heart-beat is a sufficient condition (neglecting some inessential points) of heart-sounds. Thus, no heart-sounds, no blood-pumping. But heart-sounds do not have the function of pumping blood; whereas we would be committed to this on the hypothesis that the ascription is otherwise appropriate. Moreover, since the heart-beat is sufficient for heart-sounds, we are committed to the conclusion that a function of the heart-beat is to produce heart-sounds. This is false. Clearly the difficulty here lies in the fact that heart-sounds are an accidental by-product of the heart-beat, and are no part of the cause of blood-pumping. We should not say that P and ϕ' have ϕ as their function unless they were, in some sense, part of the cause of ϕ. The difficulty here is closely connected with the famous difficulties associated with attempts to analyze the causal relation in terms of the relations of necessary and sufficient condition.

The second sort of problem is this: a function of the heart is blood-pumping, but the heart is not really necessary for blood-pumping. At least, not if by 'heart' one means the muscular chambered organ usually meant by the term. For example, the blood can be pumped by a machine. This might seem a rather trivial objection. It might be suggested (as Nagel does) that since a function of P is always a function in a system S, the relevant class of S's in which a P has the function be restricted so as to rule out such cases as blood-pumping machines. Perhaps this could be done, but I do not believe a general method of ruling them out has been described successfully. And if it were to work for the heart example, there would still be a problem. Organisms commonly have alternative means of performing the same function. My right kidney excretes urea, but if it is damaged the left one does the job; so my right kidney is not necessary for urea excretion, although that is one of its functions. Sweating aids in temperature regulation; but if I lose the ability to sweat I can make do by panting and suitable choice of behavior. And there are the many cases in which we say that P or ϕ' has the function ϕ when P or ϕ' does no more than increase the probability of ϕ under certain, perhaps rare, circumstances.

Nevertheless, there is a grain of truth in the 'necessary condition' view. The grain is extracted in the following formulation, which also escapes the second type of difficulties. (See below for the way out of the first type of difficulties.) I state it only for activities of parts of S; with obvious minor changes, it holds also for the parts of S.

Granted that the functional ascription is appropriate, then 'A function of ϕ' in S is ϕ' is true if and only if there are regularly occurring states of S and its environment in which ϕ' occurs and in which the occurrence of ϕ' causes an increase in the probability of the occurrence of ϕ. Under these circumstances we may say that ϕ' 'contributes to' the performance of ϕ.

We shall now turn to the question of what makes FA_3 inappropriate, whereas FA_2 is appropriate but merely false. Philosophers who have addressed this question have all, I think, thought that the answer must lie in some difference between, for example, vertebrate bodies and solar systems. All functional ascriptions, in this view, presuppose (or implicitly assert) that the system in question has certain properties that merely physical, chemical, and other systems lack. Nagel, for example, suggests that functional ascription 'presupposes' that "the system under consideration... is directively organized." (Nagel defines 'directive organization' carefully; I have indicated in Part I no more than we need for my subsequent arguments.) This suggestion works for FA_2 and FA_3: organisms with hearts are certainly directively organized, whereas the solar system is not. Hence the necessary presupposition is missing in the case of FA_3.

But there are difficulties. First, we do commonly ascribe functions to the parts of machines and other artifacts. These systems are not, or need not be, directively organized. Second, the functional relationship, as either I or Nagel define it, can be present in a directively organized system, but still have nothing to do with the system *qua* directively organized. FA_2 provides an example. Third, some systems are directively organized, but we do not apply functional analysis to them. An example is the ecosystem of a mountain lake. This system is directively organized with respect to the biomass ratio of predator and prey fishes. But we would not say that a function of the trout is to eat the bluegills, although this does play a role in the regulation of the ratio.

It might seem possible to avoid these difficulties by holding not merely that functional ascriptions presuppose directive organization, but that they presuppose that the ascribed function contributes to directive activity, and that, moreover, the directive activity has adaptive significance. Another possibility would be to delete reference to directive activity altogether, and define 'appropriateness' in terms of adaptive significance. In this view, functional analysis of non-living systems would be treated as a somehow parasitic or derivative procedure.

Nagel's position, and the variations on it just suggested, do provide clarification. Nevertheless, I think they narrowly miss the mark. These positions seek the distinction between FA_2 and FA_3 in some feature of the systems, some difference between e.g., animals and solar systems. My view is that the distinction lies in logical differences between the conceptual schemes we are prepared to apply to animals and to solar systems. Functional ascriptions presuppose *conceptual schemes of a certain logical character*. The ascription is inappropriate if such a scheme is missing.

In order to describe these conceptual schemes I shall first define two notions: (1) the 'net-like organization' of a system, and (2) a 'contributory system'.

(1) If we consider a function such as respiration, we find that the function is performed by a complicated set of parts that causally influence each other in complicated ways. The parts are all parts of the organism; except for the indivisible parts (if there are any) each part has parts; each part (except the whole organism) is part of another part; and finally, given any part P_j there are other parts of which P_j is not a part. (These are rather trivial remarks about the concept of a 'part'.) Every part can in principle, and in practice does, have causal influences on every other part. Of course, only some parts, and only some of the causal relations between them, need be considered in an analysis of respiration.

The part-whole relations of a system can be represented by a 'tree' diagram, such as that shown in Figure 1.

In the diagram, a and b represent parts of c; and a, b, c and d are parts of e. Such diagrams would represent perfect part-whole hierarchies if we were to define 'part-levels' in such a way that every part at a level was the same kind of part, and if every part at a level were exhaustively analyzable into parts of the same kind. The systems we analyze functionally, however, are not perfect hierarchies, if we define the parts in the way we ordinarily define them.

Suppose now we identify some activity ϕ performed by the system marked 'e' in Figure 1; and a set of causal relations between some of the parts of e which contribute to ϕ. These can be represented by the arrows in Figure 1.

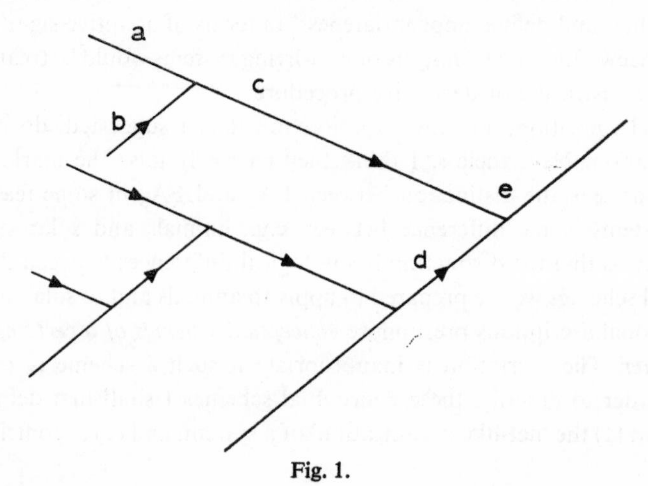

Fig. 1.

What I have in mind by 'net-like organization' could be represented by 'tree' diagrams, expect for the fact that a part P_1 can contribute to the activities of part P_2 when neither P_1 or P_2 is part of the other (for example, the diaphragm influences the lungs). I represent these causal influences by wavy lines.

Now suppose that a theory *T* leads us to identify parts of *S*, and causal relations between the parts of *S*, in such a way that as a matter of fact, the part-whole and causal relations in *S* can be represented in a net-like diagram such as Figure 2. Then I shall say that *S* possesses 'net-like organization with respect to *T*'.

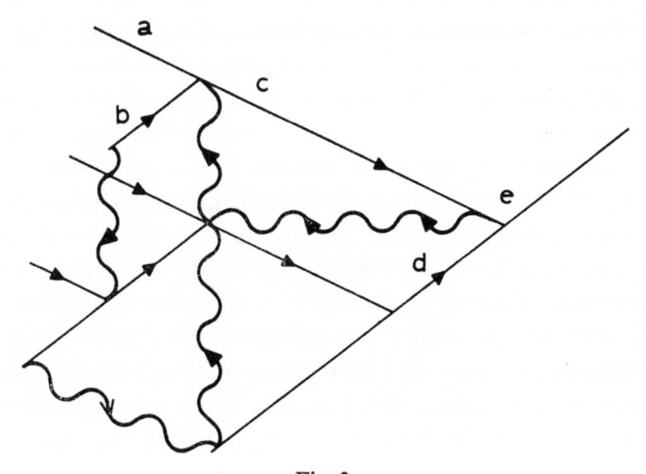

Fig. 2.

Any relatively complex material system obviously will possess net-like organization with respect to some existing theory or other. So net-like organization is possessed by systems, such as the solar system, which we refuse to subject to functional analysis.

(2) If the activities ϕ_1 in a system S_1 'contribute' (in the sense defined above) to the activities ϕ_2 of another system S_2, I shall call it a 'contributory', and we may speak of ϕ_1 as a 'contribution' to ϕ_2. S_1 need not be a part of S_2. The arrows in Figures 1 and 2, on both straight and wavy lines, represent the 'contribution' relation. Any straight line with an arrow point on it is a contributory system, and so is any part with a wavy line

leading from it. Again any complex system will contain contributory subsystems, so contributory systems are present in, for example, the solar system.

Both parts of a system with net-like organization and the contributions they make are identified with the help of a conceptual scheme. When we refer, for example, to the lungs, we do so by means of a term ('lungs') that has a certain definition, or at least a certain standard use. The definition or use places the term within our conceptual scheme. A part (or activity), such as a lung (or respiration), is a part or activity relative to the concepts we employ in identifying and describing them.

Now suppose we have identified a system S; that S exhibits activity ϕ; and that we are interested in showing how S does ϕ by ferreting out the contributions that the parts $P_1, P_2 \dots P_n$ (and their corresponding activities $\phi_1, \phi_2 \dots \phi_n$) contribute to ϕ. I want to call attention to two alternative ways of identifying the parts of S. Let 'P_j' refer, by virtue of its definition, to P_j. Granting that 'P_j' makes a contribution to ϕ, 'P_j' could be defined (in part) by reference to ϕ; or 'P_j' could be defined quite independently of any reference to ϕ. For example, the heart pumps blood (P_j contributes to ϕ); and the term 'heart', by definition, cannot be applied to any system unless it pumps blood, or is the sort of system that contributes to the pumping of blood (the term 'P_j' is defined (in part) by reference to ϕ). On the other hand, the earth does intercept passing meteorites; but the term 'the earth' is not defined (even in part) by reference to the activity of meteorite interception. In sum: it is a logical truth that hearts pump blood; it is not a logical truth, but only an empirical one, that the earth intercepts meteorites. Let us refer to the parts of systems which, like the heart, by definition contribute to an activity ϕ of a larger system as 'definitionally contributory' parts.

Moreover, the activites ϕ_i' of parts of a system, or of the whole system, may simply, as a matter of fact, contribute to an activity ϕ of a system; or, alternatively, the term 'ϕ_i'' can be so defined that ϕ_i' necessarily contributes to ϕ. The revolution of the earth about the sun contributes to the interception of meteorites, but this fact is not exploited in the definition of 'revolution'. On the other hand, (a) the courtship rituals of a bird do contribute to the selection of mates, and this contribution is part of the concept 'courtship ritual'; and (b) breathing by definition contributes to respiration. Call such activities 'definitional contributions'. They are, I think, relatively much rarer in biological theory than definitionally contributory systems.

We are now in a position to state rather briefly the major theses of Section II. First, functional ascriptions explicitly state that a part P_j or activity ϕ_i' contributes to an activity ϕ in or of system S. Second, functional ascription presupposes that S possesses a net-like organization such that (a) the strands of the net are identified by the same general conceptual scheme which is employed in the ascription itself; (b) a significant number of critical strands in the net definitionally contribute to one or more activities ϕ of the whole system S; and (c) the contribution ascribed is a contribution to activity ϕ of the whole system S, where ϕ is an activity to which a significant number of strands in the net definitionally contribute.

Let me now summarize by applying the foregoing remarks to FA_1, FA_2, and FA_3 – omitting as far as possible my own jargon.

FA_1: It is true that a function of the heart is to pump blood. The heart does pump blood; the body is a complex system of parts that by definition aid in certain activities of the whole body, such as locomotion, self-maintenance, copulation; the concepts 'heart' and 'blood' are recognizably components of the scheme we employ in describing this complex system; and blood-pumping does contribute to activities of the whole organism to which many of its organs, tissues and other parts definitionally contribute.

FA_2: The heart does produce heart-sounds; but heart-sounds do not contribute to any of the activities of the whole body which our conceptual scheme singles out as privileged benefactors of the activities of our bodily parts.

FA_3: Although the earth does intercept meteorites, we do not, I think, identify the parts of the solar system, or any of its activities, in terms of the contribution they make to activities of the whole solar system. The solar system certainly does things as a whole (for example, it revolves about the galactic center), but we have not found it useful to construct a theory which involves the definition of celestial bodies in terms of their contribution to the motions of the solar system.

Note that this analysis applies to machines and other artifacts; there is no need to regard the ascription of functions to them as in any way derivative or parasitic.

III. ELIMINABILITY

It has been suggested that teleological statements such as 'The heart beats in order to pump blood', 'The function of the heart is to pump blood', 'The heart beats for the sake of pumping blood', and so on can be eliminated in favor of non-teleological statements.

This is a very complicated question. In order to present a supportable answer, we have to pay some attention to the following questions: (1) How are we to distinguish in general between teleological and non-teleological statements?; (2) What are we to understand by the 'elimination' of statement A in favor of statement B?

Supporters of the elimination thesis usually speak of the 'translation' of teleological statement A into non-teleological statement B; but there are other sorts of relation not as stringent as translatability that might reasonably fill the bill.

I will deal with these two questions together, and in doing so, explain my own view, which is: (a) teleological statements are not translatable into non-teleological ones, but (b) every activity of a system, even those of systems which we ordinarily describe with the help of functional ascriptions, can be described in non-teleological language.

Clause (a) of my opinion is difficult to prove, in view of the lack of clarity both in our criteria for identifying teleological language and in our criteria for deciding whether B translates A. There would, no doubt, be general agreement that sentences containing the expression 'for the sake of' and 'in order that' are teleological; and similarly for those sentences I have called functional, goal-, and intention-ascriptions. But consider the following sentences, which do not fit one of these molds:

(1) Vultures break open eggs with stones.
(2) Myrtle warblers migrate in the spring into regions of abundant food.
(3) The missile swerved toward its target.
(4) He acted out of avarice.
(5) The arms of the Dean's chair are upholstered.
(6) The sight of a barracuda releases an escape reaction in anchovies.

In all of these cases there are strong lines of argument tending to show that they are teleological, if explicit functional, goal-, and intention-ascriptions are teleological. These arguments have a common pattern; they indicate that any differences between the phrasing of cases 1–6 and the phrasing of admittedly teleological sentences are matters of prose style only. For example, (1) is very much like 'Vultures use stones to break open eggs', which is very much like 'Vultures use stones in order to break open eggs'. I think these three sentences do indeed differ only stylistically, and that if one is teleological, all three are. Similarly, in case (2), the movement of birds is not a case of 'migration' unless it is movement of a regularly recurrent sort that does indeed serve a mating or food-getting function. In case (3) an object is not the 'target' of a self-guided missile unless the missile is directed toward it as a goal. In case (4), the reference to 'avarice' indicates (among other things) that the action is done for the sake of some goal. In case (5), the 'arms' of a chair are by definition parts that serve a certain function. Finally, in case (6), an 'escape reaction' is an activity whose function is the avoidance of danger.

I would suggest that there is no touchstone for identifying the teleological, and that teleological language is far more prevalent and more firmly embedded in the language of science than a casual inspection might suggest.

My conviction that teleological language cannot be translated into non-teleological language is based on the following hunch: if *A* is a teleological sentence, and if *B* translates *A*, *B* would also be teleological. I would tend to regard preservation of teleological character under translation as a condition of adequacy for any account of the translatability relation. The teleological character of a sentence is not a matter of vocabulary alone; it is a matter of the logical structure of the conceptual scheme employed in the sentence.

I have indicated in Section II what features of a conceptual scheme are presupposed by functional ascriptions. I think this account can be easily generalized to cover the features of the conceptual schemes presupposed by goal-ascriptions. We have to adopt a technical concept of goal-direction, similar in principle, if not in every detail, to those proposed by men such as Sommerhoff and Nagel. If we do this, we may regard any goal-directed activity as defining an activity ϕ of a system S which is privileged in two ways: (1) any narrowed line in the net-like organization of S which contributes to ϕ is a function in S (thus admission of goal direction extends the range of admissibility of function ascriptions); and (2) the vocabulary of success and failure is applicable to *S*, to ϕ, and to the parts (and their activities) of *S*.

All language which essentially exploits these conceptual schemes I regard as teleological language.

We may now turn to clause (b) of my thesis, that even though teleological language cannot be translated away, there is a sense in which teleological language is eliminable. The sense is this: given any single case of an activity which is describable in teleological language, that case is also describable in non-teleological language. By this I mean that every observable aspect of the activity which in fact serves as the basis for our application of teleological concepts can also be described by means of a conceptual apparatus that is not teleological in character.

This point is best clarified by means of an example. Suppose we are watching a tank in which there is a single anchovy, and that we introduce a barracuda into the tank. At first nothing happens; then, when it would be reasonable for us to suppose that the anchovy spots the barracuda, the anchovy behaves as follows: he turns sharply, swims quickly to the surface, leaps out of the water, reenters, and repeats the sequence. This whole description (call it *B*) is, let us at least suppose, in non-teleological language. And it is a description of exactly the same behavior that would serve as the basis for the teleological description (call it *A*) 'On sighting the barracuda, the anchovy engaged in an escape reaction'. Of course *B* does not translate *A*. *A* in some ways says less, and in other ways says (and presupposes) much more. *A* says less than *B* since *B* offers details not mentioned in *A*. It says more, since calling the anchovy's behavior an 'escape reaction' implies that it serves the function of escape, whereas this is not implied by *B*.

The utility of the teleological sentence *A* in relation to non-teleological *B* lies in the fact that there are contexts in which the 'less' said by *A* is hardly worth saying, and in which the 'more' said by *A* is very much worth saying. These are contexts in which we are trying to fit the single case into a certain general picture.

To see this, consider how the concept of an 'escape reaction' can be related to behavior-segments such as that described in *B*.

(1) The details described in *B* may vary considerably, and still constitute an escape reaction. For example, the same anchovy, or other

anchovies of the same species, might omit the direction change, or not break the surface. In other words, the concepts employed in the description of these various cases might not provide the concept of a 'kind' with respect to which all of these cases are the 'same kind' of cases.

(2) The *B*-type description of the behavior of anchovies might be a description of an escape reaction, while the same ('*B*-type same') behavior shown by another fish might be behavior of another kind ('*A*-type kind', such as a mating dance.)

(3) Anchovies may have more than one sort of escape reaction in their repertory. The reactions of members of a school might be entirely different. In this case even the term 'anchovy escape reaction' would not apply solely to cases like *B*.

(4) The term 'escape reaction' applies to very different sorts of behavior throughout a very wide range of taxonomic groups. All these escape reactions might have virtually nothing in common other than the common features noted in teleological language.

It should not be supposed that the term 'escape reaction' could be defined in non-teleological terms as a complex disjunction of *B*-type conditions. Such a concept could be constructed, but, even if it were in fact to apply to all and only escape reactions, it would nevertheless not be the concept of an escape reaction; for it would still be possible for a pattern of behavior not mentioned in the definition to serve the function of escape.

IV. CONCLUSION

The view of teleology sketched in the above remarks seems to me to offer a piece of candy to both the critics and guardians of teleology. The critics want to defend against a number of things: the importation of unverifiable theological or metaphysical doctrines into the sciences; the idea that goals somehow act in favor of their own realization; and the view that biological systems require for their study concepts and patterns of explanation unlike anything employed in the physical sciences. An important part of their defense has been the contention that teleological language can be eliminated without loss from the sciences. I have argued that this is true: any phenomenon that can be described in teleological language can be described otherwise.

On the other hand, eliminability does not mean translatability. I have suggested – I do not see how to prove it – that the teleological character of a sentence is so fundamental that it would be preserved under translation. Teleological character is conferred on a sentence by the manner in which it fits into a conceptual scheme designed for the description of certain classes of systems possessing net-like organization. The elimination of teleological language thus involves a conceptual shift, and involves a different method of classifying the elements of a system.

The guardians of teleology – myself among them – have insisted that teleological language is perfectly legitimate. This conclusion is plain if my account of 'appropriateness' is correct. Indeed, the development of conceptual schemes which render functional ascriptions appropriate is seen to be just a special case of a general scientific procedure: the designing of languages aimed at bringing to light those causal relations which are most interesting to us.

Pomona College, Claremont

BIBLIOGRAPHY

Beckner, Morton, *The Biological Way of Thought*, New York, 1957.
Hempel, Carl G., 'The Logic of Functional Analysis', in *Symposium on Sociological Theory*, Evanston, Ill., 1959.
Lehman, Hugh, 'Functional Explanations in Biology', *Philosophy of Science* **32** (1965).
Nagel, Ernest, *The Structure of Science*, New York, 1961.
Scheffler, Israel, 'Thoughts on Teleology', *British Journal for the Philosophy of Science* **9** (1959).
Sommerhoff, George, *Analytical Biology*, London, 1950.

LARRY WRIGHT

FUNCTIONS

The notion of function is not all there is to teleology, although it is sometimes treated as though it were. Function is not even the central, or paradigm, teleological concept. But it *is* interesting *and* important; and it is still not as well understood as it should be, considering the amount of serious scholarship devoted to it during the last decade or two. Let us hope this justifies my excursion into these murky waters.

Like nearly every word in English, 'function' is multilaterally ambiguous. Consider:

(1) $y=f(x)$/The pressure of a gas is a function of its temperature.
(2) The Apollonaut's banquet was a major state function.
(3) I simply can't function when I've got a cold.
(4) The heart functions in this way... (something about serial muscular contractions).
(5) The function of the heart is pumping blood.
(6) The function of the sweep-second hand on a watch is to make seconds easier to read.
(7) Letting in light is one function of the windows of a house.
(8) The wood box next to the fireplace currently functions as a dog's sleeping quarters.

It is interesting to notice that the word 'function' has a spectrum of meanings even within the last six illustrations, which are the only ones at all relevant to a teleologically oriented study. Numbers (3), (4), and (8) are substantially different from one another, but they are each, from a teleological point of view, peripheral cases by comparison with (5), (6), and (7), which are the usual paradigms. And even these latter three are individually distinct in some respects, but much less profoundly than the others.

Quite obviously, making some systematic sense of the logical differentiation implicit in categorizing these cases as peripheral and paradigmatic is a major task of this paper. But a clue that we are on the right track here can be found in a symptomatic grammatical distinction present in the last six illustrations: in the peripheral cases the word 'function' is itself the verb, whereas in the more central cases 'function' is a noun, used with the verb 'to be'. And since the controversy resolves around what *the function* of something *is*, the grammatical role of 'function' in (5), (6), and (7) makes them heavy favorites for the logical place of honor in this discussion.

I. SOME RUDIMENTARY DISTINCTIONS

1. *Functions v. Goals*

There seems to be a strong temptation to treat functions as representative of the set of central teleological concepts which cluster around goal-directedness. However, even a cursory examination of the usual sorts of examples reveals a very important distinction. Goal-directedness is a behavioral predicate. The *direction* is the direction of behavior. When we do speak of objects (homing torpedoes) or individuals (General MacArthur) as being goal-directed, we are speaking indirectly of their behavior. We would argue against the claim that they are goal-directed by appeal to their behavior (for example, the torpedo, or the General, did not *change course* at the appropriate time, and so forth). On the other hand, many things have *functions* (for example, chairs and windpipes) which do not behave *at all*, much less goal-directedly. And behavior can have a function without being goal-directed – for example, pacing the floor or blinking your eye. But even when goal-directed behavior has a function, very often its function is quite different from the achievement of its *goal*. For example, some fresh-water plankton diurnally vary their distance below the surface. The *goal* of this behavior is to keep light intensity in their environment relatively constant. This can be determined by experimenting with artificial light sources. The *function* of this behavior, on the other hand, is keeping constant the oxygen supply, which normally varies with sunlight intensity. There are many instances to be found in the study of organisms in which the function of a certain goal-directed activity is not some further goal of that activity, as it usually is in human behavior, but rather some natural concomitant or consequence of the immediate goal. Other examples are food-gathering, nest-making, and copulation. Clearly function and goal-directedness are not congruent concepts. There is an important sense in which they are wholly distinct. In any case, the relationship between functions and goals is a complicated and tenuous one; and becoming clearer about the nature of that relationship is one aim of this essay.

2. *A Function v. the Function*

Recent analyses of function, including all those treated here, have tended to focus on *a* function of something, by contrast with *the* function of something. This tendency is understandable; for any analysis of this sort aims at generality, and 'a function' would seem intrinsically more general than 'the function' because it avoids one obvious restriction. This generality, however, is superficial: the notion of *a* function is derivable from the notion of *the* function (more than one thing meets the criteria) just as easily as the reverse (only one thing meets the criteria). Furthermore, the notion of *a* function is much more easily confused with certain peripheral, quasi-functional ascriptions which are examined below. In short, the discussion of this paper is concerned with *a* function of *X* only in so far as it is the sort of thing which would be *the* function of *X* if *X* had no others. Accordingly, I take the definite-article formulation as paradigmatic and will deal primarily with it, adding comments in terms of the indefinite-article formulation parenthetically, where appropriate.

3. *Function v. Accident*

Very likely the central distinction of this analysis is that between the *function* of something and other things it does which are *not* its function (or one of its functions). The function of a telephone is effecting rapid, convenient communication. But there are many other things telephones do: take up space on my desk, disturb me at night, absorb and reflect light, and so forth. The function of the heart is pumping blood, not producing a thumping noise or making wiggly lines on electrocardiograms, which are also things it does. This is sometimes put as the distinction between a function, and something done merely 'by accident'. Explaining the propriety of this way of speaking – that is, making sense of the function/accident distinction – is another aim, perhaps the *primary* aim of the following analysis.

4. *Conscious v. Natural Functions*

The notion of accident will raise some interesting and important questions across another rudimentary distinction: the distinction between natural functions and consciously designed ones. Natural functions are the common organismic ones such as the function of the heart, mentioned above. Other examples are the function of the kidneys to remove metabolic wastes from the bloodstream, and the function of the lens of the human eye to focus an image on the retina. Consciously designed functions commonly (though not necessarily) involve artifacts, such as the telephone and the watch's sweep hand mentioned previously. Other examples of this type would be the function of a door knob, a headlight dimmer switch, the circumferential grooves in a pneumatic tire tread, or a police force. Richard Sorabji has argued[1] that 'designed' is too strong as a description of this category, and that less elaborate conscious effort would be adequate to give something a function of this sort. I think he is right. I have used the stronger version only to overdraw the distinction hyperbolically. In deference to his point I will drop the term 'designed' and talk of the distinction as between natural and conscious functions.

Of the two, natural functions are philosophically the more problematic. Several schools of thought, for different reasons, want to deny that there are natural functions, as opposed to conscious ones. Or, what comes to the same thing, they want to deny that natural functions are functions in anything like the same sense that conscious functions are. Some theologians want to say that the organs of organisms get their functions through God's conscious design, and hence these things *have* functions, but not natural functions *as opposed* to conscious ones. Some scientists, like B. F. Skinner, would *deny* that organs and organismic activity have functions *because* there is no conscious effort or design involved.

Now it seems to me that the notion of an organ having a function – both in everyday conversation and in biology – has no strong theological commitments. Specifically, it seems to me consistent, appropriate, and even common for an atheist to say that the function of the kidney is elimination of metabolic wastes. Furthermore, it seems clear that conscious and natural functions are functions in the same sense, despite their obvious differences. Functional ascriptions of either sort have a profoundly similar ring. Compare 'the function of that cover is to keep the distributor dry' with 'the function of the epiglottis is to keep food out of the windpipe'. It is even more difficult to detect a difference in what is being requested: 'What is the function of the human windpipe?' versus 'What is the function of a car's exhaust pipe?' Certainly no analysis should begin by supposing that the two sorts are wildly different, or that only one is really legitimate. That is a possible *conclusion* of an analysis, not a reasonable presupposition. Accordingly, the final major aim of this analysis will be to make sense of natural functions, both as functions in the same sense as consciously contrived ones, and as functions independent of any theological presuppositions – that is, independent of conscious purpose. It follows that this analysis is committed to finding a way of stating what it is to be a function – even in the conscious cases – that does not rely on an appeal to consciousness. If no formulation of this kind can be found despite an honest search, only then should we begin to take seriously the view that we actually mean something quite different by 'function' in these two contexts.

II. SOME ANALYSES OF FUNCTION

The analysis of function for which I wish to argue grew out of a detailed critical examination of several recent attempts in the literature to produce such an analysis, and it is best understood in that context. For this reason, and because it will help clarify the aims I have sketched above, I will begin by presenting the kernel of that critical examination.

The first analysis I want to consider is an early one by Morton Beckner.[2] Here Beckner contends that to say something s has function F' in system s' is to say that

> There is a set of circumstances in which: F' occurs when s' has s, AND F' does not occur when s' does not have s [p. 113].[3]

For example, 'the human heart has the function of circulating blood' means that there is a set of circumstances in which circulation occurs in humans when they have a heart, and does not when they do not. Translated into the familiar jargon, s has function F' in s' if and only if there is a set of circumstances containing s which are sufficient for the occurrence of F' and which also require s in order to be sufficient for F'. Now it is not clear whether the 'requirement' here is necessity or merely non-redundancy. If it is necessity, then under the most natural interpretation of 'circumstances' (environment), it is simply mistaken. There are *no* circumstances in which, for example, the heart is absolutely irreplaceable:

we could always pump blood in some other way. On the other hand, if the requirement here is only non-redundancy, the mistake is more subtle.

In this case Beckner's formula would hold for cases in which *s* merely *does F′*, but in which *F′* is not the function of *s*. For example, the heart is a non-redundant member of a set of conditions or circumstances which are sufficient for a throbbing noise. But making a throbbing noise is not a function of the heart, it is just something it does – accidentally. In fact, there are even dysfunctional cases which fit the formula: in some circumstances, livers are non-redundant for cirrhosis, but cirrhotic debilitation could not conceivably be the (or a) function of the liver. So this analysis fails on the functional/accidental distinction: it includes too much.

After first considering a view essentially similar to this one, John Canfield has offered a more elaborate analysis.[4] According to Canfield:

> A function of *I* (in *S*) is to do *C means I* does *C* and that *C* is done is useful to *S*. For example, '(In vertebrates) a function of the liver is to secrete bile' means 'the liver secretes bile, and that bile is secreted in vertebrates is useful to them' [p. 290].

Canfield recognizes that natural functions are the problematic ones, but he devotes his attention solely to those cases. He treats only the organs and parts of organisms studied by biology, to the exclusion of the consciously designed functions of artifacts. As a result of this emphasis, his analysis is, without modification, almost impossible to apply to conscious functions. But even with appropriate modifications, it turns out to be inadequate to the characterization of either conscious or natural function.

In the conscious cases, there is an enormous problem in identifying the system *S*, *in* which *I* is functioning, and *to* which it must be useful. The function of the sweep-second hand of a watch is to make seconds easier to read. It would be most natural to say that the system *in which* the sweep hand is functioning – by analogy with the organismic cases – is the watch itself; but it is hard to make sense of the easier reading being useful to the mechanism. On the other hand, the best candidate for the system *to which* the easier reading is useful is the person wearing the watch; but this does not seem to make sense as the system *in which* the sweep hand is functioning.

The crucial difficulty of Canfield's analysis begins to appear at this point: no matter what modifications we make in his formula to avoid the problem of identifying the system *S*, we must retain the requirement that *C* be useful. This is really the major contribution of his analysis, and to abandon it is to abandon the analysis. The difficulty with this is that, for example, in the watch case, it is clearly not necessary that easily read seconds be useful to the watch-wearer – or anyone else – in order that making seconds easier to read be the function of the sweep hand of that wearer's watch. My watch has a sweep-second hand, and I occasionally use it to time things to the degree of accuracy it allows: it is useful to me. Now suppose I were to lose interest in reading time to that degree of accuracy. Suppose my life changed radically so that nothing I ever did could require that sort of chronological precision. Would that mean the sweep hand on my particular watch no longer has the function of making seconds easier to read? Clearly not. If someone were to ask what the sweep hand's function was ('What's it do ?', 'What's it there for?') I would still have to say it made seconds easier to read, although I might yawningly append an autobiographical note about my utter lack of interest in that feature. Similarly, the function of that button on my dashboard is to activate the windshield washer, even if all it does is make the mess on the windshield worse, and hence is not useful at all. That would be its *function* even if I never took my car out of the garage – or broke the windshield.

It is natural at this point to attempt to patch up the analysis by reducing the requirement that C be useful, to the requirement that C *usually* be useful. But this will not do either, because it is easy to think of cases in which we would talk of something's having a function even though doing that thing was quite *generally* of no use to anybody. For example, a machine whose function was to count Pepsi Cola bottle caps at the city dump; or MIT's ultimate machine of a few years back, whose only function was to turn itself off. The source of the difficulty in all of these cases is that what the thing in question (watch, washer button, counting machine) was *designed* to do has been left out of the calculation. And, of course, in these cases, if something is designed to do X, then doing X is its function even if doing X is generally useless, silly, or even harmful. In fact, intention is so central here that it allows us to say the function of I is to do C, even when I cannot even *do* C. If the windshield washer switch comes from the factory defective, and is never repaired, we would still say that its *function* is to activate the washer system, which is to say: that is what it was *designed* to do.

It might appear that this commits us to the view that natural and consciously contrived functions cannot possibly be the same sort of function.

If conscious intent is what *determines* the function an artifact has got, there is no parallel in natural functions. I take this to be mistaken, and will show why later. For now it is only important to show, from this unique vantage, the nature of the most formidable obstacle to be overcome in unifying natural and conscious functions.

The argument thus far has shown that meeting Canfield's criteria is not necessary for something to be a function. It can easily be shown that meeting them is also not sufficient. We are always hearing stories about the belt buckles of the Old West or on foreign battlefields which save their wearers' lives by deflecting bullets. From several points of view that is a very useful thing for them to do. But that does not make bullet deflection the function – or even *a* function – of belt buckles. The list of such cases is endless. Artifacts do all kinds of useful things which are not their functions. Blowouts cause you to miss flights that crash. Noisy wheel bearings cause you to have the front end checked over when you are normally too lazy. The sweep hand of a watch might brush the dust off the numbers, and so forth.

All this results from the inability of Canfield's analysis to handle what we took to be one of the fundamental distinctions of function talk: accidental versus non-accidental. Something can do something useful purely by accident, but it cannot have, as its function, something it does only by accident. Something that *I* does by accident cannot be the function of *I*. The cases above allow us to begin to make some fairly clear sense of this notion of accident, at least for artifacts. Buckles stop bullets only by accident. Blowouts only accidentally keep us off doomed airplanes. Sweep hands only accidentally brush dust, if they do it at all. And this brings us back to the grammatical distinction I made at the outset when I divided the list of illustrations into 'central' and 'peripheral' ones. When something does something useful by accident rather than design, as in these examples, we signal the difference by a standard sort of 'let's pretend' talk. Instead of using the verb 'to be' or the verb 'to have', and saying the thing in question *has* such and such a function, or saying that *is* its function, we use the expression 'functioning as'. We might say the belt buckle *functioned as* a bullet shield, or the blowout *functioned as* divine intervention, or the sweep hand *functions as* a dust brush. Canfield's analysis does not embrace this distinction at all.

So far I have shown only that Canfield's formula fails to handle conscious functions. This means it is incapable of showing natural functions to be functions in the same full-blooded sense as conscious ones, which is indeed serious; but that, it might be argued, really misses the point of his analysis. Canfield is not interested in conscious functions. He would be happy just to handle natural functions. For the reasons set down above, however, I am looking for an analysis which will *unify* conscious and natural functions, and it is important to see why Canfield's analysis cannot produce that unification. Furthermore, Canfield's analysis has difficulties in handling natural functions that closely parallel the difficulties it has with conscious functions; which is just what we should expect if the two are functions in the same sense.

For example, it is absurd to say with Pangloss that the function of the human nose is to support eyeglasses. It is absurd to suggest that the support of eyeglasses is even one of its functions. The function of the nose has something to do with keeping the air we breathe (and smell) warm and dry. But supporting a pincenez, just as displaying rings and warpaint, is something the human nose does, and is useful to the system having the nose: so it fits Canfield's formula. Even the heart throb, our paradigm of non-function, fits the formula: the sound made by the heart is an enormously useful diagnostic aid, not only as to the condition of the heart, but also for certain respiratory and neurological conditions. More bizarre instances are conceivable. If surgeons began attaching cardiac pacemakers to the sixth rib of heart patients, or implanting microphones in the wrist of CIA agents, we could then say that these were useful things for the sixth rib and the wrist (respectively) to do. But that would not make pacemaker-hanging a function of the sixth rib, or microphone concealment a function of the human wrist.

There seems to be the same distinction here that we saw in conscious functions. It makes perfectly good sense to say the nose *functions as* an eyeglass support; the heart, through its thump, *functions as* a diagnostic aid; the sixth rib *functions as* a pacemaker hook in the circumstances described above. This, it seems to me, is precisely the distinction we make when we say, for example, that the sweep-second hand *functions as* a dust brush, while denying that brushing dust is one of the sweep hand's functions. And it is here that we can make sense of the notion of accident in the case of natural functions: it is merely fortuitous that the nose supports eyeglasses; it is happy chance that the heart throb is diagnostically significant; it would be the merest serendipity if the sixth rib were to be a particularly good pacemaker hook. It is (would be) only *accidental* that (if) these things turned out to be useful in these ways. Accordingly, we have already drawn a much stronger parallel between natural functions and conscious functions than Canfield's analysis will allow.

Thus far I have ignored Canfield's analysis of usefulness:

[In plants and animals other than man, that *C* is done is useful to *S* means] if, *ceteris paribus*, *C* were not done in *S*, then the probability of that *S* surviving or having descendants would be smaller than the probability of an *S* in which *C* is done surviving or having descendants [p. 292].

I have ignored it because its explicit and implicit restrictions make it even more difficult to work this analysis into the unifying one I am trying to produce. Even within its restrictions (natural functions in plants and animals other than man), however, the extended analysis fails for reasons very like the ones we have already examined. Hanging a pacemaker on the sixth rib of a cardiovascularly inept lynx would be useful to that cat in precisely Canfield's sense of 'useful': it would make it more likely that the cat would survive and/or have descendants. Obviously the same can be said for the diagnostic value of an animal's heart sounds. So usefulness – even in this very restricted sense – does not make the right function/accident distinction: some things do useful things which are not their functions, or even one of their functions.

The third analysis I wish to examine is a more recent one by Morton Beckner.[5] This analysis is particularly interesting for two reasons. First, Beckner is openly (p. 207) trying to accommodate both natural and conscious functions under one description. Second, he wants to avoid saying things like (to use his examples) 'A function of the heart is to make heart sounds' and 'A function of the Earth is to intercept passing meteorites'. So his aims are very like the ones I have argued for: to produce a unifying analysis, and one which distinguishes between functions and things done by accident. And since the heart sound is useful, and intercepting meteorites could be (perhaps already is), Beckner would probably agree in principle with the above criticism of Canfield.

Beckner's formulation is quite elaborate, so I will present it in eight distinct parts, clarify the individual parts, and then offer an illustration before going on to raise difficulties with them collectively as an analysis of the concept of function. That formulation is:

P has function *F* in *S* if and only if[6]:

(1) *P* is a part of *S* (in the normal sense of 'part').

(2) *P* contributes to *F*. (*P*'s being part of *S* makes the occurrence of *F* more likely.)

(3) *F* is an activity in or of the system *S*.

(4) *S* is structured in such a way that a significant number of its parts contribute to the activities of other parts, and of the system itself.

(5) The parts of *S* and their mutual contributions are identified by the same conceptual scheme which is employed in the statement that *P* has function *F* in system *S*.

(6) A significant number of critical parts (of *S*) and their activities definitionally contribute to one or more activities of the whole system *S*.

(7) *F* is or contributes to an activity *A* of the whole system *S*.[7]

(8) *A* is one of those activities of *S* to which a significant number of critical parts and their activities definitionally contribute.

Two points of clarification must be made at once. First, the notion of 'the same conceptual scheme' in number (5) is obscure in some respects, and the considerable attention devoted to it by Beckner does not help very much. In general all one can say is that *P*, *F*, and the other parts and activities of *S* must be *systematically* related to one another. But in practice the point is easier to make. For example, if we wish to speak of removing metabolic wastes as the function of the human kidney, the relevant conceptual scheme contains other human organs, life, and perhaps ecology in general, but not atoms, molecular bonds, and force fields. The second point concerns the 'definitional contribution' in number (6). A part (or activity) makes a definitional contribution to an activity if that contribution is part of what we mean by the word which refers to that part (or activity). For example, part of what we mean by 'heart' in a biological or medical context is 'something which pumps blood': we would allow considerable variation in structure or appearance and still call something a heart if it served that function. Beckner illustrates how all these steps work together, once again using the heart.

> It is true that a function of the heart is to pump blood. The heart does pump blood; the body is a complex system of parts that by definition aid in certain activities of the whole body, such as locomotion, self-maintenance, copulation; the concepts 'heart' and 'blood'

are recognizably components of the scheme we employ in describing this complex system; and blood-pumping does contribute to activities of the whole organism to which many of its organs, tissues and other parts definitionally contribute (p. 207].

There are several difficulties with this analysis. They appear below, roughly in order of increasing severity.

First, Beckner's problems with the system *S* are in some ways worse than Canfield's; for Beckner explicitly wants to include artifacts, and in addition he says much more definite things about the relationship among *P*, *F*, and *S*. So in this case, when we say the function of a watch's sweep hand is making seconds easier to read, we must not only find a system *of* which the sweep hand is a part, and *in* or *of* which 'making seconds easier to read' is an activity, but this activity must be or contribute to one to which a number of the system's critical parts definitionally contribute. In the case of the natural functions of the organs and other parts of organisms, the system *S* is typically a natural unit, easy to subdivide from the environment: the organism itself. But for the conscious functions of artifacts, such systems, if they can be found at all, must be hacked out of the environment rather arbitrarily. With no more of a guide than Beckner has given us, there is nothing like a guarantee that we can always find such a system. Accordingly, when our minds boggle – as I take it they do in trying to conceive of 'making seconds easier to read' being an activity at all, much less one meeting all of the other conditions of this analysis – we have to say that the analysis is at best too obscure to be applicable to such cases, and is perhaps just mistaken.

A second difficulty stems directly from the first. It is not at all clear that functions – even natural functions – have to be activities at all, let alone activities of the sort required by Beckner. Making seconds easier to read is an example, but there are many others: preventing skids in wet weather, keeping your pants up, or propping open my office door. All of these things are legitimate functions (of tire treads, belts, and doorstops, respectively); none are activities in any recognizable sense.

Thirdly, we noticed in our discussion of Canfield that something could do a useful thing by accident, in the appropriate sense of 'accident'. Similarly, a part of a system meeting all of Beckner's criteria might easily make a contribution to an activity of that system also quite by accident. For example, an internal-combustion engine is a system satisfying Beckner's criteria for *S*. If a small nut were to work itself loose and fall under the valve-adjustment screw in such a way as to adjust properly a poorly adjusted valve, it would make an accidental contribution to the smooth running of that engine. We would never call the maintenance of proper valve adjustment the *function* of the nut. If it got the adjustment right it was just an accident. But on Beckner's formulation, we would have to call that its function. The nut does keep the valve adjusted; the engine is a complex system of parts that by definition aid in certain activities of the whole body, such as generation of torque and self-maintenance (lubrication, heat dissipation); the concepts 'nut', 'valve', and 'valve adjustment' are components of the scheme we employ in describing this complex system; and proper valve adjustment does contribute to the smooth running of the (whole) engine, which is an activity to which many of the other parts of the engine definitionally contribute (flywheel, connecting rod, exhaust ports).

The final difficulty is also related to one we raised for Canfield's analysis. There we noticed that if an artifact was explicitly designed to do something, *that* usually *determines* its function, irrespective of how well or badly it does the thing it was supposed to do. An analogous point can be made here. If *X* was designed to do *Y*, then *Y* is *X*'s function regardless of what contributions *X* does in fact make or fail to make. For example, the *function* of the federal automotive safety regulations is to make driving and riding in a car safer. And this is so even if they actually have just the opposite effect, through some psychodynamic or automotive quirk.

So in spite of their enormous differences, this analysis and Canfield's fail for very similar reasons: problems with the notion of system *S*, failure to rule out some accidental cases, and general inability to account for the obvious role of design.

There have been several other interesting attempts in the recent literature to provide an analysis of function. Most notable are those by Carl Hempel,[8] Hugh Lehman,[9] Richard Sorabji,[10] Francisco Ayala,[11] and Michael Ruse.[12] The last two of these do a somewhat better job on the function/accident distinction than the ones we have examined. But other than that, a discussion of these analyses would be largely redundant on the discussions of Beckner and Canfield. So I think we have gone far enough in clarifying the issues to begin constructing an alternative analysis.

III. AN ALTERNATIVE VIEW

The treatments we have so far considered have overlooked, ignored, or at any rate failed to make, one important observation: that functional ascriptions are – intrinsically, if you will – explanatory. Merely saying of something, *X*, that it has a certain function, is to offer an important kind of explanation of *X*. The failure to consider this, or at least take it seriously, is, I think, responsible for the systematic failure of these analyses to provide an accurate account of functions.

There are two related considerations which urge this observation upon us. First, the 'in order to' in functional ascriptions is a teleological 'in order to'. Its role in functional ascriptions (the heart beats in order to circulate blood) is quite parallel to the role of 'in order to' in goal ascriptions (the rabbit is running in order to escape from the dog). Accordingly, we should expect functional ascriptions to be explanatory in something like the same way as goal ascriptions.[13] When we say that the rabbit is running in order to escape from the dog, we are explaining *why* the rabbit is running. If we say that John got up early in order to study, we are offering an explanation of his getting up early. Similarly in the functional cases. When we say that the distributor has that cover in order to keep the rain out, we are explaining *why* the distributor has that cover. And when we say the heart beats in order to pump blood, we are ordinarily taken to be offerering an explanation of why the heart beats. This last sort of case represents the most troublesome problem in the logic of function, but it must be faced squarely, and, once faced, I think its solution is fairly straightforward.

The second consideration which recommends holding out for the explanatory status of functional ascriptions is the contextual equivalence of several sorts of requests. Consider:

(1) What is the function of *X*?
(2) Why do *C*'s have *X*'s?
(3) Why do *X*'s do *Y*?

In the appropriate context, each of these is asking for the function of *X* 'What is the function of the heart?', 'Why do humans have a heart'?, 'Why does the heart beat?' All are answered by saying, 'To pump blood', in the context we are considering. Questions of the second and third sort, being 'Why?' questions, are undisguised requests for explanations. So in this context functional attributions are presumed to be explanatory. And why-form function requests are by no means bizarre or esoteric ways of asking for a function. Consider:

> Why do porcupines have sharp quills?
> Why do (some) watches have a sweep-second hand?
> Why do ducks have webbed feet?
> Why do headlight bulbs have two filaments?

These are rather ordinary ways of asking for a function. And if that is so, then it is ordinarily supposed that a function explains why each of these things is the case. The function of the quills is why porcupines *have* them, and so forth.

Moreover, the kind of explanatory role suggested by both of these considerations is not the anemic 'What's it good for?' sort of thing often imputed to functional explanations. It is rather something more substantial than that. If to specify the function of quills is to explain why porcupines *have* them, then the function must be the reason they *have* them. That is, the ascription of a function must be explanatory in a rather strong sense. To choose the weaker interpretation, as Canfield does,[14] is once again to run afoul of the function-accident distinction. For, to use his example, if 'Why do animals have livers?' is a request for a function, it cannot be rendered 'What is the liver good for?' Livers are good for many things which are not their functions, just like anything else. Noses are good for supporting eyeglasses, fountain pens are good for cleaning your fingernails, and livers are good for dinner with onions. No, the *function* of the liver is that *particular* thing it is good for which explains why animals have them.

Putting the matter in this way suggests that functional ascription-explanations are in some sense etiological, concern the causal background of the phenomenon under consideration. And this is indeed what I wish to argue: functional explanations, although plainly not causal in the usual, restricted sense, do concern how the thing with the function *got there*. Hence they *are* etiological, which is to say 'causal' in an extended sense. But this is still a very contentious view. Functional and teleological explanations are usually *contrasted with* causal ones, and we should not abandon that contrast lightly: we should be driven to it.

What drives us to this position is the specific difficulty the best-looking alternative accounts have in making the function/accident distinction. We have seen that no matter how useful it is for *X* to do *Z*, or what contribution *X*'s doing *Z* makes within a complex system,[15] these sorts of consideration are never sufficient for saying that the function of *X* is *Z*. It could still turn out that *X* did *Z* only by accident. But all of the accident counterexamples can be avoided if we include as part of the analysis something about how *X* came to be there (wherever): namely, that it is there *because it does Z* – with an etiological 'because'. The buckle, the heart, the nose, the engine nut, and so forth were not there *because* they stop bullets, throb, support glasses, adjust the valve, and all the other things which were falsely attributed as functions, respectively. Those pseudo-functions could not be called upon to explain how those things *got* there. This seems to be what was missing in each of those cases.

In other words, saying that the function of *X* is *Z* is saying at least that

(1) *X* is there *because* it does *Z*.
or
Doing *Z* is the *reason* *X* is there.
or
That *X* does *Z* is *why* *X* is there.

where 'because', 'reason', and 'why' have an etiological force. And it turns out that '*X* is there because it does *Z*',[16] with the proper understanding of 'because', 'does', and 'is there' provides us with not only a necessary condition for the standard cases of functions, but also the kernel of an adequate analysis. Let us look briefly at those key terms.

'Because' is of course to be understood in its explanatory rather than evidential sense. It is not the 'because' in 'It is hot because it is red'. More importantly, 'because' is to be taken (as it ordinarily is anyway) to be indifferent to the philosophical reasons/causes distinction. The 'because' in 'He did not go to class because he wanted to study' and in 'It exploded because it got too hot' are both etiological in the appropriate way.[17] And finally, it is worth pointing out here that in this sense '*A* because *B*' does not require that *B* be either necessary or sufficient for *A*. Racing cars have airfoils because they generate a downforce (negative lift) which augments traction. But their generation of negative lift is neither necessary nor sufficient for racing cars to have wings: they could be there merely for aesthetic reasons, or they could be forbidden by the rules. Nevertheless, if you want to know why they are there, it is because they produce negative lift. All of this comes to saying that 'because' here is to be taken in its ordinary, conversational, causal-explanatory sense.

Complications arise with respect to 'does' primarily because on the above condition 'Z is the function of X' is reasonably taken to entail 'X does Z'. Although in most cases there is no question at all about what it is for X to do Z, the matter is highly context-dependent and so perhaps I should mention an extreme case, if only as notice that we should include it. In some contexts we will allow that X does Z even though Z never occurs. For example, the button on the dashboard activates the windshield washer system (that is what it does, I can tell by the circuit diagram) even though it never has and never will. An unused organic or organismic emergency reaction might have the same status. All that seems to be required is that X be *able* to do Z under the appropriate conditions; for example, when the button is pushed or in the presence of a threat to safety.

The vagueness of 'is there' is probably what Beckner and Canfield were trying to avoid by introducing the system S into their formulations. It is much more difficult, however, to avoid the difficulties with the system S than to clarify adequately this more general place-marker. 'Is there' is straightforward and unproblematic in most contexts, but some illustrations of importantly different ways in which it can be rendered might be helpful. It can mean something like 'is where it is', as in 'keeping food out of the windpipe is the reason the epiglottis is where it is'. It can mean 'C's have them', as in 'animals have hearts because they pump blood'. Or it can mean merely 'exists (at all)', as in 'keeping snow from drifting across roads (and so forth) is why there are snow fences'.

Now, saying that (1), understood in this way, should be construed as a necessary condition for taking Z to be the function of X, is merely to put in precise terms the moral of our examination of the function/accident distinction. We saw above that the accident counterexamples could not meet this requirement. On the other hand, this condition *is* met in all of the center-of-the-page cases. This is quite easy to show in the conscious cases. When we say the function of X is Z in these cases, we are saying that at least some effort was made to get X (sweep hand, button on dashboard) where it is precisely because it does Z (whatever). Doing Z is the reason X is there. *That* is why the effort was made. The reason the sweep-second hand is there is that it makes seconds easier to read. It is there *because* it does that. Similarly, rifles have safeties because they prevent accidental discharge.

It is only slightly less obvious how natural functions can satisfy (1): We can say that the natural function of something – say, an organ in an organism – is the reason the organ is there by invoking natural selection. If an organ has been naturally differentially selected-for by virtue of something it does, we can say that the reason the organ is there is that it does that something. Hence we can say animals have kidneys *because* they eliminate metabolic wastes from the bloodstream; porcupines have quills *because* they protect them from predatory enemies; plants have chlorophyll *because* chlorophyll enables plants to accomplish photosynthesis; the heart beats *because* its beating pumps blood. And each of these can be rather mechanically put in the 'reason that' form. The reason porcupines have quills is that they protect them from predatory enemies, and so forth.

It is easy to show that this formula does not represent a sufficient condition for being a function, which is to say there is something more to be said about precisely what it is to be a function. The most easily generable set of cases to be excluded is of this kind: oxygen combines readily with hemoglobin, and that is the (etiological) reason it is found in human bloodstreams. But there is something colossally fatuous in maintaining that the function of that oxygen is to combine with hemoglobin, even though it is there because it does that. The function of the oxygen in human bloodstreams is providing energy in oxidation reactions, not combining with hemoglobin. Combining with hemoglobin is only a means to that end. This is a useful example. It points to a contrast in the notion of 'because' employed here which is easy to overlook and crucial to an elucidation of functions.

As I pointed out above, if producing energy is the function of the oxygen, then oxygen must be there (in the blood) because it produces energy. But the 'because' in 'It is there because it produces energy' is importantly different from the 'because' in 'It is there because it combines with hemoglobin'. They suggest different *sorts* of etiologies. If carbon monoxide, which we know to combine readily with hemoglobin, were suddenly to become able to produce energy by appropriate (non-lethal) reactions in our cells and, further, the atmosphere were suddenly(!) to become filled with CO, we could properly say that the reason CO was in our bloodstreams was that it combines readily with hemoglobin. We could not properly say, however, that CO was there because it produces *energy*. And that is precisely what we could say about oxygen, on purely evolutionary-etiological grounds.

All of this indicates that it is the nature of the etiology itself which determines the propriety of a functional explanation; there must be specifically functional etiologies. When we say the function of *X* is *Z* (to do *Z*) we are saying that *X* is there because it does *Z*, but with a further qualification. We are explaining how *X* came to be there, but only certain kinds of explanations of how *X* came to be there will do. The causal/functional distinction is a distinction *among* etiologies; it is not a contrast between etiologies and something else.

This distinction can be displayed using the notion of a causal consequence.[18] When we give a functional explanation of *X* by appeal to *Z* ('*X* does *Z*'), *Z* is always a consequence or result of *X*'s being there (in the sense of 'is there' sketched above).[19] So when we say that *Z* is the function of *X*, we are not only saying that *X* is there because it does *Z*, we are also saying that *Z* is (or happens as) a result or consequence of *X*'s being there. Not only is chlorophyll in plants *because* it allows them to perform photosynthesis, photosynthesis is a *consequence* of the chlorophyll's being there. Not only is the valve-adjusting screw there *because* it allows the clearance to be easily adjusted, the possibility of easy adjustment is a *consequence* of the screw's being there. Quite obviously, 'consequence of' here does not mean 'guaranteed by'. '*Z* is a consequence of *X*', very much like '*X* does *Z*' earlier, must be consistent with *Z*'s not occurring. When we say that photosynthesis is a consequence of chlorophyll, we allow that some green plants may never be exposed to light, and that all green plants may at some time or other not be exposed to light. Furthermore, this consequence relationship does not mean that whenever *Z does* occur, happen, obtain, exist, and so forth, it is as a consequence of *X*. There is room for a multiplicity of sufficient conditions, overdetermined or otherwise. Other things besides the adjusting screw may provide easy adjustment of the clearance. This (the inferential) aspect of consequence, as that notion is used here, can be roughly captured by saying that there are circumstances (of recognizable propriety) in which *X* is non-redundant for *Z*. The aspect of 'consequence' of central importance here, however, is its asymmetry. '*A* is a consequence of *B*' is in virtually every context incompatible with '*B* is a consequence of *A*'. The source of this asymmetry is difficult to specify, and I shall not try.[20] It is enough that it be clearly present in the specific cases.

Accordingly, if we understand the key terms as they have been explicated here, we can conveniently summarize this analysis as follows: The function of *X* is *Z means*

(2) (a) *X* is there because it does *Z*,
(b) *Z* is a consequence (or result) of *X*'s being there.

The first part, (a), displays the etiological form of functional ascription-explanations, and the second part, (b), describes the convolution which distinguishes functional etiologies from the rest. It is the second part of course which distinguishes the combining with hemoglobin from the producing of energy in the oxygen-respiration example. Its combining with hemoglobin is emphatically not a consequence of oxygen's being in our blood; just the reverse is true. On the other hand, its producing energy *is* a result of its being there.

The very best evidence that this analysis is on the right track is that it seems to include the entire array of standard cases we have been considering, while at the same time avoiding several very persistent classes of counterexamples. In addition to this, however, there are some more general considerations which urge this position upon us.[21] First, and perhaps most impressive, this analysis shows what it is about functions that is teleological. It provides an etiological rationale for the functional 'in order to', just as recent discussions have for other teleological concepts. The role of the consequences of *X* in its own etiology provide functional ascription-explanations with a convoluted forward orientation which precisely parallels that found by recent analyses in ascription-explanations employing the concepts goal and intention.[22] In a functional explanation, the consequences of *X*'s being there (where it is, and so forth) must be invoked to explain why *X* is there (exists, and so forth). Functional characterizations, by their very nature, license these explanatory appeals. Furthermore, as I hinted earlier, (b) is often simply implicit in the 'because' of (a). When this is so, the 'because' is the specifically teleological one sometimes identified as peculiarly appropriate in functional contexts. The peculiarly functional "because" is the normal etiological one, except that it is limited to consequences in this way. The request for an explanation as well will very often contain this implicit restriction, hence limiting the appropriate replies to something in terms of this 'because' – that is, to functional explanations. 'Why is it there?' in some contexts, and 'What does it do?' in most, unpack into 'What consequences does it have that account for its being there?'

The second general consideration which recommends this analysis is that it both accounts for the propriety of, and at the same time elucidates the notion of, natural selection. To make this clear, it is important first to say something about the unqualified notion of selection, from which natural selection is derived. According to the standard view, which I will accept for expository purposes, the paradigm cases of selection involve conscious choice, perhaps even deliberation. We can then understand other uses of 'select' and 'selection' as extensions of this use: drawing attention to specific individual *features* of the paradigm which occur in subconscious or nonconscious cases. Of course, the range of extensions arrays itself into a spectrum from more or less literal to openly metaphorical. Now, there is an important distinction within the paradigmatic, conscious cases. I can say I selected something, *X*, even though I cannot give a reason for having chosen it: I am asked to select a ball from among those on the table in front of me. I choose the blue one and am asked why I did. I may say something like 'I don't know; it just struck me, I guess'. Alternately, I could without adding much give something which has the form of a reason: 'Because it is blue. Yes, I'm sure it was the color'. In both of these cases I want to refer to the selection as 'mere discrimination', for reasons which will become apparent below. On the other hand, there are a number of contexts in which another, more elaborate reply is possible and natural. I could say something of the form 'I selected *X* because it does *Z*', where *Z* would be some possibility opened by, some advantage that would accrue from, or some other result of having (using, and so forth) *X*. 'I chose American Airlines because its five-across seating allows me to stretch out'. Or 'They selected DuPont Nomex because of the superior protection it affords in a fire'.[23] Let me refer to selection by virtue of resultant advantage of this sort as 'consequence-selection'. Plainly, it is this kind of selection, as opposed to mere discrimination, that lies behind conscious functions: the consequence *is* the function.

Equally plainly, it is specifically this kind of selection of which *natural* selection represents an extension.

But the parallel between natural selection and conscious consequence-selection is much more striking than is sometimes thought. True, the presence or absence of volition is an important difference, at least in some contexts. We might want to say that *natural* selection is really *self*-selection, nothing is *doing* the selecting; given the nature of *X*, *Z*, and the environment, *X* will *automatically* be selected. Quite so. But here the above distinction between kinds of conscious selection becomes crucial. For consequence-selection, by contrast with mere discrimination, de-emphasizes volition in just such a way as to blur its distinction from natural selection on precisely this point. Given our criteria, we might well say that *X does* select itself in conscious consequence-selection. By the very nature of *X*, *Z*, and our criteria (the implementation of which may be considered the environment), *X* will automatically be selected.[24] The cases are very close indeed.

Let us now see how this analysis squares with the desiderata we have developed. First, it is quite clearly a unifying analysis: the formula applies to natural and conscious functions indifferently. Both natural and conscious functions are functions by virtue of their being the reason the thing with the function 'is there', subject to the above restrictions. The differentiating feature is merely the *sort* of reason appropriate in either case: specifically, whether a conscious agent was involved or no. But in the functional-explanatory context which we are examining, the difference is minimal. When we explain the presence or existence of *X* by appeal to a consequence *Z*, the overriding consideration is that *Z* must be or create conditions conducive to the survival or maintenance or *X*.[25] The exact *nature* of the conditions is inessential to the possibility of this form of explanation: it can be looked upon as a matter of mere etiological detail, nothing in the essential form of the explanation. In any given case something could conceivably get a function through either sort of consideration. Accordingly, this analysis begs no theological questions. The organs of organisms could logically possibly get their functions through God's conscious design; but we can also make perfectly good sense of their functions in the absence of divine intervention. And in either case they would be functions in precisely the same sense. This of course was accomplished only by disallowing explicit mention of intent or purpose in accounting for conscious functions. Nevertheless, the above formula can account for the very close relationship between design and function which the previous analyses could not. For, excepting bizarre circumstances, in virtually all of the usual contexts, *X* was designed to do *Z* simply entails that *X* is there because it results in *Z*.

Second, this analysis makes a clear and cogent distinction between function and accident. The things *X* can be said to do by accident are the things it results in which cannot explain how it came to be there. And we have seen that this circumvents the accident counterexamples brought to bear on the other analyses. It is merely accidental that the chlorophyll in plants freshens breath. But what it does for plants when the sun shines is no accident – that is why it is there. Furthermore, in this sense, '*X* did *Z* accidentally' is obviously consistent with *X*'s doing *Z* having well-defined causal antecedents, just like the normal cases of other sorts of accident (automobile accidents, accidental meetings, and so forth). Given enough data it could even have been predictable that the belt buckle would deflect the bullet. But such deflection was still in the appropriate sense accidental: that is not why the buckle was there.

Furthermore, it is worth noting that something can get a function – either conscious or natural – *as the result of* an accident of this sort. Organismic mutations are paradigmatically accidental in this sense. But that only disqualifies an organ from functionhood for the first – or the first few – generations. If it survives by dint of its doing something, then that something becomes its function on this analysis. Similarly for artifacts. For example, if an earthquake shifted the rollers of a transistor production-line conveyor belt, causing the belt to ripple in just such a way that defective transistors would not pass over the ripple, while good transistors would, we could say that the ripple was *functioning as* a quality control sorter. But it would be incorrect to say that the ripple *had* the function of quality control sorting. It does not *have* a function at all. It is there only by accident. Sorting can, however, *become* its function if its sorting ability ever becomes a reason for preserving the ripple: if, for example, the company decides against repairing the conveyor belt *for that reason*. This accords nicely with Richard Sorabji's comment that in conscious cases, saying the function of *X* is *Z* requires at least 'that some efforts are or would if necessary be made' to obtain *Z* from *X*.[26, 27]

Third, the notion of something having more than one function is derivative. It is obtained by substituting something like 'partly because'[28] for 'because' in the formula. Brushing dust off the numbers is one of the functions of the watch's sweep-second hand if that feature is *one* of the (restricted, etiological) reasons the sweep hand is there. Similarly in the case of natural functions. If two or three things that livers do all contribute to the survival of organisms which have livers, we must appeal to all three in an evolutionary account of why those organisms have livers. Hence the liver would have more than one function in such organisms: we would have to say that each one was *a* function of the liver.

Happily, the analysis I am here proposing also accounts for the undoubted attractiveness of the other analyses we have examined. Beckner's first analysis is virtually included in this one under the rubric '*X* does *Z*'. The rest of the formula can be thought of as a qualification to avoid some rather straightforward counterexamples which Beckner himself is concerned to circumvent in his more recent attempt. Canfield's 'usefulness' is even easier to accommodate: the usefulness of something, *Z*, which *X* does is *very usually* an informative way of characterizing why *X* has survived in an evolutionary process, or the reason *X* was consciously constructed. The important point to notice is that this is only *usually* the case, not necessarily: not all useful *Z*'s can explain survival and some things are constructed to do wholly useless things. As for Beckner's most recent analysis, the complex, mutually contributory relationship among parts central to it is precisely the sort of thing often responsible for the survival and reproduction of organisms on one hand, and for the construction of complex mechanisms on the other. Again the valuable features of that analysis are incorporated in this one.

There is still one sort of case in which we clearly want to be able to speak of a function, but which offends the letter of this analysis as it stands. In several contexts, some of which we have already examined, we want to be able to say that *X* has the function *Z*, even though *X* cannot be said to do *Z*. *X* is not even *able* to do *Z* under the requisite conditions. In the cases of this sort I have already mentioned (the defective washer switch and ineffective governmental safety regulations), it has seemed necessary to italicize (emphasize, underline) the word 'function' in order to make its use plausible and appropriate. This is a logical flag: it signals that a special or peculiar contrast is being made, that the case departs from the paradigms in a systematic but intelligible way.

Accordingly, an analysis has to make sense of such a case as a variant.

On the present analysis, the italic type signals the dropping of the (usually presumed) second condition. *X* does *not* result in *Z*, although, paradoxically, doing *Z* is the reason *X* is there. Of course, in the abstract, this sounds fatuous. But we have already seen cases in which it is natural and appropriate. That *is* the reason *X* (switch, safety regulations) is there. And a slightly more defensive formulation of (2) will include them directly: a functional ascription-explanation accounts for *X*'s being there by appeal to *X*'s resulting in *Z*. These cases *do* appeal to *X*'s resulting in *Z* to explain the occurrence of *X*, even though *X* does *not* result in *Z*. So the form of the explanation is functional even in these peculiar cases.

Interestingly, this account even handles the exotic fact that these italicized functions of *X* can cease being even italicized functions without dispensing with or directly altering *X*. (Something that *X* did not do can stop being its function!) For example, if the ineffective safety regulations were superseded by another set, and merely left on the books through legislative sloth or expediency, we would no longer even say they had the (italicized) *function* of making driving less dangerous. But, of course, that would no longer be the reason they were there. The explanation would then have to appeal to legislative sloth or expediency. This is usually done with verb tenses: that *was* its function, but is not any longer; that was why it was there at one time, but is not why it is still there. A similar treatment can be given vestigial organs, such as the vermiform appendix in humans.

IV. A FINAL OBSERVATION

There are at least two sorts of plausible-looking counterexamples to this analysis. The first, which we have touched on briefly already, is the initial instance of an organ which in fact will survive because of its activities. If, e.g., the liver – with all its normal activities – had just suddenly occurred at some stage in the evolution of organisms, we could not, according to this analysis, say it had any functions upon its appearance. We would have to wait some generations to see if it survived by virtue of the things it did. If it did, only then could we say that those activities were its functions. It could plausibly be urged that this is to rule out a perfectly legitimate case of something's having a function.

The second sort of example is what Sorabji calls a "luxury function." (*op. cit.*, p. 294) His example of a luxury function is

... an organ ... which came into operation only when some lethal type of damage had occurred, e.g., a major coronary thrombosis. And suppose that the effect of this organ were to shut off sensations of pain as soon as such lethal damage occurred. This effect would not increase the chances of survival either for individual or for species.

And accordingly, its cutting off pain could not explain why the organ was there. But, Sorabji would insist, it is very natural to call this pain-elimination the function of the organ.

Now, my understanding of natural selection suggests that clear-cut examples of either of the above sorts are enormously unlikely to ever occur. And while this is not an objection to these cases as counter-examples, it does suggest that the following general logical point is relevant here.

Several interesting concepts in natural languages and in science fit the following schema.

(1) *A*, *B*, *C*, *D* and *E* together constitute a clear case of *F* (where *A*, *B*, *C*, *D*, and *E* are characteristics of something or other).
(2) *A*, *B*, *C*, *D* and *E* always or usually occur together.
(3) *E* is different from the other characteristics in that it is almost never explicitly appealed to in justifying an ascription of *F*.
(4) We commonly infer from *A*, *B*, *C*, and *D* to *F*.
(5) We commonly infer from *F*, to *E* (or to something else, *G*, which clearly (logically) presupposes *E*).

Given just this much, in the abstract, it is not clear how we might choose between the following.

(A) By '*F*' we mean *A*, *B*, *C*, *D and E*; so (4) is only inductive while $F \rightarrow E$ is deductive.
(B) What we *really* mean by '*F*' is merely *A*, *B*, *C*, and *D*; so (4) is deductive, while $F \rightarrow E$ is only inductive.

For example, in the standard cases of death, all the body's life signs and characteristic behavior cease permanently. Usually all of these occur together. But the *permanence* of the cessation is seldom if ever explicitly appealed to in judging a case of death. Usually: the life signs cease, death is pronounced. And just as usually, we infer from death to the permanence of the cessation of life-signs: we bury the body, pay off on life insurance and allow the spouse to remarry. However, it is simply not clear which inference takes the logical risk. Our first temptation is to say that the inference from death to permanence is the contingent/inductive one, but that is hasty: if someone were to 'come alive' again after a 'down' period it is not at all clear that we would be bound to pay off on his insurance and allow his spouse to remarry. And if such 'recovery' were relatively common, it would seem unreasonable not to change the criteria for issuing D.O.A. certificates in emergency wards. All of which suggests that permanence is logically part of death. On the other hand we do occasionally say things like 'he was dead for a time during the operation, but they managed to save him'; and there is always the rhetoric of Easter Sunday in the background. It is the tension between these two temptations which makes 'suspended animation' or 'limbo' a welcomed escape route: it is arbitrary to resolve the issue either way.

Often this much is misunderstood about a concept *F*. Accordingly, when one comes to a philosophical dispute about whether a given case of *A*, *B*, *C*, *D* and *not-E* is an *F*, it is sometimes sufficient merely to point out its irresolvability on what might be called formal grounds: *that* was the misunderstanding, nothing else was at issue. Examples of more-or-less classic disputes that can be treated in this manner are: whether or not a kind of perception can be *seeing* if it doesn't involve using the eyes;[29] the William James squirrel case;[30] and perhaps (though disputably) the old saw (!) about the tree falling in the forest when nobody is around to hear it.

But to say that issues of this sort are *formally* irresolvable is not to say that they cannot be resolved. Often there are reasons other than formal ones which have weight here. And it is here that a philosophical analysis can be particularly illuminating and perhaps legitimately reconstructive. It can be illuminating by pointing out the sorts of non-formal considerations relevant in a given case, and what (if any) resolution they offer. It can be reconstructive by pointing out the (net) advantage which would accrue through the general adoption of a particular resolution.[31]

I wish to argue that the luxury function and first mutation examples represent just this sort of formally irresolvable, borderline case. And they do so merely by virtue of their not being explanatory in the appropriate way. The clear cut examples of functions are obviously explanatory. And the automaticity of our inferences from function to explanation is testified to by the evident synonymy of several different requests for and formulations of functional ascriptions, some of which are palpably pregnant explanatorily (see p. 226). This is enough to justify their classification as borderline cases under the *F*-schematization above. And this, in turn, is enough to allow my analysis to escape the brunt of their force.

But a stronger position is defensible. The automaticity of our inference from function to explanation suggests that non-explanatory functional ascriptions would be systematically misleading. And indeed they are: The whole *point* of a functional ascription is explanatory. This is obviously true in scientific inquiry, but I think it is so in everyday life as well. The sole issue raised by the function/accident distinction is whether the thing is there because of something it does, or is there only by chance. And in the normal cases in which this issue is resolved in favor of chance, we feel compelled to say so, and deny a function. 'What is the function of the ripple in the conveyor belt?' 'Oh, we didn't put that there, it was the earthquake, but the thing works all right so we're letting the line run until our technicians get a look at it'. Similarly, 'what's the function of our vermiform appendix?' 'It doesn't *have* any function, though some people are glad they had one because it gave them a scar to talk about, or resulted in the early detection of a dermoid cyst which could have been trouble'.

Accordingly, there is strong reason to draw attention to the central functional atypicality of the luxury function and first-mutation cases, and to warn of the inferential trap they represent. One way of drawing such attention and providing such warning is to publicly insist that these cases be regarded as *non*-functions, just like their conceptual twins ripple and appendix.

University of California, Riverside

NOTES

[1] Richard Sorabji, 'Function', *Philosophical Quarterly* **14** (1964), 290.

[2] Morton Beckner, *The Biological Way of Thought* (New York, 1959), Ch. 6.

[3] Beckner gives an alternative formulation in which we can speak of *activities* as having functions, instead of *things*. I have abbreviated it here for convenience and clarity. The logical points are the same.

[4] John Canfield, 'Teleological Explanations in Biology', *The British Journal for the Philosophy of Science* **14** (1964).

[5] Morton Beckner, 'Function and Teleology', *Journal of the History of Biology* **2** (1969), reprinted above pp. 197–212.
[6] As before, Beckner gives an alternative formulation so that we can speak either of a thing or of an activity having a function. My treatment will be limited to things, but again the logical points are the same.
[7] Beckner seems to suggest (p. 207, top) that *F* must *be* an activity of the whole system *S*, which, of course, would conflict with part of 3. But his illustration, reproduced below, suggests the phrasing I have used here.
[8] Carl Hempel, 'The Logic of Functional Analyses', in L. Gross (ed.), *Symposium on Sociological Theory* (New York, 1959).
[9] Hugh Lehman, 'Functional Explanations in Biology', *Philosophy of Science* **32** (1965).
[10] Sorabji, *op cit.*
[11] Francisco J. Ayala, 'Teleological Explanation in Evolutionary Biology', *Philosophy of Science* **37** (1970).
[12] Michael E. Ruse, 'Function Statements in Biology', *Philosophy of Science* **38** (1971).
[13] This is not to abandon, or even modify, the previous distinction between functions and goals: the point can be made in this form only *given* the distinction. Nevertheless, support is provided for the analysis I am presenting here by the fact that the 'in order to' of goal-directedness can be afforded a parallel treatment. For that parallel treatment see my paper 'Explanation and Teleology', in the June 1972 issue of *Philosophy of Science.*
[14] Canfield, *op. cit.*, p. 295.
[15] It is sometimes urged that this sort of thing is all a teleological explanation is asserting; this is all 'why?' asks in these contexts.
[16] I take the other forms to be essentially equivalent and subject, *mutatis mutandis*, to the same explication.
[17] Of course, it follows that the notion of a *reason* offered in one of the alternative formulations is the standard conversational one as well: the reason it exploded was that is got too hot.
[18] The qualification 'causal' here serves merely to indicate that this is not the purely inferential sense of 'consequence'. I am not talking about the result or consequence of an argument – e.g., necessary conditions for the truth of a set of premises. The precise construction of 'consequence' appropriate here will become clear below.
[19] It is worth recalling here that 'is there' can only sometimes, but not usually, be rendered 'exists (at all)'. So, contrary to many accounts, what is being explained, and what *Z* is the result of, can very often *not* be characterized as 'that *X* exists' *simpliciter.*
[20] It is often claimed that the asymmetry is temporal, but there are many difficulties with this view. Douglas Gasking, in 'Causation and Recipes', *Mind* (Oct., 1955), attempts to account for it in terms of manipulability with some success. But manipulability is even less generally applicable than time order, so, as far as I know, the problem remains.
[21] The following considerations are intended primarily as support for the entire analysis considered whole. Since (a) has already been examined extensively, however, I have biased the argument slightly to emphasize (b).
[22] The primary discussions of this sort I have in mind are those in Charles Taylor's *Explanation of Behavior* and the literature to which it has given rise.
[23] Of course the advantage is not always stated explicitly; 'I chose American because of its five-across seating'. But for it to be selection of the sort described here, as opposed to mere discrimination, something like an advantage must be at least implicit.

[24] This is a version of the old problem about the tension between rationality and freedom.
[25] This formulation is at best very clumsy and misleading. The matter is much more complicated than this and cannot easily be put in a form suitable for this context. The point is adequately made merely by drawing attention to the fact that conscious and natural selection both provide consequence-etiologies.
[26] Sorabji, *op. cit.*, p. 290.
[27] Including the conveyor-belt case in this manner was clearly a mistake: it is better viewed as a derivative case with no close parallel among natural functions. As it stands it violates (F): it does *not* have a consequence-etiology, the ripple did not get there because of what it does. It can be included, as a derivative, only by taking advantage of the rich causal possibilities of an intentional context. If we can justify not eliminating the ripple by appeal to its virtue as a sorting device, then with the same utilities, there are circumstances (of availability, relative cost and so forth) in which I could justify including the ripple in the original design. So I can say that, ceteris paribus, the ripple *would* have come to be there because of what it does, if it had not gotten there fortuitously. We might call this a subjunctive consequence-etiology; but in any case it is a peculiar relative, not a paradigm.
[28] Again, it is worth pointing out that 'partly' here does not indicate that 'because', when *not* so qualified, represents a sufficient condition relationship. It merely serves to indicate that more than one thing plays an explanatorily relevant role in this particular case. More than one thing must be mentioned to answer adequately the functional 'why?' question in this context. But that answer, as usual, need not provide a sufficient condition for the occurrence of *X*.
[29] See for example p. 101 ff of J. Hospers, *An Introduction to Analytical Philosophy*, Prentice Hall, New York, 1953.
[30] Neither of James' 'definitions' of 'going round' represent anything anybody ever meant by '*A* is going round *B*'. The missing feature – represented above by '*E*' – which is usually present and not explicitly referred to is a commonly accepted fixed reference frame to which one of the parties is relatively firmly attached. Without this the issue simply is formally irresolvable.
[31] The kinds of advantage relevant here are much more narrowly circumscribed than is commonly thought. E.g., tidiness of the analysis is *not* a relevant advantage. On the other hand it would be relevant to argue that, e.g., calling something an '*F*' which does not have property *E* is systematically misleading. But even here we have some choices: we could simply insist that '*F*' is not applicable in such cases, or we could say we have to clearly distinguish *kinds* of *F*.

C. PLURALISTIC EXPLANATION

STUART A. KAUFFMAN

ARTICULATION OF PARTS EXPLANATION IN BIOLOGY AND THE RATIONAL SEARCH FOR THEM

With the realization that the grounds upon which an hypothesis comes to be formulated can be considered separately from the grounds upon which it is accepted, it has become popular among some philosophers and scientists to claim that although it may be of psychological interest to understand the genesis of hypotheses, there can be no logic of search or discovery. While it is unclear exactly what is meant by the claim that there can be no logic of search, it *is* clear that this opinion, coupled with anecdotes of Poincaré's sudden solution of a mathematical problem while stepping on a Madrid streetcar, and Kekulé's vision of a snake biting its tail, have left the aura that the generation of an hypothesis is as mysterious as a Gestalt shift in perception of a figure. Perhaps because Gestalt shifts seem to occur without a processes of reasoning, but in some sense, spontaneously, the use of such perceptual shifts as models of hypothesis formation have lent support to the claim that there can be no logic of search. To a practicing scientist, the image of the startling 'shift' and insight might seem overly flattering of the scientist's genius, and the actual generation of hypotheses seem more reasonable and less mysterious. Some of the ways in which the generation of an hypothesis is a rather reasonable affair will be discussed below in conjunction with an effort to examine some of the features of what I am calling articulation of parts explanations, as they occur in biology.

Typical explanations in biology exhibit the manner in which parts and processes articulate together to cause the system to do some particular thing. Examples include accounts of the cardiovascular system, protein synthesizing system, endocrine system, etc. I do not wish to say that articulation of parts explanations occur only in biology, or are the only form of explanation utilized in biology, merely that such explanations are prevalent in biological sciences and exhibit interesting properties which I wish to discuss.

To refer to some types of explanations as articulation of parts explanations suggests that some explanations are not. I suggest that when a thing is seen as consisting of a single part, or as a continuum, then explanations of the appropriate aspect of its behavior will not exhibit how parts work together. For example, Newton's first law of motion does not exhibit how parts of a system work together, for there is but a single part – a particle in rectilinear motion. Maxwell's field equations treat an electromagnetic field as a continuum, not as a second order infinity of points. Further, some behaviors of a given system may require an explanation by reference to the interworkings of its parts, while others of its behaviors may not. For example, the manner in which a gasoline engine falls is predictable from an account of the engine as a mass without discriminated parts.

I wish to pursue the following theses:

1. An organism may be seen as doing indefinitely many things, and may be decomposed into parts and processes in indefinitely many ways.

2. Given an adequate description of an organism as doing any particular thing, we will use that description to help us decompose the organism into particular parts and processes which articulate together to cause it to behave as described.

3. For different descriptions of what the organism is doing, we may decompose it into parts in different ways.

4. The use of an adequate description of an organism seen as doing a particular thing to guide our decomposition of it into interrelated parts and processes, and indeed part of the logic of search, is intimately connected with the sufficient conditions for the adequate description. In particular, we can use the sufficient conditions to generate a cybernetic model showing how symbolic parts might articulate together to cause a symbolic version of the described behavior.

5. We can use such a cybernetic model to help find an isomorphic causal model showing how presumptive parts and processes of the real system might articulate to cause the described behavior.

6. Since there may be more than one set of sufficient conditions for the adequate description of the behavior, more than one cybernetic model to account for the behavior may be constructed. Such different cybernetic models will not be isomorphic, and each leads us to decompose the system in a different way. Hence, not only are organisms decomposed into parts to yield articulation of parts explanations in different ways for *different* adequate descriptions of the organism seen as doing diverse

things; but different tentative decompositions can be made of the system seen as doing any one thing, through the use of the diverse sets of sufficient conditions of that adequate description.

7. A successful decomposition leads to an articulation of parts explanation of how the system does what it is seen as doing.

8. We not only use views of what a system is doing to help decompose it into parts, we use information about parts to synthesize new views of what a system is doing.

9. The descriptions of parts and processes of one decomposition need only be compatible with and not deducible from the descriptions of parts and processes of a different composition.

10. There *need* be no 'ultimate' decomposition such that all other decompositions are deducible from it, although there may be such an 'ultimate' decomposition.

1. Perhaps the first point to be made is that there is no uniquely correct view about what an organism is doing. But, in order to achieve an articulation of parts (henceforth, *A* of *P*) explanation about an organism, we must, perforce, be explaining how the parts and processes articulate together so that the system does something. To begin studying such an object, we must come to some initial view about what the system is doing; literally, a view of what is happening. Consider that an organism may be viewed as: a self-replicating system, a developing individual, a parent with similar appearing offspring, a member of an ecosystem, a system exhibiting circadian rhythms, a member of an evolving population, an open thermodynamic system maintaining a locale of low entropy, etc.

Now, not only are multiple views about what a system is doing possible, but also any system may be decomposed into parts in indefinitely many ways, and for any such part, it too can be seen as doing indefinitely many things. Our questions concern the character of the diverse possible decompositions into parts and the relations between parts within one decomposition, and between diverse decompositions.

2. I suggest that we use an adequate description of a view of what a system is doing to help guide our decomposition of the system into a particular set of parts and processes which causally articulate together to cause the system to behave as described. A view of what a system is doing sets the explanandum and also supplies criteria by which to decide whether

or not a proposed portion of the system with some of its causal consequences is to count as a part and process of the system. Specifically, a proposed part will count as a part of the system if it, together with some of its causal consequences, will fit together with the other proposed parts and processes to cause the system to behave as described. In general, it is not possible to decide that a single proposed process is to count as part of the system in isolation from decisions about the adequacy of the suppositions about the remaining parts and processes. Parts and processes are accepted more or less jointly, and with them, the adequacy of a particular articulation of parts explanation in which just those parts and processes are seen as fitting together to yield the behavior in question. Other causal consequences of these parts are then considered irrelevant. I will call such a decomposition of a system a conjugate, coherent decomposition, for it is conjugate to a particular view of what the system is doing, and coherent in that it provides an articulation of parts explanation of how the system does whatever is specified in that particular view of it.

3. Clearly, distinct views of what an organism is doing may lead us to decompose it in distinct ways. Our account of an organism as a device exhibiting circadian rhythms picks out different parts and processes from an account of the effects of chromosomal crossing over on population genetics.

4. I wish to argue that part of what might be called a logic of search involves our use of the sufficient conditions of an adequate description of a view of what an organism does to help find a cybernetic model of the phenomenon, and thence, a causal model. I will consider first a hypothetical case of tissue reaggregation, then an actual description of gastrulation in the chick, and hypotheses generated to account for these phenomena.

Hypothetical Example

Suppose we note that the cells of a sponge can be disaggregated and then allowed to reaggregate, and that, upon reaggregation they always form a particular three dimensional structure in which different specific cell types are at specific loci relative to one another. Sponges are deformed by currents in the water, and we will consider that an adequate description of this hypothetical sponge need only describe cell movement and specify which cell types have which cell types as neighbors, and in which directions, and which cells border internal and external lumens in the final structure. Note that the specification of cell types is made on the basis of some theory, here the cell theory. The observation that the cells reaggregate constitutes a view about what the system is doing, and sets us a question.

There may be more than one set of sufficient conditions for the truth of any description. By sufficient condition I here mean a description of a state of affairs such that from this description and with no further empirical information, the initial description may be deduced. I shall refer to such a description as a descriptive sufficient condition.

Let us assume that in our hypothetical aggregate, each cell type has either no cell as a neighbor in some directions, or only specific other cell types. Thus, we may assume that a description of these restricted adjacency relations, plus an account of the numbers of each type of cell, constitute descriptive sufficient conditions from which, with no further empirical information, we might deduce the initial adequate description.

We can, however, find a different set of descriptive sufficient conditions for the initial description. If the reaggregate is located in a three dimensional coordinate system with some particular cell taken as the origin, then the specification of the coordinates of each cell, and its cell type, is surely a descriptive sufficient condition of the original description, from which that original description can be deduced. I want to argue that each such set of descriptive sufficient conditions can be utilized to help find a model of how the cells manage to reaggregate, and that models derived from different sets of descriptive sufficient conditions pick out different putative parts and processes interacting in different putative articulation of parts explanations.

The power of the strategy of search I shall discuss rests on three features, two logical, one contingent. (1) If a process can be found which is causally sufficient to bring about the state of affairs described in the descriptive sufficient conditions then that process necessarily is causally sufficient to bring about the state of affairs described in the initial description. (2) Any initial description has multiple sets of descriptive sufficient conditions. (3) It is a contingent fact that very often, the ease of finding a causally sufficient process for one set of descriptive sufficient conditions is greater than for other sets of descriptive sufficient conditions. Indeed, sometimes a process to bring about a descriptive sufficient condition is suggested in a transparently obvious way by the descriptive sufficient conditions themselves. The sense of transparently obvious will be brought out below.

The first step, then, in utilizing a descriptive set of sufficient conditions is to suppose that there is a process which is causally sufficient to bring about the state of affairs referred to in the descriptive sufficient conditions. In the hypothetical aggregate we are considering, suppose we are currently using the descriptive sufficient conditions in terms of restricted adjacency relations among the different cell types. We then suppose a process which brings about just these intercell boundries. A particularly obvious choice of process is to assume that only these specific intercell boundaries form bonds between the cells. Since cells move during reaggregation, then remain properly juxtaposed, we also suppose a process which causes movement relative to one another among the cells until an allowed boundary is formed, when relative movement of the two cells stop. These hypothetical processes may now be linked, or articulated together to show how the state of affairs described initially might come about; specifically, cells move about until they come into contact with the appropriate neighbor cell type, when stable bonds are formed, relative movement stops, and the final architecture of the aggregate is generated.

At this stage the model is what I will call purely symbolic, or cybernetic. As yet, no actual causal mechanism is suggested. The cybernetic model provides a set of rules such that a set of symbols behaving as prescribed by the rules, will generate a form isomorphic to the cell assembly. The cybernetic model asserts that if causal mechanisms can be found such that only restricted bounderies form stably and relative motion of cells then ceases, then the observed behavior will occur in the aggregate. The central criterion of adequacy of such a cybernetic model is sufficiency. That is, it must be true that a set of symbolic components behaving as described by the rules in the model would, in fact, result in an aggregate isomorphic to the cell aggregate. Note also that the cybernetic model, while not yet suggesting actual causal mechanisms, exhibits the manner in which the processes of the parts of the system must articulate – namely by mediating the formation of restricted kinds of intercell boundaries. Thus, the cybernetic model, and indeed the adjacency descriptive sufficient conditions themselves, already begins to indicate the relations that are to exist among the parts and processes of the system; furthermore, the cybernetic model coupled with our current knowledge about cells begins to suggest what sorts of causal mechanisms are required. Specifically, mechanisms are required which will create specific intercell boundaries, and not allow other intercell boundaries to form. We do not know yet what particular causal mechanism achieves this, but we do know that causal consequences of putative parts of this system which do not yield specific intercell boundaries may be treated as irrelevant causal consequences of the parts of the system. Those causal consequences, that is, processes, will not be regarded as processes of the system, but as irrelevant behaviors.

The descriptive sufficient conditions of the initial adequate description speak, if you will, in the imperative mood. They are an injunction to the scientist to direct his attention to those conditions, for around them it should be possible to build a cybernetic, and later a causal, model to explain the phenomenon. It is at least partially because the sufficient conditions can be used to generate a cybernetic model that the generation of hypotheses is a rather reasonable process.

5. With the cybernetic model in hand, and background knowledge about cells, we can now search for the kind of casual processes which are likely to cause restricted boundary relations – say specific molecules on the membranes of the different cells, each of which only interlocks with the appropriate others. With the suggestion of a specific set of causal mechanisms, the model has become an hypothesis requiring verification. If verified, it will specify which processes – that is, which of the many causal consequences of each of the portions of the system – are to count as processes of the system, and which are irrelevant to this particular account. With the verification of the hypothesis we will have achieved an articulation of parts explanation of how cells manage to reaggregate. Until we are confident of the entire *A* of *P* explanation, we may remain unsure about the claim that any particular process is to be regarded as a process *of* the system.

A cybernetic model is used to help find a causal model isomorphic to it. More than one causal model might be suggested, each isomorphic to the cybernetic model and to one another. The sense of isomorphism intended may be exemplified by supposing that one has designed, on paper, an adding machine. The design is a flow chart of operations performed by as yet symbolic devices. Now an actual machine realizing this design might, for example store numbers by filling and emptying water tumblers, or filling and emptying capacitors. The different physical realizations would be isomorphic to one another and the cybernetic model by virtue of the fact that the parts and processes of one machine can be put into one correspondence with the appropriate part and processes of the other real machines or cybernetic design model. While isomorphic in this sense, the causal mechanisms by which the different physical machines realized the cybernetic design model would differ, and would be described by different causal laws. In the example of the cell aggregate, many different causal means of forming only restricted adjacency relations might be considered.

6. Different sets of descriptive sufficient conditions of the same initial adequate description can be used to generate different, non-isomorphic, cybernetic models, and in turn, non-isomorphic causal models which decompose the organism into different putative parts and processes. For example, specification of the coordinates of each cell, and its cell type, in a three dimensional coordinate system with one cell taken as origin, was a set of descriptive sufficient conditions for the initial adequate description. Utilizing this, we build a cybernetic model by supposing a process which is causally efficacious in generating the state of affairs referred to in the descriptive sufficient condition. In that descriptive sufficient condition, each cell is specified in a coordinate system. A particularly obvious choice of process suggested by this set of sufficient conditions is that there *is* a coordinate system in the reaggregate with some cell as origin, and that each cell 'knows' where it is in the coordinate system and goes to the correct place. We then suppose that an origin cell somehow elaborates a coordinate system and that cells have some means of 'knowing' their location. From this cybernetic model, we make use of knowledge about cells to suggest causal mechanisms and achieve an hypothesis.

Note that the parts and processes specified in this description, and this cybernetic model, cannot be put into one to one correspondence with the parts and processes of the adjacency cybernetic model; here, distance measuring from an origin cell is a process of the system, and formation of restricted types of boundaries between adjacent cells is an irrelevant behavior of the cells. The two cybernetic models are not isomorphic, neither are those causal models which are isomorphic to each, isomorphic to one another across cybernetic models.

7. A successful articulation of parts explanation distinguishes between irrelevant causal consequences and important causal consequences of a

(For an explanation of these figures see text on page 256.)

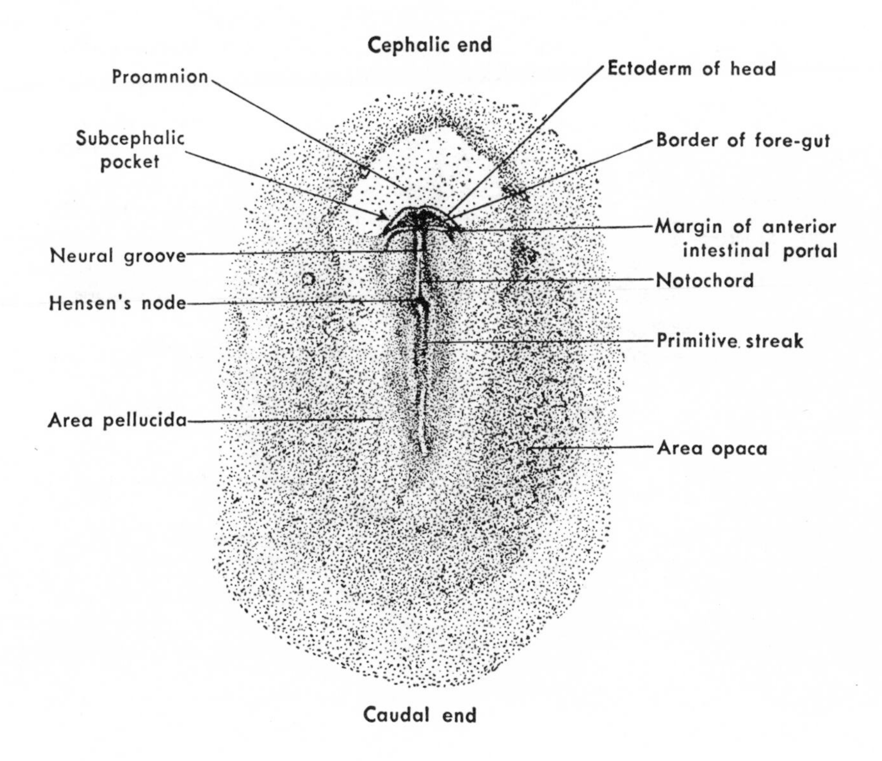

Fig. 1a. 20 hr chick embryo, dorsal view. Redrawn from Patten.

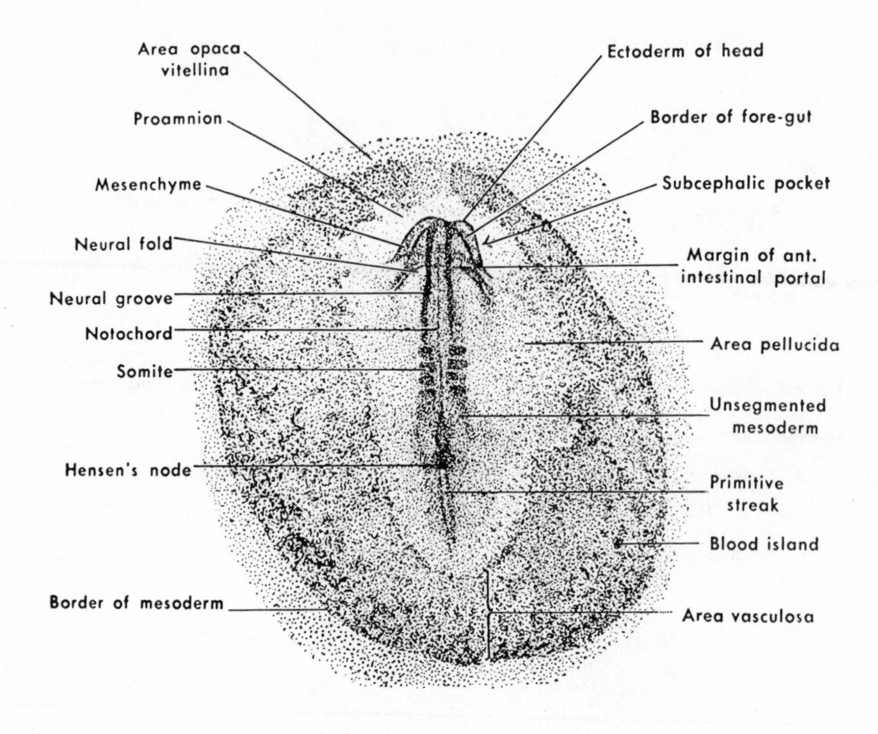

Fig. 1b. 24 hr chick embryo, dorsal view. Redrawn from Patten.

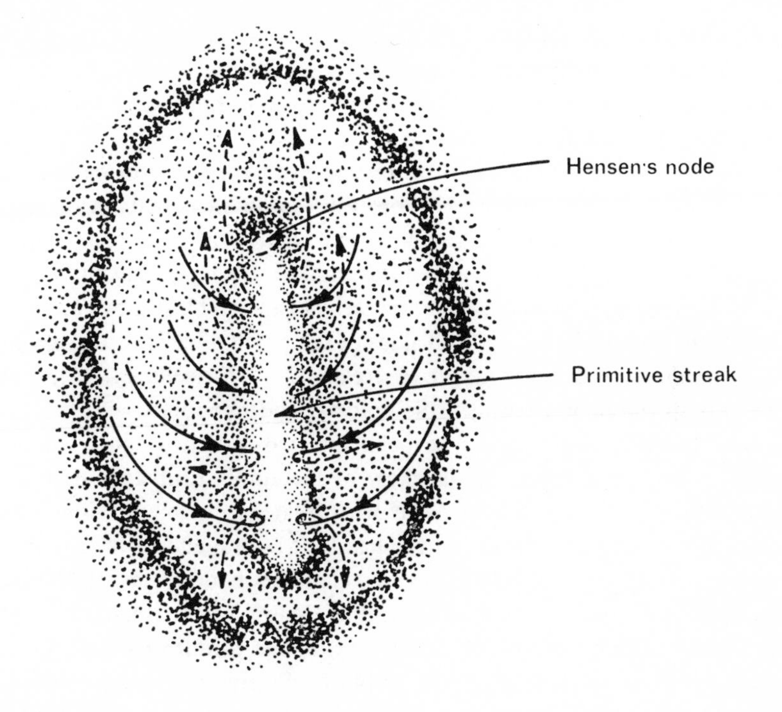

Fig. 1c. Schematic drawing of epiblast cell movement, solid arrows, and movement of mesodermal tissue, dotted arrows.

part, thus the explanation not only accounts for the behavior of the whole, it supplies a view of what it is that the parts themselves shall be seen as doing from among the indefinitely many possible things each part might be taken to be doing.

We now consider a more realistic example of hypothesis formation concerning gastrulation in chick embryogenesis (see figures). In its early stages, the chick embryo consists of a double layer, roughly oval, patch of cells on the yolk. The superficial layer is called the epiblast, the layer next to the yolk is the hypoblast. During gastrulation, cells of the epiblast move in an arc backward and medially, on both sides of the patch of cells, and condense in the midline to form the primitive streak. There, the cells sink below the epiblast to the region between the epiblast and hypoblast, and the cells spread laterally and forward between the two initial layers, forming a third, mesodermal layer. Medial flow of the epiblast, invagination, and lateral flow of the mesoderm occur with bilateral symmetry along both sides of the primitive streak. At the head, or rostral end of the primitive streak, a small mass of cells known as Hensen's node forms. With time, the primitive streak becomes shorter, and Hensen's node is carried bodily backward toward the tail, or caudal end of the primitive streak. In the region anterior to Hensen's node, epiblast cells are not moving medially, nor are cells sinking below the epiblast and migrating laterally. That is, Hensen's node is the anterior-most locus in which cells are invaginating to form the mesodermal layer, and as Hensen's node regresses backward, the primitive streak which is the region where invagination is occurring, becomes shorter. Some time after the node regresses, condensations of the mesodermal cells into aggregates to form somites occur bilaterally along both sides of the line that Hensen's node has followed during its regression; that is, on both sides of the line where the primitive streak was prior to its regression. Our problem is to explain the occurrence of bilaterally symmetrical somites in just the loci in which they occur.

Let the foregoing description be considered an adequate description. A different set of descriptive sufficient conditions might be: (1) a description of cell movements as given above. (2) The addition that somites condense only after invagination (sinking below the epiblast) and the other medial and lateral movements of gastrulation have ceased; and further, that somites condense only in those cells which were the last to invaginate to the mesoderm. This description is a sufficient condition of the initial description since it includes the claim that lateral cell movement of the mesoderm stops when invagination stops, and we can therefore deduce from this new description without further empirical information that somites form in bilaterally symmetric lines adjacent to the line of the now regressed primitive streak, as described in the initial description.

Processes capable of bringing about the state of affairs described in these new sufficient conditions will be processes capable of causing the state of affairs initially described. In addition to description of cell movement, this new set of descriptive sufficient conditions refers to: (1) formation of somites only after cessation of movement, (2) aggregation of initially disperse mesodermal cells into local aggregates; (3) the occurrence of these aggregates only among the last cells to invaginate. We need to find processes to bring about these effects. That is, we suppose processes by which (a) cessation of movement allows or causes mesodermal cells to aggregate; (b) mesodermal cells aggregate; and (c) only the last to invaginate aggregate. The descriptive sufficient conditions themselves suggest rather obvious processes. If we suppose there is some process by which mesodermal cells tend to aggregate, then movements of gastrulation might overcome the tendency to aggregate, thus cessation of movement would allow aggregation to occur. We now need a process by which there is a tendency of mesodermal cells to aggregate. A new descriptive sufficient condition for the initial description of aggregation is that originally, mesodermal cells are fairly dispersed, but move about, and when they come together, they stay together. We suppose a process causing the state of affairs described by this new descriptive sufficient condition to occur, say that mesodermal cells are sticky and attach to one another. We now need a process to cause only the last cells to invaginate to aggregate. Since we are to account for aggregation by mesodermal cells sticking together, this process must either be strongest, or most effective among the last cells to invaginate. If we suppose that mesodermal cells elaborate a diffusible 'sticky stuff', then epiblast cells will absorb sticky stuff as they migrate medially over the laterally migrating mesodermal cells. Epiblast cells which have traversed a greater number of mesodermal cells will be stickier than those which have traversed fewer. Epiblast cells arriving at the primitive streak later will be stickier than those arriving earlier, thus the last to arrive will be the stickiest, and the aggregation will occur in the proper place.

We now have a cybernetic model utilizing the three processes; that movement disrupts the formation of aggregates, that mesodermal cells tend to stick together, and that countercurrent movement of the cell layers will concentrate diffusible 'sticky stuff' near the primitive streak rather than laterally. This model can now be used to generate a causal model in which, for example, the 'sticky stuff' receives a molecular interpretation of some kind. As in the previous example, the descriptive sufficient conditions of the initial adequate description of the entire process and portions of it, have served as injunctions to direct attention in particular ways in order to build an hypothesis. And the formation of the hypothesis does indeed seem a rather reasonable affair, open to the judgement of having been done stupidly or well.

As in the previous example, differenet descriptions which are sufficient conditions for the truth of an initial description can yield radically different models. For example, somites may also be described as forming in those mesodermal cells over which Hensen's node has passed. This, coupled with a description of cell movement, is sufficient to deduce the initial description. It leads to the supposition that Hensen's node somehow causes or induces the underlying mesoderm to differentiate into somites. As in the previous example of cell reaggregation, models derived from the two different descriptive sufficient conditions are not isomorphic; each picks out parts and processes in a different way. In the last model, Hensen's node is the crucial part which induces somite formation in the underlying mesoderm; in the former model, Hensen's node is an incidental part of the mechanism having no bearing on the formation of somites. Slightly more advanced forms of the two hypotheses are currently under experimental investigation.

We need now to ask what criteria we utilize to judge whether an initial description is adequate, and what information we use to guide our choice among indefinitely many descriptive sufficient conditions and possible cybernetic and causal models.

In the initial description of the cell reaggregation, and of chick gastrulation, many things which occurred were left undescribed. For example, motions of enzymes within cells, the color of the container in which the system was maintained, and the brownian motion of ions in the medium were not described. Surely, an adequate initial description is already a highly selective description based on our background knowledge about the sort of system and process we are concerned with. The initial description is made in terms of facts and factors we have reason to believe will be the relevant facts and factors, on the basis of our current knowledge and theories. Of course, the initial description may turn out to be inadequate because our suppositions about what is relevant may be incorrect.

There are several interrelated ways in which we use background knowledge to help choose from among diverse descriptive sufficient conditions and cybernetic models those which commend themselves for further serious effort. On the basis of background information, we know a reasonable number of different processes which are, so to speak, stock in trade. For example, cells adhering to one another is such a process. These processes are simple in three senses: (1) They are well known and understood, thus intuitively simple; (2) Many plausible causal mechanisms to realize the process are known or can be supposed, indeed it is just because there are so may causal mechanisms by which cells can adhere to one another that the process 'cells adhere' is very familiar and stock in trade; (3) These processes are simple in the sense that it is by articulating together more than one of these well understood processes that we seek to explain more complex processes. For example, in the hypothetical cell reaggregate, it was permissible to suppose a process 'by which just these cell boundaries form bonds,' for specific bonds between entities is a process which is simple in the specified sense. However, in the chick gastrulation case, the supposition of a process by which cells along the line of the primitive streak, but not laterally, hapened to be the ones to aggregate, would be inacceptable, for this is just the sort of case where we want to see from *other* simple processes, how this complex process might occur.

We therefore choose among the diverse descriptive sufficient conditions those for which we think we can put together a cybernetic model articulating such simple processes. Even if we achieve an articulating cybernetic model, that model might yet be considered implausible for two quite different sorts of reasons. First, the causal mechanisms by which the cybernetic model might be supposed to be realized might be considered implausible in the system under study. Second, even if plausible causal mechanisms might be imaginable, it may be that the cybernetic model demands, in some sense, too much machinery, and it may be implausible to think the system behaves as described initially by such complex means. For example, a descriptive sufficient condition for the initial description of chick gastrulation in terms of a coordinate system is possible, and from it one can create a cybernetic model in which the chick embryo elaborates a coordinate system and cells 'know' where they are and go to the right places. Although causal mechanisms which are plausible might be imagined, the entire construction is likely to be considered an implausibly cumbersome way for a chick to go about gastrulating.

Finally, it must be added that it may not be possible to find a cybernetic model for any of the descriptive sufficient conditions of an initial description which have so far been investigated, for there is no guarantee that, with the currently acceptable set of simple processes, a cybernetic model *can* be formulated. When we are unable to find such a model, one of the most obvious strategies to take is to reexamine the phenomenon we wish to explain and alter our initial description by including new data now hoped to be relevant.

8. Having discussed ways in which we utilize a view of what a system is doing to achieve an articulation of parts decomposition into parts and processes interrelated in particular ways, we now turn to ways in which we utilize information about parts to synthesize new articulation of parts views of what the whole is doing. First, it is clear that new information about a part picked out in an old decomposition of the system can lead to new views of what the system is doing. For example, Harvey's discovery that the blood circulates led to a new view of what the heart, specified on old anatomic grounds, does. The current discoveries that cells and organisms exhibit circadian rhythms, and are, in general, temporally organized, is the focus of debate about the 'significance' of such rhythms. The efforts of this debate are to synthesize a new view of what an organism is doing – essentially to be a timekeeper of some sort. The points to notice about such new syntheses are that, in general, the new synthesis may decompose the system in a new way, cutting across parts specified in an old decomposition. For example, the endocrine system cuts across the old anatomic decompositions; the view of an organism as an open thermodynamic system maintaining low entropy cuts across the decomposition of it as a circadian system.

An important feature of achieving a new synthesis of what a system is doing is that the new view may require no new verification. It may be that true descriptions of behaviors of parts picked out on an old decomposition have accumulated, where these behaviors are irrelevant to the behaviors of the parts which linked them into a system in that old decomposition. It might then be realized that these 'irrelevant' behaviors can be interrelated to achieve a new view of what the system is doing. Since the truth of the descriptions of these irrelevant behaviors is supposed to have been adequately established already, the adequacy of the new composition may not require verification. Of course, the new composition will have *further* consequences which require verification. The 'irrelevant' behavior might have been established in two ways: It may be that, given one description of what the system is doing and a conjugate decomposition into parts and processes, the true descriptions of what the parts are doing, coupled with current theory, may entail the truth of descriptions of other 'irrelevant' causal consequences of the parts. These other causal consequences may be linked together to yield a new synthesis of what the system is doing which would pick out new features of the system and provide an *A* of *P* explanation. The truth of the explanation would be adequately established by the validity of the deductions from the true initial description of the system, and the adequacy of current theory about the stuffs of which the parts were made. On the other hand, we may merely have discovered 'irrelevant' causal consequences of parts on an old decomposition, and established the descriptions of the irrelevant causal consequences by observation, not by deduction from the old descriptions and current theory. These independently established claims about irrelevant behaviors might be linked together to achieve a new composition, or view, of what the system is doing. In either case, the adequacy of the new view does not require verification in the sense in which hypotheses are normally understood to require verification. Of course, normally new compositions often do assert claims about behaviors of parts which are not yet established, and these do require verification. Thus, achievement of a new composition *can* occur without having to reexamine the empirical world; the new composition focuses attention on new features of the system and, presumably, provides a new *A* of *P* explanation of the newly noted behavior. A new behavior is seen as calling for an explanation.

9. We have noted that different views of what a system is doing yield different decompositions of that system into parts and processes. The foregoing discussion has also indicated the ways in which the various true claims made in the diverse decompositions may be related across

decompositions. The claims must, eventually, at least be consistent with one another. However, there is no requirement that the claims made in one decomposition need ever be deducible from those made in another decomposition. That is, the claims must eventually be *compatible* with one another, but not deducible from one another. The set of causal laws sufficient to provide an *A* of *P* explanation of the system seen as doing one thing may not include causal laws necessary for the explanation of a second thing the system may be seen as doing. Thus, the various accounts may proceed independently from one another, with the restriction that they must all, eventually, be non-contradictory.

10. A final point. It may be asked whether there are not preferred views of what an organism is doing, for example, living. I am willing to grant that some views of what an organism does are currently seen as more central to the defined interests of biology, but am not willing to agree that all possible views of what an organism may be seen as doing must, in some sense, be seen as extensions of, versions of, or reducible to those preferred views. To insist, whatever an organism may be seen as doing, that that feature of it is an aspect of a particular view of the organism, say being alive, may be mistaken in each of several ways. It might be the view that only the preferred view of what the organism is doing is to be allowed. It might be the view that whatever happens in an organism is causally necessary for the organism to, e.g. live. Now this claim is almost certainly either false, or an analytic claim masquerading as an empirical one. It might be the view that if the organism is made of just those materials with those particular causal consequences accounting for how it manages to live, then these particular materials have also just the other (indefinitely many) casual consequences. Thus, if the system is to do what it does – e.g. live – then it must do these other things as well. While this version of the claim is true, it does not establish that all views of what a system is doing must be able to be brought under one view of what it is doing and deduced from claims explaining how it does that particular thing. For the truth of the claims on the different accounts need only be compatible, not deducible, from one another. Thus, while the argument above is valid, it does not establish that the diverse account of how the system does the diverse things it can be seen as doing, need be deducible from any one account. The accounts may proceed independently with the restriction that all must eventually be jointly compatible.

Thus, it seems we must admit that many of the features of organisms which we seek to explain are incidental to any particular view of it, even to 'being alive'. Articulation of parts explanations of any way be logically independent, in the sense of not being deducible from many or all the others. There is no reason to restrict the features of organisms we will seek to explain. We can insist that we achieve articulation of parts explanations rather than compilations of true descriptions which fail to articulate. Finally, while we can and do form new inclusive views of what an organism is doing which joins several earlier views in new and rewarding ways, there seems to be no reason whatsoever to insist that all possible views of what an organism is doing and all possible true articulation of parts explanations, need ever be brought under some overarching, ultimate view of what the organism is 'really' doing. For there are, indeed, indefinitely many things which an organism can legitimately be seen as doing, and there is every reason to expect biologists to pursue an ever widening ring of puzzles to explain.

ACKNOWLEDGEMENT

The author wishes to thank Drs. Dudley Shapere and David Hull for helpful criticism.

University of Chicago

PART IV

EVOLUTION

LAWRENCE B. SLOBODKIN

THE STRATEGY OF EVOLUTION*

There is an embarrassing lack of surprises in certain aspects of evolutionary theory. For example, if a bird has only seawater to drink, it is not surprising, given our knowledge of birds in general, that some mechanism must exist for eliminating salt if the bird is to survive. It is not possible to state *a priori* what the mechanism is for any particular bird. It is therefore intellectually pleasing that gulls excrete strong brine through a nasal gland (Schmidt-Nielsen, 1960). The novelty of the arrangement provides the same kind of intellectual satisfaction in the twentieth century that I imagine the information in the Bridgewater treatises provided in the nineteenth (Whewell, 1833), or Henderson's *Fitness of the Environment* (1913) to the teachers of our teachers. Nevertheless, physiological and even behavioral adaptations, no matter how apt, seem only what ought to be expected from our knowledge of natural selection, survival of the fittest, and other catch words.

In the literature of biology, discussions of evolution consist of either historical statements, whether phylogenetic or genetic, and of physiological or morphological descriptions. Despite this reticence on the part of biologists, an enormous amount of bad philosophy, psychology, and, on occasion, political theory, has been initiated by a consideration of evolutionary phenomena. There is apparently an almost irresistible temptation to try to find some point, purpose, or goal in the evolutionary process. In some sense, the narrowly biological discussions of evolution are unsatisfying, so that para-biologists feel called upon to provide a theatrically acceptable finale.

Even among biologists, goal-oriented evolutionary statements are made in private conversation. We tend to speak in generally teleological terms as if the confrontation between organisms and their environment were some kind of game in which the successful organism has somehow outwitted a powerful but not necessarily malicious opponent. This manner of speaking does not do violence to normal standards of mechanistic science since it is an axiom of mechanistic biology that, in fact or in principle, any teleological statement in biology is replaceable by a suitably non-teleological one. The non-teleological replacement statement may always be made of two parts, first, a physiological or biochemical description of the events occurring in the organism confronted with the situation of interest, second, a statement about the effect of these events on the probability of survival of the organism.

The process of taking organisms from one environment and placing them in another can be thought of, from the quasi-teleological viewpoint of a game analogy, as confronting them with a problem to solve. More usually, the situation would be described in physiological terms. The actual responses of the organisms to this new situation are quite independent of the observer's viewpoint, as are the data recorded from the experiment. The conclusions drawn from the experiment, the experiments undertaken, and the kind of data actually recorded are, however, very much dependent on the observer's viewpoint.

I will present one set of experimental examples and will interpret these examples as if they constituted a kind of game between the organisms and their environment. I will then attempt to make statements about the game analogy in evolutionary theory and about the 'purpose' of evolution.

I accept as an initial axiom (as did Darwin and Wallace) that, barring drastic physical changes or the introduction of new species, the number and kind of animals found in any region of the world tend to remain essentially constant from year to year.

It is also a valid experimental platitude that if a few fertile organisms are taken from either the field or a stock culture and placed in a suitably salubrious but spatially confined environment it is vanishingly unlikely that their general physiological state will not change. In particular, their abundance in the new environment will come to some new level which will thereafter characterize the experimental container. We will refer to this relatively constant population as the 'steady state population'.

The fact of introducing them to a new container is a violent alteration of the environment of the animals in the inoculum. If the inoculum is sufficiently small, the number of animals will increase. When careful examination of animals in increasing populations has been made it has been found that the animals are growing more rapidly as individuals, are reproducing more rapidly, have a higher fat content, and by most reasonable criteria are healthier individuals than animals of the same kind taken from a steady state population in the same environment (Smith, 1963). The degree and kind of physiological changes that occur in the individual animals of a population during the transition period between the inoculation and the attainment of steady state vary from species to species but they always do occur. This is tautological, since the steady state population can be defined by an identity of birth and death rates while death rates are significantly lower than birth rates in an increasing population. We conclude that there is generally an evolutionary advantage to having a relatively larger number of less healthy animals than to having a smaller number of more healthy ones. There is some balance struck for any particular species in any particular environment between optimal physiological condition of the individual and optimal abundance. The mechanism of achieving this balance differs from species to species.

Recent experimental studies in my laboratory permit a tentative analysis of the adaptive significance of this balance in two species of hydra. The hydra are morphologically simple animals with no hard skeletal structure at all. Essentially, they consist of a hollow bag of tissue with tentacles growing around the lip of the bag. The tissue layers are relatively thin. The body wall thickness, as well as the size and shape of the animals, can be altered within fairly wide limits by differential contraction and relaxation of muscle-like cells. In still water, for example, the tentacles may be extended as fine filaments 20 cm or longer while the same animal when shaken will shrink to a mound not more than 3 mm across. This drastic alteration in the morphology of a single individual is fairly common among the soft-bodied coelenterates.

If either green or brown hydra are inoculated in small numbers into an experimental container with an abundant but constant daily food ration, they will initially grow larger as individuals and produce buds. As the number of actively feeding mouths increases, the food available for each mouth is reduced. As this reduction occurs, the budding rate falls off and eventually ceases and the body size of the individual animals diminishes. *Hydra oligactis*, *Hydra littoralis*, *Hydra pseudoligactis* (the brown hydra in my laboratory) or *Hydra viridis* and *Chlorohydra viridissima* (green hydra with symbiotic algae embedded in their body cells) all overlap broadly in their size ranges so that any hydra in any of these species can be brought to the same body size as any other by appropriate feeding. In the process of population growth, each population will increase numerically and the animals in it will come to a fixed body size, considerably smaller than their genetic potential. Griffing (personal communication) has shown that in *H. oligactis* and *H. pseudoligactis* there is a direct relation between body weight, budding rate, and feeding rate. This seems valid for all hydra. The size of animals in a steady state population is therefore that size at which the budding rate is just sufficient to compensate for the mortality rate.

While the precise body weight to which each of these various species come at the population steady state is not known for all of them, it is clear that a threefold difference in steady state body size does occur between *Hydra littoralis* and *Ch. viridissima* (Slobodkin, 1964).

If animals with flexible body size tend to one particular size, it may be inferred that being either larger or smaller is disadvantageous. That is, it should be possible to define a set of circumstances under which large body size would be favorable for the organisms and a contrasting set of circumstances under which small body size would be favorable. In the case of the hydra, for example, we can suggest that large body size is advantageous if the individual food particles are relatively large so that too small an animal cannot catch them and hold them. Large body size might also have an advantage if the food supply were extremely sporadic, since during starvation the hydra diminish in body size with time. Large body size is a direct insurance against starvation.

Small body size has a conceivable advantage if the food supply consists of numerous small particles evenly distributed in time. The total feeding surface per unit of living bulk to be maintained is greater in a population of small animals. It is also conceivable that, if the population is threatened in some sense by environmental processes that destroy individual animals regardless of their size, many small animals have an adaptive advantage over a few large ones. Note, I leave adaptive advantage undefined.

All of the species of hydra listed above can be maintained on a ration of brine shrimp nauplii (Artemia sp.). *H. littoralis*, *H. oligactis* and *H. pseudoligactis*, and *H. viridis* can also be maintained on daily rations of young Daphnia. Two species of Daphnia were available, *D. pulex* with newborn animals weighing 3–5 micrograms each and *D. magna* with newborn animals weighing 9–11 micrograms each. The weight of a single Artemia nuplius is around 2 micrograms. By 'maintaining the hydra on the food organism' I am here referring to keeping individual animals.

Populations of all these species can be maintained on Artemia and *D. pulex* and all of them except the green hydra can maintain populations fed on *D. magna*. For the brown hydra, the number of animals maintained on a given number of *D. magna* per daily ration is somewhat larger than can be maintained by the same number of *D. pulex*, but the increase is not quite proportional to the difference in the energetic content of the two species of Daphnia. Within each type of food supply, the size of the population that can be maintained is linearly proportional to the amount of food supplied.

When green hydra populations are fed on *Daphnia magna* there is an initial increase in abundance followed by a decline below the level of a population fed on a corresponding number of *Daphnia pulex*. After approximately fifty days, food consumption stops in the green hydra populations on *D. magna*. The animals are simply too small to catch *D. magna*. If no other food supply is made available, the green hydra population eventually starves to death. If these tiny green animals are put on a diet of Artemia nauplii they can be increased in size to the point that they can again capture and feed on Daphnia. So long as they are consistently well fed on *D. magna*, the green hydra maintain their body size and bud freely. We have, however, never seen a bud from a green hydra feed on *D. magna* young as its first meal.

The inadequacy of *D. magna* as a diet for populations of *Hydra viridis* is, therefore, not a biochemical deficiency of the hydra nor is it related to the biochemistry of the Daphnia except in so far as this relates to the size of the Daphnia, nor is it due to the inability of the green hydra to grow to a sufficiently large size to eat *D. magna*. It is rather that, in the process of adjusting to a new environment, the green hydra produce very small animals rather than a smaller number of large ones, as if the green hydra are playing the strategy of being numerous and small while the brown ones play the contrasting strategy of being relatively less numerous but larger. It should be noted again that these commitments to specific sizes are not permanently binding in the sense that suitably starved individual brown animals become as small as individual green ones and suitably fed green ones become as large as brown ones.

We have seen a situation in which small size is disastrous for green hydra populations. Clearly we are required to show a correspondingly advantageous situation or be faced with the problem of explaining why green hydra exist at all. When *Hydra littoralis* and *Chlorohydra viridissima* are simultaneously inoculated into a laboratory culture, fed on Artemia nauplii and maintained under a fluorescent lamp, both species increase in number of animals, the brown less rapidly than the green (Slobodkin, 1964; Stiven, 1962). When the number of hydra are approximately twice the number of Artemia per day provided in the food ration, the brown hydra begin to decrease while the green continue to become more numerous. In approximately forty days, the brown hydra have been completely eliminated and the green hydra approach a rather vacillatory steady state.

In any two species system, the available energy must be divided between the two species. By becoming extremely numerous the small green hydra could capture an overwhelmingly large share of the small Artemia nauplii, starving out the brown hydra. This was demonstrated by removing a given fraction of the increase of the two populations daily. This type of removal acts as an increment to the death rate and reduces the number of animals in the residual population. The greater the fraction of the increase removed from the two populations, the longer the persistence of the brown hydra. When 90% of the increase was removed daily, the combined number of hydra of the two species in the population was not too different from what might be found in a control population of only brown hydra, and the two species persisted indefinitely. The competitive outcome between brown and green hydra is not in any obvious way related to any mutual interaction on the biochemical level but is explicable in terms of the relation between number and size of the animals and their food supply.

So far, we have seen that the green hydra have the advantage of greater food-catching surface for their population as a whole. This proves of value when the food supply consists of small particles, but is dangerous if the food supply consists of large particles. The brown hydra have the advantage of individual large size, increasing the likelihood that each individual animal will acquire some food and decreasing the probability of starving to death between meals.

How can the green hydra counteract the risks of small size? If hydra get smaller and smaller in the absence of food, we might expect that they would eventually get so small as to be unable to eat any metazoan. As a matter of fact, sufficient duration of starvation will give exactly this result. The danger is considerably reduced however by the fact that the green hydra can derive a certain amount of nutrition from the photosynthetic processes of their symbiotic algae.

The significance of the photosynthetic processes of the algae for the green hydra has been demonstrated in three ways. First, we have already indicated that brown hydra are eliminated by green hydra when grown. in the same container and fed on Artemia under a fluorescent light. When the experiment is performed under a black cloth wrapper, both species persist (Slobodkin, 1964). Also, when *Hydra viridis* from a stock culture were starved for a twenty-eight day period, the total weight of the hydra was one-third the initial weight for animals kept in cloth wrappers but essentially identical with the initial weight for animals kept in twenty-four hours of fluorescent light. (This experiment is now being repeated at various temperatures and light periodicities and intensities and must be considered cautiously.)

The third demonstration of the role of the algae in green hydra also provides a rough estimate of the relative amount of energy provided by algae. This involved a series of experiments to determine 'ecological efficiency'. This concept is defined and discussed in fair detail elsewhere (Slobodkin, 1962). Suffice it for the present purpose that ecological efficiency is a fraction with a numerator which is the energy that a predator can get from a population and with a denominator which is the energy consumed by the prey population as food. The evaluation is made at steady state conditions. Whenever this has been evaluated, a maximum value of approximately 10–15% has been found, except in the case of green hydra populations in twenty-four hours of light. If the denominator is taken as the Artemia eaten by the hydra, then the ecological efficiency is approximately 40%. From this I infer that approximately three-fourths of the energy supply of green hydra is derived from the photosynthetic activity of their algal symbionts.

The story of size control in hydra is not yet complete by any means. Preliminary experiments indicate a rather delicate adjustment between the budding mechanisms of the hydra and the rate of photosynthetic activity of the symbionts, for example. We have gone far enough, however, to get the impression that the hydra seem to act as if they are playing some kind of game with their environment, making certain adjustments only when there is reasonable assurance that the risks associated with these adjustments have been covered. This case is simply an example.

Almost any set of ecological phenomena can be analyzed the same way.

In what sense does this manner of speaking contribute insights that would not be available from a more conventional biochemical or physiological approach?

Consider the analogies that are possible between evolutionary processes and changes on the one hand and games on the other. For example, imagine a group of non-chess playing observers at a tournament of mute chess players. It would be possible for such a group of observers to determine the rules of the game fairly readily by simply observing the players. Regardless of superficial differences such as size, shape, and color, there are functional identities between all pawns, all knights, all rooks, etc. There are very limited categories of moves that each of these can make regardless of their relative position on the board or association with each other. Pawns move only forward, polarization of the bishops' activity is at a 45 degree angle to the polarization of the rooks', there is a quasi-helical pattern to the activities of the knights and so on. Occasionally peculiarities might be noted such as 'castling', '*en passant* capture', and 'queening of pawns' that might be somewhat more difficult to characterize fully.

Once the rules have been determined, it is possible to consider that the game of chess is now 'solved' in some sense and go on to the next game or problem. The more complicated the rules of a game, the greater the sense of having solved the game when the rules are determined. To know the rules of a game makes it possible to play the game in the sense that a person manipulating objects on the play area contrary to the rules is not playing the game at all.

All the players of any particular game do play by the same rules. All of them in that sense understand the game. Not all of them win, and this is why games are interesting. One of the standards of quality that can be applied to a game, aside from the intrinsic aesthetics of the manipulative procedures and the social rewards of proficiency, is the subtlety of the relation between playing by the rules and winning. That is, a game in which playing by a set of rules ensures victory is a bore, no matter how complex the rules. The various games involving building a path of one or another color across a field (such as John, Hex, or Bridgit) are dull in this sense. Somewhat better games are those in which the rules or the objects involved embody chance or unknown elements. Card games, dicing games, and contests with a biological element (as races, baseball, golf, etc.) are better games. Betting on their outcome is a worthwhile activity. Some games of this sort are amenable to the construction of interesting game theories in the mathematical sense.

The more delicate board games like chess and Go can be considered either from the standpoint of chance or from the standpoint of strategy. Such games are played one move at a time. At each player's turn, there is available to him a very large class of bad moves and, if it is an interesting game, a very small class of good moves. In the game of Go, the class of good moves is defined in part by a series of adages or proverbs: 'Don't make empty triangles', 'Don't peep at the bamboo joint', etc. (Segoe, 1960). To a degree, the same type of adage is used in teaching chess: 'In the early game don't move the same piece twice', 'Avoid pawns in line' etc. (Mason, 1958). These adages are, in principle, replaceable by a table in which are catalogued all the moves in a very large series of games, the configuration of the board when these moves were made, and information as to the eventual outcome of each game. If the first player makes a particular move, this alters the configuration of the board. This configuration is associated with a certain probability of each player winning or losing. Within the class of moves possible in the new configuration there is a very small subclass which alters the configuration so as to increase maximally the probability of the second player winning. Note that the concept of probability in relation to this type of game does not imply chance events in the game but rather a sampling that has been made of past games. The probabilistic argument as to how to play would, however, work in any case. For example, the poker adage 'Never draw to an inside straight', has a statistical justification of a formally similar kind.

Teachers of chess and Go present the adages to their pupils and rationalize them in terms of future events that might occur in the game. 'If you do this then when he does that...'. They do not make any attempt to equip the student's memory with a probability table for each possible move, although everyone concedes that this is, in principle, possible.

No attempt at all is made to deduce the appropriate moves from the permissible ones. That is, the fact that a knight is permitted to move in a particular way is only in a very trivial way the reason for making a particular knight move at a particular juncture of the game.

The above information is enough to demonstrate that simply saying that evolutionary processes are analogous to a game is meaningless since there are many significantly different kinds of games.

Let us now consider the ways in which the evolutionary process does or does not resemble a game. It is fairly safe to say that all organisms are obeying the same rules in that they have the same or very similar biochemical processes occurring in them. It is possible to define life in terms of these processes in the same way that one may define chess as the placing and moving of pieces on a board in accord with the rules of chess.

In a game, playing by the rules is irrelevant to winning or losing. So, being alive, in the sense of following the biochemical rules that define life, may be irrelevant to 'winning or losing' the evolutionary game.

That sentence must be examined very carefully and accepted with all due caution. The biochemical processes occurring in the mammoth, mastodon, or in the saber-toothed tiger were, as far as we can tell, not different in any major way from those occurring in the Indian elephant, African elephant, or pussy cat. One trio is gone, the other is not. It is conceivable that these species became extinct for lack of an enzyme, but it does not seem likely. It has been suggested that simply the accidents of history determine extinction or survival. This position has a misleading plausibility to it. It is exactly the conclusion that a naive observer might come to after deducing the rules of chess and being confronted with the fact that half the players playing at any given moment in time are destined to lose (excluding drawn games).

A rulebook in almost any game is short, definite, and written in declarative sentences. A book on the strategy of the same game is typically much longer, consisting of highly restricted subjunctive statements. Nevertheless, a player who carefully studies a book on strategy becomes much more likely to win than a player who only knows the rules. Is there a definite empirical strategy of evolution? I would like to suggest that the kind of statements I made earlier with reference to the advantages and disadvantages of various body sizes in hydra were in fact elementary examples of evolutionary strategy. The animals that are now alive are successful players at the evolutionary game in that, while we have not yet considered how one can win in evolution, extinction represents a kind of losing.

If we, as articulate observers, can determine what the stakes are in the game, we ought to be able to expand our knowledge of its strategy by examining the properties of the successful players whether or not they are articulate or 'intelligent'. Note that we are already fairly clear on the idea that the difference between good and bad strategy need not be biochemical in any interesting way.

In the formal mathematical theory of games, the kind of payoff or penalties that the participants in the game receive must be specified. There is a quantifiable reward for winning and some kind of measurable penalty for losing. Lewontin (1961) discussed the evolutionary process from this formal game theoretical standpoint. Can we assign a quantifiable payoff to the evolutionary game?

This question can be answered indirectly. Huizinga (1950), in an historical essay on the role of games in human culture, pointed out the simple, but highly significant fact that games are always played either on a playing field or playing board or some other specified spatial region. Within this region, the only rules that apply are the rules of the game. At the termination of any game, there is a distribution of rewards and penalties. The players then leave the play area. The value of the reward does not lie in the play area itself but in the non-play world in which the play area is embedded. In other words, for a game to be worth the candle there must be a way of cashing in the winnings and this implies a place must exist other than the gaming board itself.

Animals have no such place. If evolution is a game at all, it is the kind of game that Kafka or Sartre might have created. The only payoff is in the continuation of the game. It is in one sense possible to build a formal game theory, using some suitably defined probability of survival of the population as a whole as the payoff. While this may be logically sound, it runs into considerable operational difficulties.

How would one play a purely existential game in which everything is subordinated to persistence? Neither efficiency, nor complexity, nor power, nor even destruction of one's opponent, can serve as a goal in itself. What is the best strategy in such a game? If we imagine that a player knows only what he himself played at his last turn, his obvious best play is to repeat what he did last time. If the general configuration of the game has changed from one turn to the next, this move may improve his probability of persistence or may reduce it. That is, doing the same thing at each turn, in a game of changing configurations, becomes a kind of random walk.

If we consider that the state of a player's fortunes can be specified by an arbitrarily long series of variables and that, at some initial time, each of these variables has a certain value, unless all the players do exactly as they did before, any player that persists in the same move round after round is most probably doomed. Even if all the players stick to their earlier moves, it still is not necessarily the case that any one player will persist in the game. After sufficient time has elapsed, however, each of the surviving players would have locked in on a constant move or a closed cycle of moves. This type of system can persist only if there is no extraneous source of variance being introduced into the configuration of the game. Unless all such extraneous new variance is damped out as rapidly as it is generated by some kind of homeostasis in the system, the configuration of the system will change as a random walk process.

The process of homeostasis, anywhere it occurs, involves adjustments which tend to keep some property of a system constant despite ambient changes. These adjustments, however, involve changes in other properties of the system whose constancy is not of prime significance. (A proper heating system adjusts the rate of fuel consumption so as to keep the temperature of the house constant.) What kinds of things do organisms hold constant, and what kinds of things do they alter for this constancy? Notice we have arrived at an empirical question with a fairly high degree of intrinsic interest. Much of this interest, as in most empirical questions, arises from the context. This context is a somewhat strange one in its deviations from the usual pattern of scientific argument.

Before we attempt to answer the empirical question it seems advisable to recapitulate the argument which made it matter.

I described portions of an experimental analysis of the regulation of size in several species of hydra. The biochemical and physiological size control mechanisms, to the degree that they are understood, were not really helpful. An essentially teleological approach, asking 'What are the advantages and dangers of this or that size?' did seem reasonably fruitful, as if the hydra were making intelligent moves in a game with their environment or, at least, if I were a hydra making intelligent moves in a kind of game with the environment, it would seem reasonable to do what the brown hydra do, given the attributes of brown hydra and to do what green hydra do, given the attributes of green hydra. We then considered the analogy between games and evolution and found that, if the analogy holds at all, it is only with a very restricted kind of game. This restriction of the analogy led us to the statement that the only point of evolution is persistence. This is equivalent to the theatrically unsatisfying end that Darwin came to.

Returning now to the question of what is actually held constant in a constant population by the homeostatic mechanisms of evolution, we are forced by the argument into an apparently paradoxical position. It is undeniable that 'fitness' in some sense is either held constant or maximized in the evolutionary process. Unfortunately, the word 'fitness' has been used to mean everything from physiological vigor to reproductive capacity (Birch *et al.*, 1963). If we accepted any of these as the thing which the homeostatic mechanisms of organisms are 'trying' to hold constant, we would also be saying that this was the payoff for the evolutionary game. We can get around this difficulty if we consider that homeostatic ability itself is the thing which evolutionary homeostasis is 'trying' to hold constant. The apparent paradoxes generated by all other definitions of 'fitness' now disappear. Even the concept of 'reproductive value' increase (Fisher, 1930) takes on the character of a special case. There is the danger that, in eliminating the paradoxes, we have eliminated all content and questions. This is not the case, although the form of evolutionary questions must, of necessity, now alter, as will be discussed later.

This conclusion is not new. Bateson (1963) points out that there seems to be a hierarchy of adjustment mechanisms all acting to preserve flexibility to environmental changes. The time required for each of these various adjustment mechanisms to respond to an alteration in the environment is inversely related to the range of environmental changes in which the mechanism in question will respond. All of the responses are in such a direction as to maximize the ability of the evolutionary unit (as defined in Thoday, 1953) to respond to subsequent environmental change. On the shortest time scale and narrowest environmental range are the rapid physiological changes comparable to those that occur over a diurnal cycle in man. Individual organisms may become acclimatized to a shift in environmental circumstances. In addition, certain kinds of alterations in environment result in numerical alterations in the number of organisms in the population which bring about changes in the state of the organisms themselves.

Should an environmental change be an essentially permanent one, the evolutionary unit will adjust by gene frequency changes to restore the flexibility of the acclimatization mechanisms just as they restored the flexibility of the short range physiological adjustment mechanisms. Presumably, any genetic change will reduce the capacity for further genotypic change. Bateson speculates that flexibility of the genotype would itself be selected for. This would also be anticipated from our argument as to the significance of homeostatic ability *per se*.

The effect of adjusting each of the higher levels in this hierarchy of feedback devices is to restore the flexibility of response to the next lower level. Bateson says "... there must be an *economics of somatic flexibility* and this economics must, in the long run, be coercive on the evolutionary process.... The organism or species would benefit (in survival terms) by genotypic change that would *simulate* Lamarckian inheritance.... Such a change would confer a bonus of somatic flexibility and would therefore have marked survival value." (Italics his) This is, in effect, identical with the conclusion derived from consideration of the existential game.

Bateson is not the only author to derive approximately this conclusion. Thoday (1953) states that survival probability is the only appropriate measure of fitness. Waddington (1957) says "... we are in fact suggesting that all natural selection is in fact a selection for the ability of the organism to adapt itself... in the environment in which it finds itself." Lewontin (1957) indicates clearly that he believes the central problem of evolution to be the development of a theory of homeostasis which maintains constant the survival potential of the evolutionary unit.

With the possible exception of Thoday and Bateson, authors writing about this problem seem to back away from the conclusion that the homeostatic devices of the evolutionary unit are themselves being maintained by homeostatic mechanisms. They tend to substitute more easily measured physiological properties as the thing whose constancy is of prime evolutionary significance. Waddington (1957) condemns Thoday for insisting on an almost unmeasurable concept as an evolutionary goal and feels that this is a misuse of the term homeostasis.

Waddington's concept of developmental homeostasis is somewhat different from the notion of flexibility discussed here (Waddington, 1957). He is referring to the homeostatic maintenance of constancy in development so that a fairly broad spectrum of environmental alterations will be compensated for by the organism during the developmental process. From our present standpoint, this is most significant as implying that the terminal resultant morphology of this process is the morphology of maximal homeostatic capacity (i.e., flexibility) for the environment in question. That is, if developmental homeostasis in the sense of Waddington is highly developed, it implies that the terminal morphology is of major significance in fitness while, if it is not, there is the implication that other methods are available for coping with environmental variability. This is of interest in that it indicates the existence of a class of meaningful ecological questions that can be formulated and in part answered from knowledge of the organisms themselves, having reference to the organisms' environment without involving direct environmental measurement.

The contention that all of the adjustment mechanisms of organisms and populations function to maintain the flexibility of the evolutionary unit may only be valid for biological systems although Ashby's homeostat operates in the same general way, but without the hierarchical property (Ashby, 1952).

No insurmountable logical problem is generated by holding homeostatic ability constant nor are normal assumptions of the physical-chemical bases of life violated.

It might be argued that all biological statements other than those explicitly stated in terms of chemical reactions are of necessity invalid. This position would deny that strategy matters in chess. It is a kind of fundamentalist mechanism and represents an almost complete circle in the history of biology.

In the late nineteenth century conflict between mechanism and vitalism, one of the strongest general arguments against vitalists was that their insistence on the intrinsic impossibility of biological analysis closed a series of intellectual pathways while a mechanist or 'agnostic' position permitted these pathways to be explored.

To close the door to enquiry in the name of mechanism would seem a tragedy.

Lerner's concept of genetic homeostasis (Lerner, 1954; Wallace, 1963) is analogous to Waddington's developmental homeostasis in that both are concerned with the mechanisms by which those components of fitness which are appropriate for long term average environmental conditions avoid altering too rapidly to accommodate to short term fluctuations. In the absence of mechanisms of this type, changes in the environment of a species can only be met by direct changes in gene frequencies. As a matter of fact, for some morphological features this seems to be the case. Human eye color or blood type, however complex the actual genetic or biochemical mechanism of their determination, are fixed within an extremely narrow range for the duration of the individual's life and cannot be altered without selection. Others, for example the human waistline circumference, while related to genotype are notoriously flexible during the life of an individual.

The expressivity of the genes controlling different morphological properties is very different. That this is not a property of specific kinds of genes can be demonstrated by the differences that exist between species in precisely which properties are variable during the life of an animal and which are not. In the context of maintenance of homeostatic ability of the evolutionary unit, a morphological property controlled by genes of low expressivity would be assumed to have its primary significance with reference to an environmental variable which is likely to fluctuate at intervals that are relatively short compared to the generation time of the organisms in question.

Levins (1961) has discussed this problem in an interesting way. He says

> ... Natural selection in a slowly changing environment in general permits a species to follow the changes and improve its fitness. But if the environment changes too rapidly, a response to selection may be disastrous. Suppose that a population of butterflies has one summer and one winter generation per year. If there are genetic differences in winter and summer viability, the winter generation will be modified by selection to produce a summer generation even more adapted than their parents to winter conditions, but less well adapted to the summer environment in which they live. We have designated this the Epaminondas effect after the small boy who always did the right thing for the previous situation. Here the optimal situation would be one in which the same genotype produces either winter or summer forms depending on say temperature of early development.

In summary, from a consideration of the relation between physiological, ecological, and genetic mechanisms by which populations adjust to their environments it seems likely that only flexibility or homeostatic ability is always maintained at a relatively high level by the evolutionary process. All other features of all organisms are expendable to this end. The response of an evolutionary unit to any environmental change seems to involve the simultaneous initiation of a series of physiological, ecological, and genetic changes all acting to restore the ability of the evolutionary unit to respond to subsequent environmental changes. These various flexibility-restoring mechanisms occur at different rates so that certain of them are essentially not influenced by short term fluctuations in the environment.

It also seems likely that a major mechanism of flexibility maintenance is that the more slowly reacting homeostatic devices tend to restore the flexibility of response of the more rapidly responding ones. This seems to occur by a readjustment of the norm of the more rapidly acting mechanism, as if one had a slow response thermostat that would make seasonal adjustments in the setting of a rapidly responding thermostat. The more slowly responding mechanisms are in some sense 'deeper' properties of the organisms, involving more widespread adjustments of various properties of the individuals and the population (cf. Bateson, 1963).

Gene frequency changes in this scheme would appear as last resort responses to essentially permanent environmental changes. The process of speciation is a special case of gene frequency changes in allopatric populations.

The adjustment mechanisms and their hierarchical interactions can be demonstrated in part for many organisms, and in some cases it can be shown that flexibility is in fact restored by the adjustment processes. Nevertheless, stating the case as strongly as I have done here involves frank speculation.

While it may be speculative it is not merely a semantic exercise. This can be shown in two ways.

First, empirical predictions are generated. For example, it would be predicted from this scheme that short-lived organisms would tend to have genetic systems involving higher expressivity than related long-lived organisms. It would also be predicted that speciation and extinction of populations would always be accompanied by relatively rapid and permanent environmental changes.

Second, alternative speculations involving the goals of the evolutionary process are denied. If the mechanism I have outlined is valid, then the absence of morally and philosophically satisfying conclusions about the ultimate meaning of evolutionary history are not weaknesses of biological theory nor are they due to a shortage of biological facts. They are simply wrong. Evolution, in fact, is simply a consequence of the general homeostatic ability of organisms combined with the biochemical properties of genetic material.

ACKNOWLEDGMENTS

The experiments with hydra were supported by the National Science Foundation, Environmental Biology Program (NSFGB-1595). I have benefited from discussion with the students and faculty of the University of Michigan ecology group, particularly Drs. Smith and Hairston, and with Anatol Rapoport and Karl Guthe. The starvation experiments with hydra were suggested by a remark of Graciella Candelas of Rio Piedras.

State University of New York, Stony Brook

NOTE

* This article is an expansion of a Friday evening Lecture given at Marine Biological Laboratory, Woods Hole, Mass., July 19, 1963.

BIBLIOGRAPHY

Ashby, W. R.: 1952, *Design for a Brain.* New York, John Wiley and Sons Inc. ix +260 pp.

Bateson, G.: 1963, 'The Role of Somatic Change in Evolution', *Evolution* **17**, 529–539.

Birch, L. C., Dobzhansky, Th., Elliot, P. O., and Lewontin, R. C.: 1963, 'Relative Fitness of Geographic Races of *Drosophila serrata*, *Evolution* **17**, 72–83.

Fisher, R. A.: 1930, *The Genetical Theory of Natural Selection*, Oxford, The Clarendon Press, xiv + 272 pp.

Henderson, L. J.: 1913, *The Fitness of the Environment: an Inquiry into the Biological Significance of the Properties of Matter*, New York, Macmillan Company, xv+317 pp.

Huizinga, J.: 1950, *Homo Ludens; a Study of the Play Element in Culture*, Boston, Beacon Press, 220 pp.

Lerner, I. M.: 1954, *Genetic Homeostasis*, London, Oliver and Boyd, vii + 134 pp.

Levins, R.: 1961, 'Mendelian Species as Adaptive Systems', *General Systems* **6**, 33–39.

Lewontin, R. C.: 1957, 'The Adaptations of Populations to Varying Environments', *Cold Spring Harbor Symposia on Quantitative Biology* **22**, 395–408.

Lewontin, R. C.: 1961, 'Evolution and the Theory of Games', *J. Theoret. Biol.* **1**, 382–403, reprinted below, pp. 286–311.

Mason, J.: 1958, *The Art of Chess*, New York, Dover Publications Inc., xii + 373 pp. Note, pp. 173–184.

Schmidt-Nielsen, K.: 1960, 'The Salt-Secreting Gland of Marine Birds', *Circulation* **21**, 955–967.

Segoe, K.: 1960, *Go Proverbs Illustrated*, Tokyo, Japanese GO Association, 262 pp.

Slobodkin, L. B.: 1962, 'Energy in Animal Ecology', *Advances in Ecol. Res.* **1**, 69–101.

Slobodkin, L. B.: 1964, 'Experimental Populations of Hydrida', *J. Animal Ecology* **33** (Supplement), 131–148.

Smith, F. E.: 1963, 'Population Dynamics in *Daphnia magna* and a New Model for Population Growth', *Ecology* **44**, 651–663.

Stiven, A. E.: 1962, 'The Effect of Temperature and Feeding on the Intrinsic Rate of Increase of Three Species of Hydra', *Ecology* **43**, 325–328.
Thoday, J. M.: 1953, 'Components of Fitness', *Symp. Soc. Exp. Biol.* **7**, 96–113.
Waddington, C. H.: 1957, *The Strategy of the Genes*, New York, The Macmillan Company, ix + 262 pp.
Wallace, B.: 1963, 'The Annual Invitation Lecture, Genetic Diversity, Genetic Uniformity, and Heterosis', *Canadian J. Genetics and Cytology* **5**, 239–253.
Whewell. W.: 1833, *Astronomy and General Physics Considered with Reference to Natural Theology*, Philadelphia, Carey, Lea and Blanchard, 284 pp.

R. C. LEWONTIN

EVOLUTION AND THE THEORY OF GAMES

I. THE PRESENT STATE OF EVOLUTIONARY THEORY

The modern theory of evolutionary dynamics is founded upon the remarkable insights of R. A. Fisher and Sewall Wright and set forth in the *loci classici The Genetical Theory of Natural Selection* (1930) and 'Evolution in Mendelian Populations' (1931). By the time of the publication of Wright's paper in 1931 all of the theory of population genetics, as it is presently understood, was established. It is a sign of the extraordinary power of these early formulations, that nothing of equal significance has been added to the theory of population genetics in the thirty years that have passed since that time. Yet we cannot take this period to mean that we now have an adequate theory of evolutionary dynamics. On the contrary, the theory of population genetics, as complete as it may be in itself, fails to deal with many problems of primary importance for an understanding of evolution.

The structure of population genetic theory may be briefly summed up as follows. Given an assemblage of organisms defined as a genetic population, given the breeding structure of that population, the frequencies of various alleles, the phenotypes of the various genotypes, the statistical nature of the environment, the amounts of recombination between loci, the mutation rates and migration rates, then it is possible in theory to predict the genetic structure of the population at some future time. Such a prediction may be unambiguous, as in deterministic theories, or else a statement about the probability that the population will be in a given state, as in stochastic theories. But whether deterministic or stochastic, these theories are restricted to predictions about changes in the genotype composition of a *given population* with *given* forces determining the changes of gene frequency. Population genetics is not genetics *of* populations, but genetics *in* populations. It is the genetics of phyletic change, of the gradual replacement of one set of alleles by another within a phylad genetically continuous in time and space.

A complete theory of evolution, however, must address itself to other problems besides those of phyletic change within populations. Despite the great amount written about speciation, there is as yet no mathematical theory of species formation nor are there rigorous formulations of the process of phyletic extinction. Yet speciation and extinction stand together with phyletic change as the main features of evolution. What theory will account for the origin and rise of individual homeostasis as an evolutionary mode? Have we a theory of the waxings and wanings of sexuality, of the variations in amount of recombination, of rates of mutation, of all those factors which enter as parameters in the formulations of population genetics? Except for intuitive theories, the answer to these questions remains, 'No'. For example, the genetic conditions which favor species formation have recently been discussed by Mayr (1959), Carson (1959) and Wallace (1959) and each, while arguing cogently, comes to a different conclusion. Another example is the problem of the relative amounts of genetic variability to be expected in marginal populations as opposed to central populations. Dobzhansky (1955) and Carson (1955, 1959) hold essentially opposing views on this question and a third, closer to that of Carson, is given by Lewontin (1957). It is not the purpose of this introduction to examine these questions in detail, but rather to point out the diversity of conclusions arrived at by evolutionists. This diversity arises not from disagreement about facts, but from the lack of a rigorous theoretical framework for solving problems in which characteristics of a population *as a whole* are the basic parameters. Disagreements of this nature do not arise in population genetics, *sensu strictu*, because of the great power of population genetic theory.

II. SPECIFICATION OF THE PROBLEM

The problem that faces the theorist of evolution, then, is to account for differences between populations and species in their degrees of inbreeding and outbreeding, in amount of recombination, in mutation rates, in degree of somatic plasticity as opposed to genetic flexibility, in dispersal rates, in short, in all those characteristics of populations which govern adaptation. The solution of this problem requires two things. First, *some measure of the success or degree of adaptation of a population must be specified.* Thus, it must be possible, in some unambiguous way, to state that a population with free recombination has a success value X, while one with no recombination has a success value Y, when all other conditions are specified. This is analogous to the measure of Darwinian fitness, or adaptive value of a genotype, which is defined in population genetics as the relative probability that an egg of that genotype will produce an egg in the next generation.

Second, given a measure of success or adaptation, *it must be possible to construct an adaptive calculus* which will predict the kinds of genetic structures that populations will evolve under given environmental conditions. In population genetics the adaptive calculus is built upon the basic rules of mendelism and the mathematical techniques are the standard ones of algebra, the calculus and the theory of stochastic processes. When the population as a whole is the unit of enquiry, however, some new basis must be found. Although the results of population genetic theory will play some part, it is possible that the calculus of population evolution will be a radically different mathematical technique, perhaps one that is yet to be invented by mathematicians. It is the purpose of this article to propose one possible mode of attack, provided by the theory of games developed by von Neuman and Morgenstern (1944).

III. OUTLINES OF A RELEVANT GAME THEORY

Game theory is not well known to biologists so that some of its general outlines must be presented here. What follows is not a strictly conventional description of the theory of games and some new notions have been introduced. There are several reasons for this. First, game theory has its roots in the social sciences and therefore much of the terminology carries connotations of purposive behavior. The notions of a 'player', 'choosing a strategy', 'preferring an outcome' are foreign to modern mechanistic biology and so must be either discarded or very carefully redefined. Second, some of the axioms of game theory flow directly from ideas of purposive behavior. The very definition of the important concept of the 'utility' of an outcome depends tautologically on human preferences. Third, many of the criteria of optimum solutions to game theoretical problems are inapplicable because they appeal intuitively to human preferences. This last objection is of some help to the evolutionist, because he may reject many of the suggested solutions and narrow down the field of choice to what may be considered evolutionarily important solutions. Finally, the processes of evolution are sufficiently different from human choice processes, that some new ideas must be introduced into the classical picture of game theory.

In what follows, any divergence from the standard theory of games will be noted when this divergence is significant.

A. *A Game in Extensive Form*

We may imagine one or more players engaged in a game according to a set of rules. Besides the actions of the player or players there are physical conditions, say the order of the cards in a shuffled pack, which determine the events of the game. The game proceeds in discrete steps called *moves* and at each move one or more players performs an *act* A_i from a denumerable set of available acts $\{A\}$. The model of discrete moves and acts simplifies the description of the game, but a further extension to *time-continuous* games and games in which $\{A\}$ is a continuous set can and must be made for some biological problems.

In the most general game the set of available acts $\{A\}$ may be different for each player and moreover may change at each move in a way dependent upon the act performed at a previous move. For example, in bridge the cards available to a player in any given trick are determined by the cards he has played in previous tricks.

At each move the physical conditions extraneous to the players' acts may also change. The set of *states of nature* $\{N\}$ is formally identical with a set of acts, $\{A\}$ and in this sense nature may be considered simply another player with its own set of possible acts.

After each move has been completed there may be a result which we shall call a *local outcome* (this term does not exist in standard game theory) which depends upon the Act A_i performed by each player and the state of nature N_i. Whether or not a local outcome does result after every move depends upon the definition of an outcome for each game. In bridge again, the taking of a trick is a local outcome and this will occur only after four moves, rather than one. In addition to local outcomes there will be a *terminal outcome* which may be either a special local outcome signaling the end of the game or else some function of the local outcome which is not itself a local outcome. In bridge the terminal outcome is the total number of tricks taken by a pair of players, or more generally a function of that total and the total contracted for by the bidding. It is important to notice that the local outcome has an intricate relationship with previous and future outcomes and acts. The set of local outcomes $\{o\}_t$ at any time may depend upon past local outcomes, *even though those past outcomes do not affect the act* performed by a player at time *t*. Thus, a player may decide to bet half his remaining resources at every toss of a coin. After any given toss, t, the outcome in terms of money lost or gained will be correlated

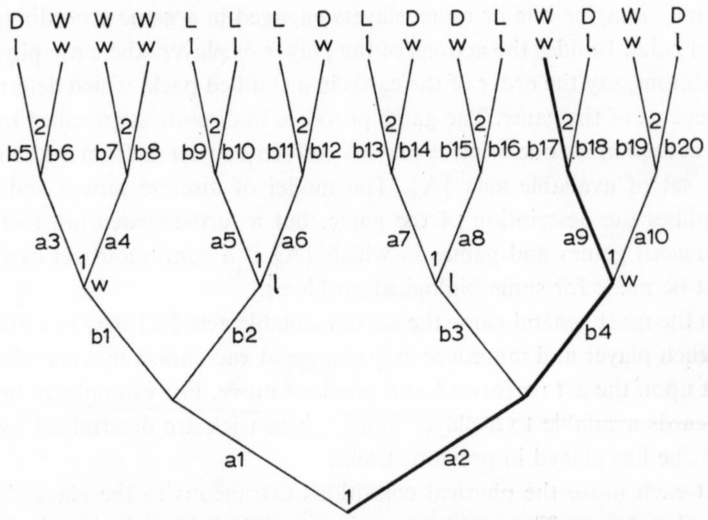

Fig. 1. A game in extensive form. The lower case letters l and w indicate local outcomes after each two moves. The upper case letters indicate terminal outcomes D (draw), W (win) and L (lose). The letters a and b with subscripts are alternative acts for players 1 and 2 respectively. The heavy line is one possible game with player 1 winning.

with all previous values of o but these will not in any way have affected the betting pattern of the player. This kind of dependence is of great importance for evolutionary problems. Suppose that population size is the outcome of interest. Population size N_t will depend upon N_{t-1} and the various factors which determine the rate of increase of a population. Yet the genetic structure of a population, which determines its rate of increase, may be totally insensitive to population size.

All of these concepts can be summarized as in Figure 1, a so-called game tree. It is assumed that there are two players (or one player and states of nature) and that their moves alternate strictly. The number at each branch point is the player number, the letters along each branch are the possible acts available to each player. There is a local outcome after each cycle of two moves and a terminal outcome, win, lose, or draw. Both outcomes are expressed relative to player 1. This figure is not the diagram of a particular set of plays, but of all possible sets given the structure of the game. Any particular play is represented by following one set of branches to the top of the tree. The heavy line shows one such play with player 1 winning. The terminal outcome is defined as the majority of local outcomes.

B. *The Normal Form of a Game*

The enumeration of a game in extensive form with all its moves and local outcomes can be reduced to a simpler form, the *normal form*. By tracing any particular path along the game tree a set of sequential acts for a given player can be enumerated. In Figure 1, for example, the heavy line includes acts a_2 and a_9 for player 1 and b_4, b_{17} for player 2. The set $\{a_2, a_9\}$ is a compound act for player 1 and is called the player's *strategy*. For each player in a game there will be a set of strategies, each strategy corresponding to one of the possible sets of sequential acts. The set of strategies S_i for one player taken together with the set of strategies S_j for the other player will determine a set of terminal outcomes O_{ij}. These can then be put in the form of Table I, the normal form of a game. Thus, it can be said that if player 1 adopts strategy 1 and player 2 strategy 3 the terminal outcome of the game will be O_{13}.

It is not necessary to encompass the entire game in the normal form. Any subset of acts performed in a sequence of moves will form a strategy

TABLE I

The normal form of a game. There are two players with 3 and 4 strategies respectively. The O_{ij} are terminal outcomes of a given strategy pair.

		Player 2			
		S_1	S_2	S_3	S_4
	S_1	O_{11}	O_{12}	O_{13}	O_{14}
Player 1	S_2	O_{21}	O_{22}	O_{23}	O_{24}
	S_3	O_{31}	O_{32}	O_{33}	O_{34}

and to each of these strategy pairs S'_i, S_j will correspond an outcome O'_{ij} suitably defined. In the case that there is only a single act comprising a strategy S' the outcome O' will be a *local outcome*, while for larger subsets it will be of the nature of a terminal outcome. That is: O'_{ij} may be some arbitrary function of the local outcomes included in that portion of the game. This notion of a strategy S' encompassing only a subset of sequential acts is not usual in game theory since it is the final outcome of a game which is usually considered. However, for our purposes, it is important to know 'who is ahead' at any particular time more often than it is to know 'who won'. At any particular moment in evolution, populations are in the process of playing the game so that much of our data comes from only partly completed matches. The entire question of the importance of local outcomes and terminal outcomes will be discussed below under the heading of pay-off functions. What is of importance here is that the normal form allows a game extended in time to be represented as a single step game, the analysis of which is much simpler. The complexity of the model has not been lessened by this procedure, however. In place of a series of simple acts and local outcomes we have a smaller set of complex strategies, and terminal outcomes.

IV. THE GAME AGAINST NATURE

In place of one of the players in the two person game of Table I, we may put the physical universe. In such a model, the *game against nature*, we have a single player adopting some course of action, a strategy, in the face of some state of nature N_j. Depending upon the state of nature, N_j, each strategy of the player will result in a specific outcome O_{ij}. As an example we may take a population of self-fertilized plants which may be homozygous for an allele *a* or *A*. Individual homozygous *aa* produce 5 seeds in dry weather and 10 seeds in wet weather. On the other hand *AA* plants have the reverse norm of reaction. Homozygosity for *A* or for *a* can be considered as alternative strategies for the population. Then the game theoretical form of this situation could be represented in one of the two forms shown in Table II. Table IIA is the normal form of an extended game of *t* generations duration. In such a case the states of nature are not simply *wet* or *dry* but a complete specification of all the possible sequences of wet and dry seasons. Assuming that the populations in question are growing exponentially, the order of seasons is not relevant, but only the number of wet and dry seasons. There are then $t+1$ states of nature (0 wet and t dry, 1 wet and $t-1$ dry,... t wet and 0 dry). The terminal outcomes expressed as total population size, are shown in the body of Table IIA.

Table IIB gives the local outcomes of a *one-move-game* form of the same situation. There are then only two states of nature, wet and dry, and two strategies. The entire sequence of t generations is then a sequence of games of type 2b. The game shown in Table IIB is called a *component game* of a *sequentially compounded game.*

A. *Sequentially Compounded Games*

The sequential compounding of one-move normal games is of very direct interest in the solution of evolutionary problems, so that a description in some detail is worthwhile.

TABLE II

Two normal forms of a game against nature. Homozygotes *AA* and *aa* have intrinsic rates of increase 5 and 10 in wet weather and the reverse in dry. Outcomes are expressed as population size. Table IIA shows terminal outcomes after t seasons, Table IIB local outcomes after 1 season.

A

		States of nature					
		0 t	1 $t-1$	2... $t-2$...	r... $t-r$...	t 0	wet dry
Strategies	*AA*	10^t	$10^{t-1}5^1$	$10^{t-2}5^2$	$10^{t-r}5^r$	5^t	
	aa	5^t	$10^1 5^{t-1}$	$10^2 5^{t-3}$	$10^r 5^{t-r}$	10^t	

B

		States of nature	
		wet	dry
Strategies	*AA*	5	10
	aa	10	5

In general, each component game in a sequentially compounded game will have a different set of strategies, states of nature and local outcomes. Even in the simple situation of Table IIB, the outcomes in the second generation are not the same as in the first and more important *depend* upon the *outcome of the first*. Suppose, for example, the population had adopted the strategy *AA* and the weather had been wet in the first generation. Then there would be 5 plants in second generation and the component game for this second generation would have the local outcomes shown in Table IIIA. On the other hand, had the weather been dry in the first generation there would be 10 plants in the second generation, so that the second component game would appear as in Table IIIB. Usually the dependence will be far more complex than this. In a population segregating

TABLE III

Second component games of a sequentially compounded game shown in Table IIB. Table IIIA is the second game on the assumption that at the first game strategy *AA* was chosen and the weather was wet. Table IIIB is the second component game if the weather had been dry in the first season.

A

		States of nature	
		wet	dry
Strategy	*AA*	25	50
	aa	50	25

B

		States of nature	
		wet	dry
Strategies	*AA*	50	100
	aa	100	50

for both A and a, there will be changes in gene frequency in successive generations so that the strategy set will also change in a dependent manner.

In general let Γ^k be the kth alternative component game with a set of strategies $\{S\}^k$ and states of nature $\{N\}^k$. The local outcome of a strategy S_i^k and state of nature N_j^k for this kth component game will be represented as

$$o_{ij}^k = a_{ij}^k \quad \text{and} \quad (P_{ij}^{k0}\Gamma^0, P_{ij}^{k1}\Gamma_{ij}^{k1}, \ldots, P_{ij}^{kn}\Gamma^n).$$

The meaning of this rather formidable expression is simple. If component Γ^k is played with strategy S_i^k and state of nature N_j^k actually occurring, then the outcome will be made up of two parts. First there will be an absolute local outcome a_{ij}^k (increase of the population, say). In addition there will be a probability P_{ij}^{k0} of playing component game Γ^0 at the next generation, a probability P_{ij}^{k1} of playing game Γ^1 at the next generation and so on up to a probability P_{ij}^{kn} that the second component game played will be Γ^n. To be completely general we will define Γ^0 as a special *terminal game* consisting of a single outcome and of no further play. Thus P_{ij}^{k0} is the probability that play stops (the population becomes extinct, for example) and that at that moment there is some terminal outcome O. If two species are in competition a terminal game would occur when one of the species became extinct and the terminal outcome with respect to the victor might be a sudden expansion to fill the space formerly preempted by the loser.

There are a number of variations of sequentially compounded games depending upon assumptions about the number of component games, Γ, and the existence of a terminal game, and the existence of absolute local outcomes. These are variously called recursive, stochastic, attrition, and survival games, but the names are misleading in a biological context and they will not be discussed in detail in this introductory essay. Further details can be found in Luce and Raiffa (1957).

V. THE NOTION OF AN OPTIMAL STRATEGY

Up to this point we have sketched a formalism for describing interactions between populations and states of nature. This formalism must now be put to use for solving some of the problems alluded to in the introduction. For the sake of concreteness, let us consider a clonally reproducing plant occupying an area in which there are both wet and dry localities. In the simplest case, we may consider three genetic strategies available to this plant: the population consists entirely of *AA* genotypes, entirely of *aa* genotypes, or entirely *Aa* genotypes. Table IV shows the outcomes in

TABLE IV

Hypothetical seed set for three genotypes of a clonally reproducing plant in two states of nature

	N1 wet	N2 dry
AA	15	2
Aa	8	7
aa	4	10

terms of seed set per plant in wet and dry localities. In terms used by Dobzhansky and Levene (1955), the homozygotes are 'narrow specialists' with a high seed set in one environment and a small seed set in the other, while the heterozygotes are 'jacks-of-all-trades' with an intermediate seed set in both environments. We may now ask "which of the three strategies is *best* for the species?" The answer to this question requires two separate decisions. First, what is the relation between outcome values and the values of some success measure of the population? It is necessary to choose a *utility* associated with each outcome, and what is most important, utilities are not necessarily the same as outcomes. It is not true that in all cases a population setting 10 seeds per plant is 'better off' than one setting 5 seeds per plant, even if all other things are equal. It is entirely possible that high seed set may result in such overcrowding at germination as to damage the next generation. The problem of choosing a proper numerical utility associated with a given outcome is a biological rather than a mathematical one. We shall discuss this point in a later section. For the moment let us assume that such a relationship between a given outcome and a utility can be made. In the present state of game theory it is necessary to postulate that the *utility function* M defined over a set of alternative outcomes is *linear*. What is meant by a linear utility function is that if outcome O_1 is given utility M_1 and outcome O_2 is assigned utility M_2, then any mixture of these outcomes in proportion p and $(1-p)$ will have a utility defined as

$$M = pM_1 + (1-p)\,M_2$$

Note that the linearity of the utility function is *not* a condition imposed upon the relation of utilities to outcomes. The outcomes 5, 10 and 15 can be assigned the utilities 7, 43 and 12, say, perfectly within the definition of utility. It will certainly be necessary to relax this stricture for some biological problems, but such a relaxation will require certain radical changes in present methods of solution for game theoretical problems.

Having assigned a numerical linear utility function M to the set of alternatives the second decision must be made: What criterion defines the best or *optimal strategy*? In the example of Table IV assume that the utilities are as in Table V. Will the population be 'better off' if it adopts strategy *AA*, *aa* or *Aa*? If it were sure that the state of nature was 'wet' then strategy *AA* would have the highest utility, while *aa* would be best if the state of nature was always dry. On the other hand if the states of nature differed in different localities the answer is not clear. If the probability distribution of the states of nature were known, the problem would be solved by definition of the utility function M. Since the utility function M is linear, a *utility index* can be calculated for each strategy. Let p_j be the probability of the state of nature N_j and let u_{ij} be the utility associated

TABLE V

A set of arbitrary utilities illustrating the principle of the admissible set. The admissible set here is (S1, S2, S3). These three strategies are those of Table IV.

	N1	N2	*optimality*
S1	8	0	maximax
S2	3	4	maximin
S3	2	7	insufficient reason
S4	2	5	inadmissible
S5	0	3	inadmissible

with outcome o_{ij}. Then the utility index of strategy S_i is

$$M(S_i)=p_1u_{i1}+p_2u_{i2}+\cdots+=\sum_{j-1}^{n} p_ju_{ij}.$$

That is $M(S_i)$ is the *average utility* of strategy S_i and best strategy, by definition of utility, is the one with the highest value of $M(S_i)$. In Table V let the probability of a wet locality be $\frac{1}{4}$ and that of a dry locality $\frac{3}{4}$. Then

$$M(AA)=\tfrac{1}{4}(8)+\tfrac{3}{4}(0)=2.00$$
$$M(Aa)=\tfrac{1}{4}(3)+\tfrac{3}{4}(4)=3.75$$
$$M(aa)=\tfrac{1}{4}(2)+\tfrac{3}{4}(7)=5.75$$

and *aa* is the optimal strategy.

In general, then, there is no serious problem of defining an optimum strategy when an *a priori* distribution of states of nature or of the second player's strategies can be specified. The optimal strategy, *by definition of linear utility functions*, is the strategy giving the highest average utility.

VI. UNCERTAIN STATES OF NATURE

The real use of game theory becomes clear in situations for which there is no *a priori* distribution of states of nature specifiable. This is certainly the usual, if not universal, case. The empirical definition of probability depends upon large (virtually infinite) numbers of repetitions of given events. In fact the events of nature are so complex that they are seldom if ever repeated in the lifetime of a population or species. With respect to living organisms nature is *capricious* rather than *random*. What is the probability of a flood, a succession of dry years, of competition with a new invading species of a particular sort? In general such probabilities cannot be specified (or even defined). Since probabilities cannot be specified for different states of nature, no average utility can be calculated so that other criteria of optimality are needed.

A. *Admissible Strategies*

Following the notation of Luce and Raiffa (1957) we will say that for two strategies S and S′

(a) S is *equivalent* to S′ ($S \sim S'$) if they have identical utilities for every state of nature.

(b) S *strongly dominates* S′ ($S > S'$) if the utilities of S are greater than for S′ for *every* state of nature.

(c) S *weakly dominates* S′ (S $\gtrsim$ S′) if S has a higher utility than S′ for at least one state of nature and never has a lower utility than S′.

Then the set of *admissible* strategies â* is the set such that *no* strategy S in â weakly dominates any other strategy S′ in â. Whatever criterion of optimality is used, the optimal strategy, must be in the admissible set S*. That is, no reasonable criterion of optimality needs to consider a strategy which is not better than a second strategy in *some* state of nature. Table V makes this notion of an admissible set clear. S5 is first rejected because it is strongly dominated by S2, S3 and S4. It is the worst in every state of nature. S4 is also to be rejected because it never has a higher utility than S2 or S3. It is weakly dominated by S2 and S3. The admissible set is then $\{S_1, S_2, S_3\}$ because none of these is weakly dominated by any other.

B. *The Principle of Insufficient Reason*

It is sometimes stated as a principle of probability theory that if there is no *a priori* reason for assigning specific probabilities to alternative outcomes of a process, they should all be regarded as equally likely. Using such a principle each state of the possible state of nature would be assigned a probability of $1/n$. With this uniform probability distribution an expected utility can be calculated and the optimal strategy determined. The shortcomings of such a criterion are obvious and have been extensively reviewed in the probability literature. One example will suffice to concretize these objections. If two indistinguishable coins are tossed, there will be three distinguishable outcomes: 2 heads, 2 tails, 1 head and 1 tail. The principle of insufficient reason would result in a probability of $\frac{1}{3}$ for each of these events. If the coins have different dates or one were scratched so that they could now be distinguished, there will be four distinguishable outcomes: head, head; tail, tail; head, tail; tail, head; and each has a probability of $\frac{1}{4}$ by insufficient reason. But this is absurd for either the outcome 'two heads' has a probability $\frac{1}{3}$ or a probability $\frac{1}{4}$ and that probability cannot depend upon whether or not the coins are distinguishable.

The counter-intuitive result of the principle of insufficient reason does not, however, rule it out as a method of solution for organisms. It may very well turn out that some species will be found to employ strategies which are optimal under this criterion. If various combinations of temperature, humidity, soil type, other organisms, and so on, really represent distinguishable environments, then over long periods of time only a small fraction of the possible environmental combinations will occur and each will occur essentially once. Even lumping environments that are fairly similar as indistinguishable in order to produce a probability distribution will result in a platykurtic function very similar to a uniform distribution. Within the range of extreme environments, then, an *a priori* uniform distribution of states of nature may provide a close approximation to an optimal strategy. In Table V strategy S3 is optimal under this criterion.

C. *Strategies Based on Extreme Utilities*

If no *a priori* probabilities can be set for the state of nature or for an opponent's choice of strategy, then the complete set of utilities will not really provide any criterion for optimality. In such a case we turn to the extreme values of utilities, the greatest and the least utility that a given strategy could yield. One such criterion is to choose that strategy whose *greatest* possible utility under any alternative state of nature is larger than for any other strategy. This is the *maximax* optimality criterion. In Table V strategy S1 is the maximax strategy since it has a utility of 8 under one of the states of nature, which is higher than any utility for any other strategy.

Under what conditions may we expect populations to adopt such a one-sided strategy? One possibility is in a species expanding into an unoccupied territory or a territory in which competitors are relatively weak. In unfavorable environments the invader would have a small rate of increase or perhaps even lose in competition with the incumbent group. In favorable conditions, however, there would be an explosive increase in the population and the setting up of many new populations by migration. Even though some of these populations may be extinguished in unfavorable environments, the maximax strategy would assure in the end successful spread. Such a maximax strategy would be especially important if rapid colonization were important to eventual success since rapid colonization and occupation of the open niche might preclude later successful invasions of other species.

A second optimum strategy based on extreme utilities is one based on a weighting of the maximum and minimum utilities for each strategy. Let α be the weight assigned to the minimum utility of a given strategy m_i and $1-\alpha$ the weight assigned to the maximum utility of that strategy, M_i.

Then an index due to Luce and Raiffe (1957) is

$$H(\alpha)=\alpha m_i+(1-\alpha)\,M_i$$

and that strategy is optimal which has the highest $H(\alpha)$. Such an index makes it possible to take into account both favorable and unfavorable states of nature, but the precise criterion for the choice of α is not clear and seems to imply some *a priori* distribution of the best and worst conditions. When $\alpha=0$ we have the *maximax* criterion discussed above. The most interesting and probably the most useful case for evolutionary problems is the assignment of $\alpha=1$ so that only the minimum utility is considered. This leads to the *maximin* criterion of optimality.

D. *The Maximin Criterion of Optimality*

When α is set to unity, that strategy is optimum whose lowest utility over all states of nature is larger than for any other strategy. That is, the population is guaranteed a utility m_i at the very least. In Table V the maximin strategy is S2. I propose that this maximin strategy is the one which will be found in most cases for reasons which will be discussed below.

For a given matrix of utilities the maximin strategy or strategies can be seen by examination. However, it is possible to increase the *security level* of a game (the maximin utility) by extending the strategies to include so-called *mixed strategies*. In a mixed strategy each of the pure strategy alternatives is assigned a probability p_i and it is assumed that the game is played over and over again with strategy S_i being chosen p_i of the time. The biological analogue of a mixed strategy is a species with many local populations, a proportion p_i of these populations adopting a strategy S_i. The problem is then to find the values of p_i which will maximize the security level of the population. There is a simple method of solution of this problem which *guarantees an average utility to the population at least as high as the maximin pure strategy, irrespective of the distribution of states of nature*. This procedure, because it applies irrespective of the probability of various states of nature, is a very powerful one.

We shall demonstrate the method of solution graphically, although from the graphical solution it is obvious that what is really involved is the solution of a series of simultaneous linear equations. Figure 2 shows such a solution. The abscissa represents proportions of two strategies S_i and S_k from 0.00 S_i, 1.00 S_k to 1.00 S_i, 0 S_k. On the ordinate are utility values.

Choosing a particular pair of strategies S_i, S_k, the utilities of pure S_i and pure S_k for a given state of nature N_j will be represented by a point on the left margin and right margin of the diagram respectively. Because utility is defined linearly, the utility of any mixture of S_i and S_k will be the value on the ordinate of the straight line joining the values of the two pure strategy utilities m_i and m_k. A similar line is drawn for every state of nature with respec to these two strategies. In Figure 2 the two dashed lines

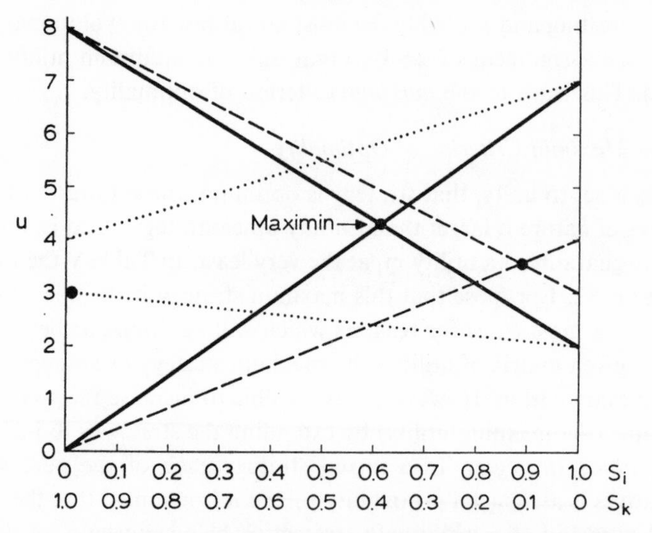

Fig. 2. Graphical solution of the maximin problem for strategies S1, S2 and S3 of Table V. The abscissa is in varying proportions of S_i and S_k. The ordinate is in utility units. Points of intersection represent the maximin utility and the appropriate weighting of S_i and S_k for a given strategy pair S_i, S_k. Dashed lines are for the pair S2, S1, dotted lines for the pair S3, S2 and solid lines for pair S1, S3.

are those for the two states of nature N1 and N2 in Table V with respect to strategies S2 and S1. The point of intersection of the lines provides the maximin mixed strategy (0.11 S1, 0.89 S2) and the security level of this mixture, 3.56. A similar set of lines can be drawn for every pair of strategies S_i, S_k and a maximin strategy for that combination found in the same way. The pair of solid lines in Figure 2 is for strategies S3 and S1 while the dotted lines are for strategies S3 and S2. Note that the S3, S2 lines do not intersect so that the maximin strategy for this pair is pure S2 with a security level of 3. To determine the maximin mixture of all three strategies the results of the pairs can now be compared. The mixed strategy (0.385 S1, 0.615 S3) gives the best security level, 4.31 of any pair. Moreover, any weight given to S2 will lower this security level since substitution of S2 for either S1 or S3 always results in a lower security level *even though* S2 *is the pure maximin strategy*. Thus, no weight should be given this strategy in the optimal strategy mixture and the optimal mixed strategy remains (0.385 S1, 0.615 S3).

In the case that there are more than two states of nature, there will be several mixed utility lines for each strategy pair, as shown in Figure 3. The maximin strategy mixture is that marked by the heavy dot. The heavy line marks off the *admissible boundary* of solutions and the heavy dot is that point on the admissible boundary which has the highest security level.

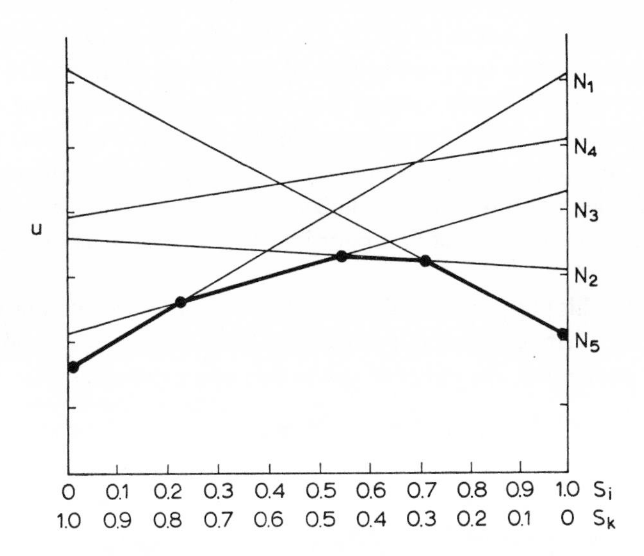

Fig. 3. An example of the linear graphic solution of a maximin problem with more than two states of nature. The dark line is the admissible boundary of solution with the maximin mixed strategy indicated by the large dot.

VII. AN EXAMPLE OF AN EVOLUTIONARY PROBLEM

The artificial example of Table V presents an excellent opportunity to demonstrate the application of game theory to an evolutionary problem.

It is fairly generally accepted that in diploid sexually-reproducing organisms, homozygotes are more specialized in their adaptive properties than heterozygotes. A heterozygote will have about equal fitness in a great variety of environments, while a homozygote will have a low fitness in most environments but a much higher fitness in some environment to which it is specifically adapted. Consider Table VI showing the utility in two states of nature of a homozygote *AA*, a heterozygote *Aa*, and the other homozygote *aa*. The utilities are taken from Table V and we have already demonstrated that although the heterozygote is the maximin *pure* strategy, a greater security level can be obtained by a mixed strategy in which about 40% of the population of the species are homozygous *AA* and 60% homozygous *aa*. A population which was heterozygous *Aa* completely (clonal reproduction) would lower the security level of the entire species. But we may carry the analysis a step further. Nothing has been said about a polymorphic population in which *AA*, *Aa*, and *aa* are all present in a segregating mixture. It is quite possible that polymorphism for each population would be a better strategy than the mixed one suggested above. It is important to note that a segregating population is adopting a *pure* strategy, *not a mixed* one in our sense. If segregating populations are allowed, then there are a great number of added pure strategies S4... S*n* each representing some set of proportions of *AA*, *Aa* and *aa* in a population.

To solve this problem we will assume that there is no facilitation among genotypes. That is, if there is a proportion X_1 of *AA*, X_2 of *Aa* and X_3 of *aa* in a population, the utility of this polymorphic population is

$$m = X_1 m_{AA} + X_2 m_{Aa} + X_3 m_{aa}$$

for any state of nature. Thus for states of nature N_1 and N_2 respectively:

$$m_1 = 2X_1 + 3X_2 + 8X_3$$
$$m_2 = 7X_1 + 4X_2 + 0X_3 .$$

The values of X_1, X_2 and X_3 will be determined by the gene frequency of *A*, say q, and the degree of inbreeding F. In general:

$$X_1 = q^2(1 - F) + Fq$$
$$X_2 = 2q(1 - q)(1 - F)$$
$$X_3 = (1 - q)^2(1 - F) + F(1 - q)$$

Table VI enumerates seven examples (S_4–S_9) with the following characteristics:

	F	q
S_4	0	0.9
S_5	0	0.5
S_6	0	0.1
S_7	0.9	0.5
S_9	0.9	0.1
S_{10}	1.0	0.615

For any value of F there is a continuum of strategies corresponding to a continuum of gene frequencies from 0 to 1. It is obviously not possible to try all possible two-way comparisons of these strategies to find the optimum mixture.

It can be shown, however, that the solution to the paired comparisons (the points of intersections of the utility lines) form a set bounded by a

TABLE VI

A set of monomorphic and polymorphic strategies and their associated utilities in two states of nature. Strategies S1, S2, and S3 are the homogeneous populations *AA*, *Aa* and *aa*. Strategies S4–S9 are segregating populations with different gene frequencies and inbreeding coefficients. S10 is an optimal mixture of *AA* and *aa* homozygotes

	N1	N2
S1	8	0
S2	3	4
S3	2	7
S4	7.04	0.79
S5	4.00	3.75
S6	2.24	6.39
S7	7.36	0.71
S8	4.90	3.53
S9	2.56	6.31
S10	4.31	4.31

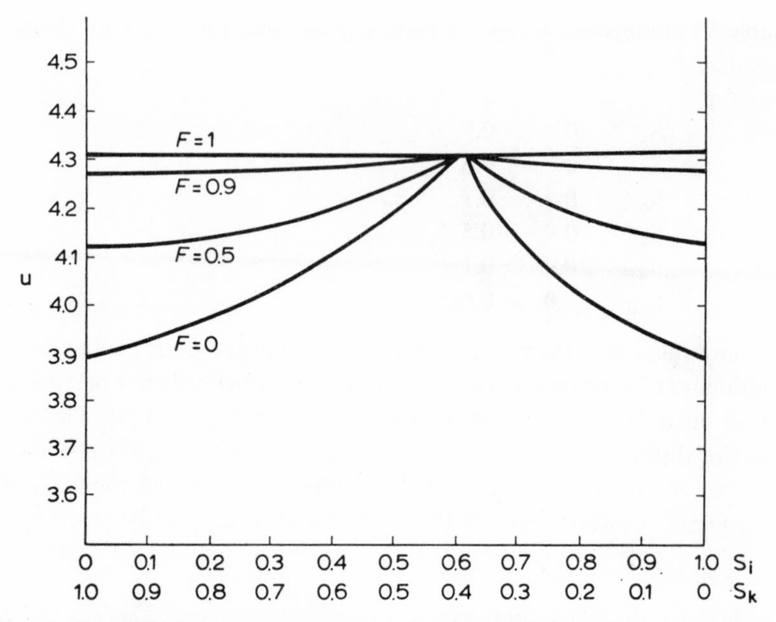

Fig. 4. The upper bounds of solutions to mixed strategies when the genotypes in Table v are assumed to be segregating in populations. Four bounds are shown corresponding to four levels of inbreeding, $F=0, 0.5, 0.9$ and 1.0. The point at which they all meet is the maximin solution discussed in the text.

curve as shown in Figure 4. Each curve in Figure 4 is the upper bound of the set of solutions for a different value of inbreeding from $F=0$ to $F=1$. Each curve is in the form of two cusps with a singularity at $S_i=0.615$, $S_k=0.385$. The left hand limb of each curve happens to correspond to a mixture of S_i=homozygous *aa* and S_j segregating. The right hand limb of each corresponds to S_i segregating and S_j=homozygous *AA*. The singularity for each curve then has the identical meaning: a mixture of 0.615 *aa* populations and 0.385 *AA* populations with *no* segregating populations. Since the singularity is the highest point with a utility of 4.31, the optimal strategy is one in which there is a mixture of homozygous *AA* and homozygous *aa* populations in proportions 0.385 and 0.615 respectively.

The line for complete inbreeding, $F=1$, is particularly interesting. It is a horizontal straight line with the utility equal to 4.31 at every point. There is then no unique maximum solution for completely inbred populations. However, all the strategy mixtures on this line have the following properties:

(1) There are no heterozygotes. This follows from the complete inbreeding, $F=1$.

(2) The overall frequency of *aa* is 0.615 and of *AA* is 0.385 for every point on the line.

The second property is simply a reflection of the fact that completely inbred populations really consist of sub-populations each homozygous for *AA* or *aa*.

Thus the optimal strategy for this species consists in possessing only two genotypes *aa* and *AA* in the proportions 0.615 to 0.385. This can be accomplished by having all populations homogeneous within themselves, 40% being *AA* population and 60% *aa* populations. It can equally well be accomplished by inbreeding so that each population is a mixture of *AA* and *aa* homozygotes with an overall proportion of 40% *AA* to 60% *aa* in the species as a whole.

In biological terms, the situation of Table VI favors *genetic isolation* between *AA* and *aa* genotypes. This means either speciation or complete selfing within one species. Which of these two alternatives is truly optimal depends upon the form of the game for other loci. Complete selfing leads to homozygosity of all the genes in the genome and this may not be optimal. Speciation, on the other hand, creates an optimal condition for the new *species pair*, without forcing homozygosity at other loci.

VIII. THE CHOICE OF A UTILITY

I have deferred until last the question of an appropriate choice of a utility function because it is closely related to the question of the criterion of optimality. Population geneticists and ecologists are not in agreement about the parameter or parameters which measure the 'success' of a population.

For many population geneticists the average fitness of a population is the appropriate choice. Dobzhansky (1951), Cain and Sheppard (1954) and Li (1955) have discussed this point thoroughly and the result of that discussion may be quickly summed up. The average fitness of a population, $\bar{W}$, does tend to a local maximum under the forces of intra-populational selection in most simple cases. However, $\bar{W}$ is defined only within a population and does not, in itself, provide a standard of comparison between populations. All other things being equal, a higher $\bar{W}$ means greater 'success', but, as always, other things are never equal. In some sense a population riddled with lethal genes is not as 'successful' as one free of them since more of its zygotes are being wasted in the elimination of these genes. A large genetic load *is* a disadvantage to a population, but it is not the whole story. With some kinds of selection an increase in $\bar{W}$ within a population may actually lower the competitive ability of that population with respect to others of lower $\bar{W}$.

Among population ecologists the rate of increase, r, is popular as a measure of success. It is necessary to distinguish two meanings of r. First, there is the rate of increase of a population calculated as the birth rate minus the death rate at any instant of time. An r defined in this way is sensitive to population density and as MacArthur (1960) points out, r is identically zero for all populations which are stable in size, irrespective of that size. Birch (1960) uses the parameter r_0 (although not always), the rate of increase of a population *in its logarithmic growth phase*. The chief disadvantage to this measure is that it is *independent* of population size and measures only the returning power of a decimated population or the rate of increase in newly founded colonies. While these two phases in the life of a population are important, they cannot be the whole measure of success. A great advantage of r_0 is that it will be increased by natural selection since an individual who leaves more offspring leaves more genes. (This is not necessarily so if there is inter-family selection.) Any parameter under the control of a direct mechanism like natural selection has distinct advantages as a measure of population success.

Thoday (1953) has made the intriguing proposal that the *probability of survival* of a population or other evolutionary unit, is the best measure of fitness of that unit. The difficulty of such a parameter is that, as envisioned by Thoday, it is virtually impossible to measure. Thoday considered the probability of survival to some distant, unspecified, time, but using the techniques of stochastic game theory, this concept may become more manageable. Consider a population playing a stochastic game extended in time. At each step or move, there is a probability P_{i0} that the population will contain no individuals at time $t+1$, given that it contained i individuals at time t. These probabilities will be different for different strategies and states of nature and, in fact, correspond to the probability of playing the terminal game Γ^0 referred to in a previous section. This extinction probability will be a function of both $\bar{W}$ and r and thus includes both genetic and ecological parameters. I propose, then, that ($P_n = 1 - P_{i0}$), the probability of survival, is a parameter which it will be of advantage for a population to maximize over the whole stochastic game. Since states of nature are uncertain, however, this maximization must take the form of a maximin criterion. That is, the population's optimal strategy is to keep the *local probability* of survival as high as possible under the worst combination of states of nature. The application of the maximin principle to a stochastic game has not been discussed here, but methods are well-known. What is essential here is that if P_n is accepted as a measure of utility, then a maximin solution seems called for.

IX. THE ADOPTION OF STRATEGIES

The final element needed in a game theoretical approach to evolution is a *mechanism*. For intra-population genetics, the maximization of $\bar{W}$ is a good principle because, in fact, the processes of natural selection provide the mechanism for this maximization. Does a similar mechanism exist for '*maximinization*' of the probability of survival? The answer to this question depends entirely upon the importance of population and species extinction. The notion of natural selection is tautological in that it simply states that those individuals with the highest probability of survival are most fit. A similar tautology holds for population and species. Species which survive are by definition more fit than those which expire. Thus we expect that surviving groups will be those which have, *simply by chance*, acquired optimal strategies. This is completely analogous to the Darwinian notion of natural selection of chance variation on the individual level. But what makes natural selection go is the fact that every organism is mortal and the rate of individual extinction is very high. The rate of natural selection within a population is limited by the rate of individual extinction (barring exponentially increasing populations, which are rare). By the same token the effectiveness of inter-deme selection among strategies is bounded by the rate of extinction of demes.

The commonness of population extinction is subject to much discussion. Andrewartha and Birch (1954) believe the rate of extinction of demes to be high while others, notably Nicholson (1957), hold the opposite view. It is extremely difficult to get data on extinction rates of species or higher taxonomic categories because of the confusion of phyletic extinction with taxonomic extinction, the changing of a taxonomic name due to phyletic evolution. Really intensive examination of certain aspects of the fossil record on the lines of Simpson's elucidation of the Equids would provide the necessary information. If the horse were typical, which is unlikely, the rate of extinction of distinct phyla would be quite high. Living Equid phyla represent less than 1% of all known phyla since the Eocene.

On the side of experimentation much can be done. Although the time spans involved in selection among populations is obviously much greater than among individuals, it does not follow that the course of intrapopulation selection cannot be followed in the laboratory. For example large numbers of experimental bottle or vial populations can be kept simultaneously with organisms like *Tribolium* or *Drosophila*. If different populations were allowed different pure strategies (homozygosity for different alleles, polymorphism, different amounts of recombination, etc.) the extinction of these populations could be followed in fluctuating environments. Mimicking of natural fluctuation in temperature, say, would be relatively easy.

In short, experimentation and observation will reveal the kinds of strategies that promote the longevity of populations and species and thus define *biologically* the meaning of an optimal strategy.

Harvard University

BIBLIOGRAPHY

Andrewartha, H. G. and Birch, L. C.: 1954, *The Distribution and Abundance of Animals*, University of Chicago Press.
Birch, L. C.: 1960, *Amer. Nat.* **94**, 5.
Carson, H. L.: 1955, *Cold Spr. Harb. Symp. quant. Biol.* **20**, 276.
Carson, H. L.: 1959, *Cold Spr. Harb. Symp. quant. Biol.* **24**, 87.
Cain, A. J. and Sheppard, P. M.: 1954, *Amer. Nat.* **88**, 321.
Dobzhansky, Th.: 1951, *Genetics and the Origin of Species*, 3rd Edition, Columbia University Press.
Dobzhansky, Th.: 1955, *Cold Spr. Harb. Symp. quant. Biol.* **20**, 1.
Dobzhansky, Th. and Levene, H.: 1955, *Genetics* **40**, 797.
Fisher, R. A.: 1930, *The Genetical Theory of Natural Selection*, Clarendon Press, Oxford.
Lewontin, R. C.: 1957, *Cold Spr. Harb. Symp. quant. Biol.* **22**, 395.
Li, C. C.: 1955, *Amer. Nat.* **89**, 281.
Luce, R. D. and Raiffa, H.: 1957, *Games and Decisions*, John Wiley & Sons, New York.

MacArthur, R. H.: 1960, *Amer. Nat.* **94**, 313.
Mayr, E.: 1959, *Cold Spr. Harb. Symp. quant. Biol.* **24**, 153.
Nicholson, A. J.: 1957, *Cold Spr. Harb. Symp. quant. Biol.* **22**, 1.
Thoday, J. M.: 1953, *Symp. Soc. exp. Biol.* **7**, 96.
von Neumann, J. and Morgenstern, O.: 1944, *Theory of Games and Economic Behavior*, 1st Edition. Princeton University Press.
Wallace, B.: 1959, *Cold Spr. Harb. quant. Biol.* **24**, 193.
Wright, S.: 1931, *Genetics* **16**, 97.

FRANCISCO J. AYALA

BIOLOGY AS AN AUTONOMOUS SCIENCE

The goal of science is the systematic organization of knowledge about the universe on the basis of explanatory principles that are genuinely testable. The starting point of science is the formulation of statements about objectively observable phenomena. Common sense knowledge also provides information about the world. The distinction between science and common sense knowledge is based upon the joint presence in science of at least three distinctive characteristics. First, science seeks to organize knowledge in a systematic way by exhibiting patterns of relations among statements concerning facts which may not appear obviously as mutually related.

The information obtained in the course of ordinary experience about the universe is frequently accurate, but it seldom provides any explanation of why the facts are as alleged. It is the second distinctive characteristic of science that it strives to provide explanations of why the observed events do in fact occur. Science attempts to discover and to formulate the conditions under which the observed facts and their mutual relationships exist.

Thirdly, the explanatory hypotheses provided by science must be genuinely testable, and therefore subject to the possibility of rejection. It is sometimes asserted that scientific explanatory hypotheses should allow one to formulate predictions about their subject matter which can be verified by further observation and experiment. However, in certain fields of scientific knowledge, as in those fields concerned with historical questions, prediction is considerably restricted by the nature of the subject matter itself. The criterion of testability can then be satisfied by requiring that scientific explanations have precise logical consequences which can be verified or falsified by observation and experiment. The word 'precise' is essential in the previous sentence. To provide genuine verification, the logical consequences of the proposed explanatory hypotheses must not be compatible with alternate hypotheses.

It is the concern of science to formulate theories, that is, to discover

M. Grene and E. Mendelsohn (eds.), Topics in the Philosophy of Biology, 312–29.

patterns of relations among vast kinds of phenomena in such a way that a small number of principles can explain a large number of propositions concerning these phenomena. In fact, science develops by discovering new relationships which show that observational statements and theories that had hitherto appeared as independent are in fact connected and can be integrated into a more comprehensive theory. Thus, the Mendelian principles of inheritance can explain, about many different kinds of organisms, observations which appear as *prima facie* unrelated, like the proportions in which characters are transmitted from parents to offspring, the discontinuous nature of many traits of organisms, and why in outbreeding sexual organisms no two individuals are likely to be genetically identical even when the number of individuals in the species is very large. Knowledge about the formation of the sex cells and about the behavior of chromosomes was eventually shown to be connected with the Mendelian principles, and contributed to the explanation of additional facts, for example, why certain traits are inherited independently from each other while other traits are transmitted together more frequently than not. Further discoveries have contributed to the formulation of a unified theory of inheritance which explains many other diverse observations, including the discreteness of natural species, the adaptive nature of organisms and their features, and paleontological observations concerning the evolution of organisms.

The connection among theories has sometimes been established by showing that the principles of a certain theory or branch of science can be explained by the principles of another theory or science shown to have greater generality. The less general branch of science, called the secondary science, is said to have been *reduced* to the more general or primary science. A typical example is the reduction of Thermodynamics to Statistical Mechanics.[1] The reduction of one branch of science to another simplifies and unifies science.

Reduction of one theory or branch of science to another has repeatedly occurred in the history of science. During the last hundred years, several branches of Physics and Astronomy have been to a considerable extent unified by their reduction to a few theories of great generality like Quantum Mechanics and Relativity. A large sector of Chemistry has been reduced to Physics after it was discovered that the valence of an element bears a simple relation to the number of electrons in the outer orbit of the atom. The impressive success of these and other reductions has led in certain circles to the conviction that the ideal of science is to reduce all natural sciences, including biology, to a comprehensive theory that will provide a common set of principles of maximum generality capable of explaining all our observations about the material universe.

To evaluate the validity of such claims, I will briefly examine some of the necessary conditions for the reduction of one theory to another. I will, then, attempt to show that, at the present stage of development of the two sciences, the reduction of biology to physics cannot be effected. I will further claim, in the second part of this paper, that there are patterns of explanation which are indispensable in biology while they do not occur in the physical sciences. These are teleological explanations which apply to organisms and only to them in the natural world, and that cannot be reformulated in non-teleological form without loss of explanatory content.

I. CONDITIONS FOR REDUCTION

In general, reduction can be defined in the present context as "the explanation of a theory or a set of experimental laws established in an area of inquiry, by a theory usually though not invariably formulated for some other domain." [2] Nagel has stated the two formal conditions that must be satisfied to effect the reduction of one science to another. First, all the experimental laws and theories of the secondary science must be shown to be logical consequences of the theoretical constructs of the primary science. This has been called by Nagel the *condition of derivability*.

Generally, the experimental laws formulated in a certain branch of science will contain terms which are specific to that area of inquiry. If the laws of the secondary science contain some terms that do not occur in the primary science, logical derivation of its laws from the primary science will not be *prima facie* possible. No term can appear in the conclusion of a formal demonstration unless the term appears also in the premises. To make reduction possible it is then necessary to establish suitable connections between the terms of the secondary science and those used in the primary science. This may be called the *condition of connectability*. It can be satisfied by a redefinition of the terms of the secondary science using terms of the primary science. For example, to effect the reduction of genetics to physical science such concepts as gene, chromosome, etc., must be redefined in physicochemical terms such as atom, molecule, electrical charge, hydrogen bond, deoxyribonucleic acid, etc.

The problem of reduction is sometimes formulated as whether the properties of a certain kind of objects, for instance organisms, can be explained as a function of the properties of another such group of objects, like the organism's physical components organized in certain ways. This formulation of the question is spurious and cannot lead to a satisfactory answer. Indeed it is not clear what is meant by the 'properties' of a certain object which enters as a part or component of some other object. If *all* the properties are included, it appears that reduction could always be accomplished, and it is in fact a trivial issue. Among the properties of a certain object one will list the properties which it has when it is a component of the larger whole. To use a simple example, one may list among the properties of hydrogen that of combining in a certain way with oxygen to form water, a substance which possesses certain specified properties. The properties of water will then be included among the properties of oxygen and hydrogen.

The reduction of one science to another is not a matter of deriving the *properties* of a kind of objects from the properties of some other group of objects. It is rather a matter of deriving a set of *propositions* from another such set. It is a question about the possibility of deriving the experimental laws of the secondary science as the logical consequences of the theoretical laws of the primary science. Scientific laws and theories consist of propositions about the world, and the question of reduction can only be settled by the concrete investigation of the logical consequences of such propositions, and not by discussion of the properties or the nature of things.

From the previous observation it follows that the question of reduction can only be solved by a specific reference to the actual stage of development of the two disciplines involved. Certain parts of chemistry were reduced to physics after the modern theories of atomic structure were developed some fifty years ago, but the reduction could not have been accomplished before such development. If the reduction of one science to another is not possible at the present stage of development of the two disciplines, it is empirically meaningless to ask whether reduction will be possible at some further time, since the question can only be answered dogmatically or in terms of metaphysical preconceptions.

Simpson has suggested that the unification of the various natural sciences be sought not "through principles that apply to all phenomena but through phenomena to which all principles apply." Science, according to Simpson, can truly become unified in biology, since the principles of all natural sciences can be applied to the phenomena of life.[3] To be sure, the theoretical laws of physics and chemistry apply to the physicochemical phenomena occurring in organisms. Besides, there are biological theories that explain observations concerning the living world but have no application to non-living matter. To conclude therefore that biology stands at the center of all science is true as far as it goes, but it is trivial and constitutes no progress in scientific understanding that I can discern.

The goal of the reductionistic program is not, as Simpson seems to believe, to establish a "body of theory that might ultimately be *completely* general in the sense of applying to *all* material phenomena," nor a "search for a least common denominator in science." It is rather a quest for a comprehensive theory that would *explain* all phenomena – the living as well as the inanimate world – with an economy of stated laws and a corresponding increase in our understanding of the world. Whether such an ideal can be accomplished is a different issue that I shall consider presently.

II. THE REDUCTION OF BIOLOGY TO PHYSICAL SCIENCE

The question of the reducibility of biology to physicochemistry has been raised again in the last decade particularly in connection with the spectacular successes accomplished in certain areas of biology. In genetics, research at the molecular level has contributed to establish the chemical structure of the hereditary material, to decipher the genetic code, and to provide some understanding of the mechanisms of gene action. Brilliant achievements have also been obtained in neurophysiology and other fields of biology. Some authors have claimed that the understanding of all biological phenomena in physicochemical terms is not only possible but the task of the immediate future. It is thus proclaimed that the only worthy and truly 'scientific' biological research is what is called in recent jargon 'molecular biology', that is, the attempt to explain biological phenomena in terms of the underlying physicochemical components and processes.

It is easy to dispose of two extreme positions which seem equally unprofitable. On one end of the spectrum there are substantive vitalists who defend the irreducibility of biology to physical science because living phenomena are the effect of a non-material principle which is variously called vital force, entelechy, élan vital, radial energy, or the like. A non-material principle cannot be subject to scientific observation nor lead to genuinely testable scientific hypotheses.[4]

At the other end of the spectrum stand those who claim that reduction of biology to physicochemistry is in fact possible at present. In the current stage of scientific development, a majority of biological concepts, such as cell, organ, species, ecosystem, etc., cannot be formulated in physicochemical terms. Nor is there at present any class of statements belonging to physics and chemistry from which every biological law could be logically derived. In other words, neither the condition of connectability nor the condition of derivability – two necessary formal conditions of reduction – are satisfied at the present stage of development of physical and biological knowledge.

Two intermediate positions have also appeared in the recent literature. First, a reductionist position maintains that although the reduction of biology to physics cannot be effected at present, it is possible in principle. The factual reduction is made contingent upon further progress in the biological or in the physical sciences, or in both. Secondly, certain anti-reductionist authors claim that reduction is not possible in principle because organisms are not merely assemblages of atoms and molecules, nor even of organs and tissues standing in merely external relation to one another. Organisms are alleged to be 'wholes' that must be studied as wholes and not as the 'sum' of isolable parts.[5]

Although biological laws are not in general derivable from any available theory of physics and chemistry, the reductionists claim that such accomplishment will be possible in the future. Such assertions are frequently based on metaphysical preconceptions about the nature of the material world. In any case it cannot be convincingly argued empirically, since it is only a statement of faith about the possibility of some future event. It must be noted, however, that advances in various areas of molecular biology are continuously extending the, as yet, exiguous realm of biological phenomena that can be explained in terms of physicochemical concepts and laws.

As for the anti-reductionist position that maintains that organisms and their properties cannot be understood as mere 'sums' of their parts, I have already stated that it rests on an unsatisfactory formulation of the problem. The question of reduction is whether *propositions* concerning organisms can be logically derived from physicochemical laws, and not whether the *properties* of organisms can be explained as the result of the properties of their physical components. It should perhaps be added that the phenomenon of so-called 'emergent' properties occurs also in the non-living world. Water is formed by the union of two atoms of hydrogen with one atom of oxygen, but water exhibits properties which are not the immediately apparent consequence of the properties of the two gases, hydrogen and oxygen. Another simple example can be taken from the field of thermodynamics. A gas has a temperature although the individual molecules of the gas cannot be said to possess a temperature.[6]

The reduction of biology to physicochemistry admittedly cannot be effected at the present stage of scientific knowledge. Whether the reduction will be possible in the future is an empirically meaningless question. A majority of biological problems cannot be as yet approached at the molecular level. Biological research must then continue at the different levels of integration of the living world, according to the laws and theories developed for each order of complexity. The study of the molecular structure of organisms must proceed hand in hand with research at the levels of the cell, the organ, the individual, the population, the species, the community, and the ecosystem. These levels of integration are not isolated from each other. Laws formulated at one level of complexity illuminate the other levels, both lower and higher, and suggest additional research strategies.[7] It is perhaps worth pointing out that in fact biological laws discovered at a higher level of organization have more frequently contributed to guide research at the lower level than vice versa. To mention but one example, the Mendelian theory of inheritance preceded the identification of the chemical composition and structure of the genetic material, made possible these discoveries, and has not become superogatory because of the latter.

III. THE NOTION OF TELEOLOGY

I will now proceed to discuss the role of teleological explanations in biology. I shall attempt to show that teleological explanations constitute patterns of explanation that apply to organisms while they do not apply to any other kind of objects in the natural world. I shall further claim that, although teleological explanations are compatible with causal accounts, they cannot be reformulated in non-teleological form without loss of explanatory content. Consequently, I shall conclude that teleological explanations cannot be dispensed with in biology, and are therefore distinctive of biology as a natural science.

The concept of teleology is in general disrepute in modern science. More frequently than not it is considered to be a mark of superstition, or at least a vestige of the non-empirical, a prioristic approach to natural phenomena characteristic of the prescientific era. The main reason for this discredit is that the notion of teleology is equated with the belief that future events – the goals or end-products of processes – are active agents in their own realization. In evolutionary biology, teleological explanations are understood to imply the belief that there is a planning agent external to the world, or a force immanent to the organisms, directing the evolutionary process toward the production of specified kinds of organisms. The nature and diversity of organisms are, then, explained teleologically in such a view as the goals or ends-in-view intended from the beginning by the Creator, or implicit in the nature of the first organisms.

Biological evolution can however be explained without recourse to a Creator or a planning agent external to the organisms themselves. There is no evidence either of any vital force or immanent energy directing the evolutionary process toward the production of specified kinds of organisms. The evidence of the fossil record is against any directing force, external or immanent, leading the evolutionary process toward specified goals. Teleology in the stated sense is, then, appropriately rejected in biology as a category of explanation.

In *The Origin of the Species* Darwin accumulated an impressive number of observations supporting the evolutionary origin of living organisms. Moreover, and perhaps most importantly, he provided a causal explanation of evolutionary processes – the theory of natural selection. The principle of natural selection makes it possible to give a natural explanation of the adaptation of organisms to their environments. Darwin recognized, and accepted without reservation, that organisms are adapted to their environments, and that their parts are adapted to the functions they serve. Penguins are adapted to live in the cold, the wings of birds are made to fly, and the eye is made to see. Darwin accepted the facts of adaptation, and then provided a natural explanation for the facts. One of his greatest accomplishments was to bring the teleological aspects of nature into the realm of science. He substituted a scientific teleology for a theological one. The teleology of nature could now be explained, at least in principle, as the result of natural laws manifested in natural processes, without recourse to an external Creator or to spiritual or non-material forces. At that point biology came into maturity as a science.

The concept of teleology can be defined without implying that future events are active agents in their own realization nor that the end-results of a process are consciously intended as goals. The notion of teleology arose most probably as a result of man's reflection on the circumstances connected with his own voluntary actions. The anticipated outcome of his actions can be envisaged by man as the goal or purpose toward which he directs his activity. Human actions can be said to be purposeful when they are intentionally directed toward the fulfilment of a goal.

The plan or purpose of the human agent may frequently be inferred from the actions he performs. That is, his actions can be seen to be purposefully or teleologically ordained toward the fulfilment of a goal. In this sense the concept of teleology can be extended, and has been extended, to describe actions, objects or processes which exhibit an orientation toward a certain goal or end-state. No requirement is necessarily implied that the objects or processes tend consciously toward their specified end-states, nor that there is any external agent directing the process or the object toward its end-state or goal. In this generic sense, teleological explanations are those explanations where the presence of an object or a process in a system is explained by exhibiting its connection with a specific state or property of the system to whose existence or maintenance the object or process contributes. Teleological explanations require that the object or process contribute to the existence of a certain state or property of the system. Moreover, and this is the essential component of the concept, teleological explanations imply that such contribution is the explanatory reason for the presence of the process or object in the system. Accordingly, it is appropriate to give a teleological explanation of the operation of the kidney in regulating the concentration of salt in the blood, or of the structure of the hand of man obviously adapted for grasping. But it makes no sense to explain teleologically the motions of a planet or a chemical reaction. In general, as will be shown presently, teleological explanations are appropriate to account for the existence of adaptations in organisms while they are neither necessary nor appropriate in the realm of non-living matter.

There are at least three categories of biological phenomena where teleological explanations are appropriate, although the distinction between the categories need not always be clearly defined. These three classes of teleological phenomena are established according to the mode of relationship between the structure or process and the property or end-state that accounts for its presence. Other classifications of teleological phenomena are possible according to other principles of distinction. A second classification will be suggested later.

(1) When the end-state or goal is consciously anticipated by the agent. This is purposeful activity and it occurs in man and probably, although in a lesser degree, in other animals. I am acting teleologically when I buy an airplane ticket to fly to Mexico City. A cheetah hunting a zebra has at least the appearance of purposeful behavior. However, as I have said above, there is no need to explain the existence of organisms and their adaptations as the result of the consciously intended activity of a Creator. There is a purposeful activity in the living world, at least in man; but the existence of the living world, including man, need not be explained as the result of purposeful behavior. When some critics reject the notion of teleology from the natural sciences, they have in mind exclusively this category of teleology.

(2) Self-regulating or teleonomic systems, when there exists a mechanism that enables the system to reach or to maintain a specific property in spite of environmental fluctuations. The regulation of body temperature in mammals is a teleological mechanism of this kind. In general, the homeostatic reactions of organisms belong to this category of teleological phenomena. Two types of homeostasis are usually distinguished by biologists – physiological and developmental homeostasis, although intermediate and additional types do exist.[8] Physiological homeostatic reactions enable the organism to maintain certain physiological steady states in spite of environmental shocks. The regulation of the composition of the blood by the kidneys, or the hypertrophy of muscle in case of strenuous use, are examples of this type of homeostasis.

Developmental homeostasis refers to the regulation of the different paths that an organism may follow in its progression from zygote to adult.

The development of a chicken from an egg is a typical example of developmental homeostasis. The process can be influenced by the environment in various ways, but the characteristics of the adult individual, at least within a certain range, are largely predetermined in the fertilized egg. Aristotle, Saint Augustine, and other ancient and medieval philosophers, took developmental homeostasis as the paradigm of all teleological mechanisms. According to Saint Augustine, God did not create directly all living species of organisms, but these were implicit in the primeval forms created by Him. The existing species arose by a natural 'unfolding' of the potentialities implicit in the primeval forms or 'seeds' created by God.

Self-regulating systems or servo-mechanisms built by man belong in this second category of teleological phenomena. A simple example of such servo-mechanisms is a thermostat unit that maintains a specified room temperature by turning on and off the source of heat. Self-regulating mechanisms of this kind, living or man-made, are controlled by a feedback system of information.

(3) Structures anatomically and physiologically constituted to perform a certain function. The hand of man is made for grasping, and his eye for vision. Tools and certain types of machines made by man are teleological in this third sense. A watch, for instance, is made to tell time, and a faucet to draw water. The distinction between the (3) and (2) categories of teleological systems is sometimes blurred. Thus, the human eye is able to regulate itself within a certain range to the conditions of brightness and distance so as to perfoɪm its function more effectively.

IV. TELEOLOGY AND ADAPTATION

Teleological mechanisms and structures in organisms are biological adaptations. They have arisen as a result of the process of natural selection. Natural selection is a mechanistic process defined in genetic and statistical terms as differential reproduction. Some genes and genetic combinations are transmitted to the following generations on the average more frequently than their alternates. Such genetic units will become more common, and their alternates less common, in every subsequent generation.

The genetic variants arise by the random processes of genetic mutation and recombination. Genetic variants increase in frequency and may eventually become fixed in the population if they happen to be advantangeous as adaptations in the organisms which carry them, since such organisms are likely to leave more descendants than those lacking such variants. If a genetic variant is harmful or less adaptive than its alternates, it will be eliminated from the population. The biological adaptations of the organisms to their environments are, then, the result of natural selection, which is nevertheless a mechanistic and impersonal process.

The adaptations of organisms – whether organs, homeostatic mechanisms, or patterns of behavior – are explained teleologically in that their existence is ultimately accounted for in terms of their contribution to the reproductive fitness of the species. A feature of an organism that increases its reproductive fitness will be selectively favored. Given enough generations it will extend to all the members of the population.

Patterns of behavior, such as the migratory habits of certain birds or the web-spinning of spiders, have developed because they favored the reproductive success of their possessors in the environments where the population lived. Similarly, natural selection can account for the existence of homeostatic mechanisms. Some living processes can be operative only within a certain range of conditions. If the environmental conditions oscillate frequently beyond the functional range of the process, natural selection will favor self-regulating mechanisms that maintain the system within the functional range. In man, death results if the body temperature is allowed to rise or fall by more than a few degrees above or below normal. Body temperature is regulated by dissipating heat in warm environments through perspiration and dilation of the blood vessels in the skin. In cool weather the loss of heat is minimized, and additional heat is produced by increased activity and shivering. Finally, the adaptation of an organ or structure to its function is also explained teleologically in that its presence is accounted for in terms of the contribution it makes to reproductive success in the population. The vertebrate eye arose because genetic mutations responsible for its development occurred, and were gradually combined in progressively more efficient patterns, the successive changes increasing the reproductive fitness of their possessors in the environments in which they lived.

There are in all organisms two levels of teleology that may be labeled *specific* and *generic*. There usually exists a specific and proximate end for every feature of an animal or plant. The existence of the feature is explained in terms of the function or property that it serves. This function or property can be said to be the specific or proximate end of the feature. There is also an ultimate goal to which all features contribute or have contributed in the past – reproductive success. The generic or ultimate end to which all features and their functions contribute is increased reproductive efficiency. The presence of the functions themselves – and therefore of the features which serve them – is ultimately explained by their contribution to the reproductive fitness of the organisms in which they exist. In this sense the ultimate source of explanation in biology is the principle of natural selection.

Natural selection can be said to be a teleological process in a *causal* sense. Natural selection is not an entity but a purely mechanistic process. But natural selection can be said to be teleological in the sense that it produces and maintains end-directed organs, when the functions served by them contribute to the reproductive efficiency of the organism.

The process of natural selection is not at all teleological in a different sense. Natural selection is not in any way directing toward the production of specific kinds of organisms or toward organisms having certain specific properties. The over-all process of evolution cannot be said to be teleological in the sense of proceeding toward certain specified goals, preconceived or not. The only nonrandom process in evolution is natural selection understood as differential reproduction. Natural selection is a purely mechanistic process and it is opportunistic.[9] The final result of natural selection for any species may be extinction, as shown by the fossil record, if the species fails to cope with environmental change.

The presence of organs, processes, and patterns of behavior can be explained teleologically by exhibiting their contribution to the reproductive fitness of the organisms in which they occur. This need not imply that reproductive fitness is a consciously intended goal. Such intent must in fact be denied, except in the case of the voluntary behavior of man. In teleological explanations the end-state or goal is not to be understood as the efficient cause of the object or process that it explains. The end-state is causally – and in general temporally also – posterior.

V. INTERNAL AND EXTERNAL TELEOLOGY

Three categories of teleological phenomena have been distinguished above, according to the nature of the relationship existing between the object or mechanism and the function or property that it serves. Another classification of teleology may be suggested attending to the process or agency giving origin to the teleological system. The end-directedness of living organisms and their features may be said to be *internal* teleology, while that of man-made tools and servo-mechanisms may be called *external* teleology. It might also be appropriate to refer to these two kinds of teleology as *natural* and *artificial*, but the other two terms, 'internal' and 'external', have already been used.[10]

Internal teleological systems are accounted for by natural selection which is a strictly mechanistic process. External teleological systems are the products of the human mind, or more generally, are the result of purposeful activity consciously intending specified ends. An automobile, a wrench, and a thermostat are teleological systems in the external sense; their parts and mechanisms have been produced to serve certain functions intended by man. Organisms and their parts are teleological systems in the internal sense; their end-directedness is the result of the mechanistic process of natural selection. Organisms are the only kind of systems exhibiting internal teleology. In fact they are the only class of *natural* systems that exhibit teleology. Among the natural sciences, then, only biology, which is the study of organisms, requires teleology as a category of explanation.

Organisms do not in general possess external teleology. As I have said above, the existing kinds of organisms and their properties can be explained without recourse to a Creator or planning agent directing the evolutionary process toward the production of such organisms. The evidence from paleontology, genetics, and other evolutionary sciences is also against the existence of any immanent force or vital principle directing evolution toward the production of specified kinds of organisms.

VI. TELEOLOGICAL EXPLANATIONS IN BIOLOGY

Teleological explanations are fully compatible with causal accounts. "Indeed, a teleological explanation can always be transformed into a causal one."[11] Consider a typical teleological statement in biology, "The function of gills in fishes is respiration." This statement is a telescoped argument the content of which can be unraveled approximately as follows: Fish respire; if fish have no gills, they do not respire; therefore fish have gills. According to Nagel, the difference between a teleological explanation and a non-teleological one is, then, one of emphasis rather than of asserted content. A teleological explanation directs our attention to "the *consequences* for a given system of a constituent part or process." The equivalent non-teleological formulation focuses attention on "some of the *conditions*... under which the system persists in its characteristic organization and activities."[12]

Although a teleological explanation can be reformulated in a non-teleological one, the teleological explanation connotes something more than the equivalent non-teleological one. In the first place, a teleological explanation implies that the system under consideration is directively organized. For that reason teleological explanations are appropriate in biology and in the domain of cybernetics but make no sense when used in the physical sciences to describe phenomena like the fall of a stone. Teleological explanations imply, while non-teleological ones do not, that there exists a means-to-end relationship in the systems under description.

Besides connoting that the system under consideration is directively organized, teleological explanations also account for the existence of specific functions in the system and more generally for the existence of the directive organization itself. The teleological explanation accounts for the presence in an organism of a certain feature, say the gills, because it contributes to the performance or maintenance of a certain function, respiration. In addition it implies that the function exists because it contributes to the reproductive fitness of the organism. In the non-teleological translation given above, the major premise states that 'fish respire'. Such formulation assumes the presence of a specified function, respiration, but it does not account for its existence. The teleological explanation does in fact account for the presence of the function itself by implying or stating explicitly that the function in question contributes to the reproductive fitness of the organism in which it exists. Finally, the teleological explanation gives the reason why the system is directively organized. The apparent purposefulness of the ends-to-means relationship existing in organisms is a result of the process of natural selection which favors the development of any organization that increases the reproductive fitness of the organisms.

If the above reasoning is correct, the use of teleological explanations in biology is not only acceptable but indeed indispensable. Organisms are systems directively organized. Parts of organisms serve specific functions that, generally, contribute to the ultimate end of reproductive survival. One question biologists ask about organic structures and activities is 'What for?' That is, 'What is the function or role of such a structure or such a process?' The answer to this question must be formulated in teleological language. Only teleological explanations connote the important fact that plants and animals are directively organized systems.

It has been argued by some authors that the distinction between systems that are goal directed and those which are not is highly vague. The classification of certain systems as teleological is allegedly rather arbitrary. A chemical buffer, an elastic solid or a pendulum at rest are examples of physical systems that appear to be goal directed. I suggest using the criterion of utility to determine whether an entity is teleological or not. The criterion of utility can be applied to both internal and external teleological systems. Utility in an organism is defined in reference to the survival and reproduction of the organism itself. A feature of a system will be teleological in the sense of internal teleology if the feature has utility for the system in which it exists and if such utility explains the presence of the feature in the system. Operationally, then, a structure or process of an organism is teleological if it can be shown to contribute to the reproductive efficiency of the organism itself, and if such contribution accounts for the existence of the structure or process.

In external teleology utility is defined in reference to the author of the system. Man-made tools or mechanisms are teleological with external teleology if they have been designed to serve a specified purpose, which therefore explains their existence and properties. If the criterion of utility cannot be applied, a system is not teleological. Chemical buffers, elastic solids, and a pendulum at rest are not teleological systems.

The utility of features of organisms is with respect to the individual or the species in which they exist at any given time. It does not include usefulness to any other organisms. The elaborate plumage and display is a teleological feature of the peacock because it serves the peacock in its attempt to find a mate. The beautiful display is not teleologically directed toward pleasing man's aesthetic sense. That it pleases the human eye is accidental, because it does not contribute to the reproductive fitness of the peacock (except, of course, in the case of artificial selection by man).

The criterion of utility introduces needed objectivity in the determination of what biological mechanisms are end-directed. Provincial human interests should be avoided when using teleological explanations, as Nagel says. But he selects the wrong example when he observes that "the development of corn seeds into corn plants is sometimes said to be natural, while their transformation into the flesh of birds or men is asserted to be merely accidental."[13] The adaptations of corn seeds have developed to serve the function of corn reproduction, not to become a palatable food for birds or men. The role of wild corn as food is accidental, and cannot be considered a biological function of the corn seed in the teleological sense.

Some features of organisms are not useful by themselves. They have arisen as concomitant or incidental consequences of other features that are adaptive or useful. In some cases, features which are not adaptive in origin may become useful at a later time. For example, the sound produced by the beating of the heart has become adaptive for modern man since it helps the physician to diagnose the condition of health of the patient. The origin of such features is not explained teleologically, although their preservation might be so explained in certain cases.

Features of organisms may be present because they were useful to the organisms in the past, although they are no longer adaptive. Vestigial organs, like the vermiform appendix of man, are features of this kind. If they are neutral to reproductive fitness these features may remain in the population indefinitely. The origin of such organs and features, although not their preservation, is accounted for in teleological terms.

To conclude, I will summarize the second part of this paper. Teleological explanations are appropriate to describe, and account for the existence of, teleological systems and the directively organized structures, mechanisms, and patterns of behavior which these systems exhibit. Organisms are the only natural systems exhibiting teleology; in fact they are the only class of systems possessing internal teleology. Teleological explanations are not appropriate in the physical sciences, while they are appropriate, and indeed indispensable, in biology which is the scientific study of organisms. Teleological explanations, then, are distinctive of biology among all the natural sciences.[14]

University of California, Davis

NOTES

[1] E. Nagel, *The Structure of Science*, New York: Harcourt, Brace and World, 1961, pp. 338–345.

[2] *Ibid.* p. 338; see also pp. 336–397.

[3] G. G. Simpson, *This View of Life*, New York: Harcourt, Brace and World, 1964, p. 107. According to J. G. Kemeny (*A Philosopher Looks at Science*, Van Nostrand, 1959, pp. 215–216) the most likely solution of the question of the reduction of biology to physics is that a new theory will be found, covering both fields, in new terms. Inanimate nature will appear as the simplest extreme case of this theory. In that case, one would say that physics was reduced to biology and not biology to physics.

[4] Except for the general conclusion that biological phenomena will never be satisfactorily explained by mechanistic principles.

[5] E. S. Russell, *The Interpretation of Development and Heredity*, Oxford, 1930; see also E. Mayr, 'Cause and Effect in Biology', *Cause and Effect*, D. Lerner (ed.) New York: Free Press, 1965, pp. 33–50.

[6] The temperature of the gas is identical by definition with the mean kinetic energy of the molecules.

[7] Th. Dobzhansky, 'Biology, Molecular and Organismic', *The Graduate Journal* VII(1), 1965, pp. 11–25.

[8] For instance, the maintenance of a genetic polymorphism in a population due to heterosis can be considered a homeostatic mechanism acting at the population level.

[9] F. J. Ayala, 'Teleological Explanations in Evolutionary Biology', *Philosophy of Science*, **37** (1970), 1–15.

[10] T. A. Goudge, *The Ascent of Life*, Toronto: Univ. of Toronto Press, 1961, p. 193.

[11] E. Nagel, 'Types of Causal Explanation in Science', *Cause and Effect*, D. Lerner (ed.), New York: Free Press, 1965, p. 25.

[12] E. Nagel, *The Structure of Science*, p. 405; see also F. J. Ayala, ref. in note 9 above.

[13] E. Nagel, *The Structure of Science*, p. 424.

[14] The author is a recipient of a PHS Research Career Development Award from the National Institute of General Medical Sciences.

RONALD MUNSON

BIOLOGICAL ADAPTATION

I

The vagueness typical of many traditional biological concepts continues to be a matter of concern to biologists and philosophers. Even though the results are not always generally accepted, efforts to explicate and clarify such fundamental concepts as organism, species, structure, and function have all contributed to the development of a more critical biology. With this general end in view, the specific purpose of this paper is to analyze the concept of adaptation as it appears within the context of biology and, in particular, in the synthetic theory of evolution.[1]

Sentences which contain the terms 'adaptation', 'adaptive', or 'adapted' will be called 'adaptational sentences', and it will be shown that no adaptational sentence need involve reference to any purpose, final cause, or other non-empirical notion in order to be meaningful. Such sentences as 'The gray color of the British moth is adaptive in a light environment', it will be argued, can in principle be reformulated as equivalent sentences which have an empirical content, and contain no teleological terms. It will also be argued that, while certain sentences which contain adaptational terms can be regarded prima facie as teleological explanations, when such sentences are given an empirical content, they have no explanatory worth. Rather, adaptational sentences in general are ways of expressing data which require explanation. Throughout this paper, actual biological usage and practice will be taken to provide the standard of adequacy and correctness.

It has long been recognized by biologists that sentences whose meaning depends upon the notion of adaptation can be regarded as implicitly teleological. (see [16], p. 10) As Pittendrigh remarks, the most common definitions of 'adaptation' "all connote that aura of design, purpose or end-directedness which has, since Aristotle, seemed to characterize the living thing." ([18], p. 392) Simpson and Beck, among others, observe that "adaptations are the apparently goal-directed features of living things that constantly impress us with the notion that organisms do have purposes." ([21], p. 445) Thus the reason for the implicit teleological character of adaptational sentences is simply that to say a trait is adaptive (or is an adaptation)[2] is apparently to claim that it is related in some way or other to an end of the organism or species which possesses it, i.e. to say it serves a purpose. In evolutionary theorizing, no matter whether the trait is morphological, behavioral, or physiological, its 'final purpose' is always taken to be the survival of the individual organism or the species, though it may also have an intermediate or 'functional purpose'. Therefore, when we say 'This animal is adapted' we are saying, to quote Burnett and Eisner, "This animal is adapted for survival" ([1], p. 33), and when we say 'This is an adaptive trait of the species', we are saying "This trait is adaptive for the survival of the species."[3]

Granted that this is so, then sentences which mention traits as adaptive lend themselves to being restated as ones which are explicitly teleological. For example, "In a light environment, the gray color of the British moth is an adaptive trait" might be reformulated as "In a light environment, the British moth is gray in order to survive (for the purpose of survival)." The color, that is, is related to survival as an end. Inasmuch as such sentences might be employed to account for the possession of the trait by the species which they mention, they might be viewed as a species of teleological explanation. Hempel and Oppenheim, in their brief discussion of teleological explanations, have also called attention to this (See [10], p. 329).

Because adaptational sentences seem to be meaningful only by virtue of an implicit commitment to purpose, many biologists have recommended a rejection of such terms as 'adaptive' and 'adaptation' from the vocabulary of biology (see [12], p. 412; [18], p. 393). This advice is rarely followed in practice, however. Moreover, such a recommendation assumes, perhaps without warrant, that the only alternative to teleology is to declare adaptational sentences unacceptable. Consequently, if adaptational language is to be retained and the possibility of introducing non-empirical notions into biology by its use is to be avoided, some analysis of adaptational sentences must be given which shows that their meaning does not necessarily depend upon a reference to purpose. What is more, before any estimation of the explanatory worth of adaptational sentences can be made, their meaning must be made clear. Though, as pointed out, some may be regarded as prima facie teleological explanations, it does not follow that once they have been provided with an empirical content they will be explanatory.

The fact that some adaptational sentences which mention a trait as adaptive differ in an important way from the example discussed above deserves a brief comment. Sentences like "The fangs of the wolf are an adaptation to environmental requirements for offense and defense" ([20], p. 144), and "The size and surface contours of the molars and pre-molars in *Equus* are adaptations to grazing" ([18], p. 391), are instances of a very special kind of adaptational sentence. In a sense such sentences are doubly teleological in import, for they make explicit reference to the purpose (function) served by a trait (e.g. offense and defense, grazing) and implicit reference to a further purpose which is served by the satisfactory fulfillment of the first (e.g. species preservation). As already indicated, sentences which make no explicit reference to ends or purposes but merely state that a trait is adaptive for, say, a species are understood as asserting that the trait serves the (final) purpose of preserving the species. That it does this by satisfying some environmental requirement or condition is assumed but unstated. (In fact, the condition, which may be the functional end of the trait, may be unknown. Thus inquiry into the 'adaptive significance' of a trait has the object of determining not only that a trait is associated with the evolutionary success of a species, but also the role which the trait plays in its continuation.) Similarly, to say a species is adapted to its environment is neither to mention traits nor ends, but it is to presuppose that the species possesses traits which, in relation to environmental factors, do something (perform functions) which serve the purpose of preserving the species. (Thus a biologist may know, from a study of the fossil record for example, that a species was adapted during some given time period, but he may not know what traits rendered it such nor why they did.)

It will later be shown that sentences which are doubly teleological are best interpreted as ones which state both that a trait is adaptive and that it has a certain function. Such sentences, that is, are adaptational-functional compounds. But this is a claim which can be established only after adaptation and function have been distinguished and after other kinds of adaptational sentences have been identified. In any case, the possession of a functional component ought not to be regarded as an essential characteristic of adaptational sentences. For an analysis of adaptational sentences to be complete, at least so far as they are employed in evolutionary theorizing, sentences which assert merely that a trait is adaptive and that organisms and species are adapted or better adapted to their environments must also be considered.

The following cases[4] illustrate adaptational sentences of the sort with which we must be concerned:

(1) The speckled-gray color of some moths of the species *Biston betularia* is adaptive in a light environment.

(2) The granular coloration of the lizard *Uma* is adapted to desert conditions.

(3) Some perissodactyls were adapted to conditions in the Tertiary.

(4) The Brazilian fish *Monocerihus polycanthus* is adapted to streams which contain abundant vegetation.

(5) The camel is better adapted to desert conditions than the horse.

(6) Some moths of the species *Biston betularia* are better adapted to their environment than others of the same species.

Although each of the above sentences is adaptational in the sense specified, it should be noticed that they may differ from one another in important ways. More will be said about the significance of their differences in the proper place, but for the moment it is enough merely to call attention to some of them. Sentences (1) and (2) are the only ones which mention adaptive traits. They assert, not that an organism or species is adapted, but that a trait[5] or property is adaptive for an organism or species. Sentences (3) and (4) assert the adaptation of organisms, either of at least one individual or a species.[6] The last sentences make comparative claims about species and about classes of individuals within the same species.

Content aside, it might be thought that except for differences in logical form there is no need to distinguish between (1) and (2), (3) and (4), or (5) and (6), for when biologists speak of a trait as adaptive and an organism as adapted or better adapted, the reference is not (necessarily) to particular organisms. Rather, the reference is to a class of organisms which have the trait or to a class of organisms which are adapted. This logical point is well-taken and is not being denied, but it must not be construed to mean that saying a trait is adaptive for the individual organisms which possess it (i.e. the class) is the same as saying it is adaptive for a species, nor that saying individual organisms are adapted is the same as saying the species is adapted. The reason this must not be done resides in the matter of fact that a trait may be adaptive for the class of individuals which possess it but not for the species and, though some individuals may be better adapted than others of the same species, their adaptation does not necessarily render the species adapted. (Why this is so will be discussed later.) Whether or not the adaptation of individuals and of species coincides, and whether or not traits which are adaptive for the class of individual organisms which have them are also adaptive for the species, are empirical questions. Accordingly, the distinctions among the members of the three sets of sentences may also be distinctions of use in referring. We must be told or learn from the context, for example, that (1) asserts that speckled-gray individuals of the species possess a trait which is adaptive for them (the class) in the specified environment. In a different context, (1) might be construed as asserting that the trait possessed by some members of the species is adaptive for the species. Sentences like (2) are generally taken to assert that a trait is adaptive for a species, but they might also be interpreted as asserting that the trait is adaptive for the class of individuals of the species (which might be all the members of the species during a given time) which possesses it. Similar differences hold between the members of the other two groups. (For example, a species may be better adapted than another without its members being better adapted than those of the other.) Thus an adequate account of the meaning of adaptational sentences must be one which makes it possible to assign an empirical content to sentences like those above, while recognizing and preserving the differences indicated.

The remainder of this paper will be an attempt to find answers to the two general questions which have been raised. First, how can we provide adaptational sentences with an empirical content and thereby show that an implicit teleological reference is not necessary in order for them to be meaningful? In light of the above illustrations and comment on the use of adaptational language, an adequate answer to this question must also be an answer to the following subsidiary questions: (a) What does it mean to say of some trait T of an organism (at least one individual) O, that T is adaptive for O? (b) What does it mean to say that T is adaptive for a species? (c) What does it mean to say that an organism (at least one individual) is adapted and that a species is adapted? (d) Finally, what does it mean to say that one organism (at least one, the class possessing a certain trait) is better adapted than another and that a species is better adapted than some other? The second general question is more straightforward: if adaptational sentences can be interpreted in principle as non-teleological ones, can such sentences, as interpreted, be employed for explanation?

Before considering how these question may be answered, two points should be made clear. One is rather obvious, the other less so. First, traits which are characterized as adaptive and organisms or species which are said to be adapted are always asserted to be such only relative to specific kinds of environment or environmental factors.[7] Thus the light color of most desert animals is adaptive in consideration of the environmental peculiarities of the landscape, the climate, and so on. In a forest region, such a color might not be adaptive. Dark colors might also be adaptive for desert life, though not in relation to the same environmental factors. It is, however, possible to know that both traits are adaptive in the same general environment (e.g. the desert) without knowing in reference to exactly what features they are adaptive. Similarly, a trait may be recognized as adaptive for a class of organisms or a species in a particular kind of environment though it is not known in relation to what environmental characteristics the trait has adaptive significance (i.e. the functional role of the trait and its connection with survival). ([5], p. 33) A class of organisms and a species may be determined to be adapted (or better adapted than some other) in a specified environment, though it is not known what traits render them so. (These points were mentioned earlier, and more will be said about them below.) Finally, a trait is always characterized as adaptive only relative to organisms or species of a determinate kind, and a trait adaptive for one kind might not be for some other, even in the same environment.

Second, adaptational sentences are properly used in biology only in reference to functional or teleonomic systems. The reference is direct when a system itself (e.g. a particular organism) is said to be adapted, indirect when a species is said to be adapted or when a trait is characterized as adaptive for individuals or species. Given this peculiarity of usage, on first view it might be thought that the problem of finding an analysis of adaptational sentences is the same as providing an analysis of sentences which employ functional terms. Thus an account of the meaning of sentences of the form 'The function of trait T in system S is G' would also be an account (indirectly) of '*T* is adaptive for *S*'. Since several analyses of functional sentences have been offered (e.g. [9]; [17], pp. 401ff.), there would be no justification for a separate consideration of adaptational sentences. Such a view is not acceptable, however. It results from an identification of adaptation and function, due to the fact that 'adaptation' is sometimes employed to refer to the state of a functional system which results when a function is served. For example, the function of the heart is to circulate the blood; when circulation is disturbed by variation in the external or internal state conditions and then restored to its usual efficiency, the result is said to be an adaptation of the system. So, it might be claimed to say a trait is adaptive is only to say it has a function, a role to play in maintaining a functional system in a certain state, and adaptation is identified with the maintenance or restoration of a state of the system.

Much could be said about such a claim, but in order to get on with the tasks at hand, it is only necessary to show that, although functional and adaptational sentences are both teleological, adaptation and function are not identical. That they are not can be shown by at least three considerations. First is the fact that not all functions or items which serve them can be regarded as adaptive. For example, the function of the large antlers of the Irish Elk might be said to be to secure protection from predators and mating rivals. If 'functional system' is broadly enough construed to make it legitimate to consider an organism and its environment such a system, then it is true that this function might render individuals of the species adapted, i.e. maintain them in certain conditions. Yet there is good reason to believe that the antlers were inadaptive for the species and contributed to their extinction. Apart from maintaining the state of some particular system, then, to say that a trait serves a function or that a function is served leaves open the question as to whether or not the trait or function are adaptive for the species. To put the point another way, if adaptation and function were identical, sentences of the form 'The function of *T* in systems of kind *S* was *G*, and *T* was not adaptive for the species' would be contradictory. Yet they are not, but can be confirmed or disconfirmed by the available evidence. Second, if adaptation and function were the same, then if a sentence which mentions a trait as adaptive is true, then a sentence which mentions the same trait as having a function ought to be true. That is, adaptational and functional sentences about the same trait ought to have the same truth-value in every case. That they do not is shown by the fact that 'Heart-sounds in man are adaptive, for they permit the detection of heart dysfunction' may well be true, whereas 'The function of the heart is to produce heart-sounds' is patently false. Thus not all adaptations (i.e. adaptive traits) are functions. Third, and perhaps most important, evolutionists sometimes conclude that a trait is (or was) adaptive, even when its function is unknown. As Simpson points out ([19], p. 167), there is good reason to believe that the claws of ungulates were adaptive, even though there is no agreement about what their function was. Thus, the evidence upon which adaptational sentences which mention traits are established is not the same as that required to establish functional sentences, which would not be the case if adaptation and function were the same.

None of these considerations is intended to deny that function and adaptation are related. At least in some (non-evolutionary) contexts, to say that an item serves a function is the same as saying that it is adaptive for the system. Moreover, the presence of items which serve certain functions, and that certain functions are served (e.g. digestion of grass, circulation), are often characterized as traits adaptive for individuals and species. All that is being denied is that an account of the meaning of functional sentences obviates the need of giving an account of adaptational sentences.

II

Hogben observed almost forty years ago that "the word adaptation is frequently used by biologists without a very clear agreement as to its content." ([11], p. 103) It is true that, for the most, biologists still leave such terms as 'adapted' and 'adaptive trait' undefined, yet the situation characterized by Hogben has altered. Though explicit definitions are rarely given, there is ordinarily agreement among evolutionists in case-by-case decisions about what traits are adaptive and whether or not a species is adapted. I wish to now indicate the ground for this agreement and show that it permits us to endow adaptational sentences with an empirical content.

The fact that biologists designate as 'adaptive' only those traits which they also characterize as 'useful' or 'advantageous' (cf. [19], p. 161; [4], p. 406; [1], p. 33; [21], p. 444; [14], p. 16) to the organisms or species which possess them suggests a way of expressing the meaning of adaptational sentences which secures the desired end of avoiding the introduction of non-empirical notions. In short, judging from the use of adaptational language in biology, it appears that to say a trait is adaptive is to assert just that the trait is advantageous to the organism or species which is mentioned as possessing it under stipulated environmental conditions. On this interpretation, then, adaptational sentences of the form

Trait *T* of organism *O* is adaptive in environment *E*

may be restated as equivalent sentences of the form

O has *T* and *T* is advantageous to *O* in *E*.

Accordingly, to call a trait 'adaptive' is to assert only that its possession is advantageous, and it is correspondingly inconsistent to characterize a trait both as adaptive and as disadvantageous to its possessors in that environment.

'The fangs of a wolf are adaptive', on this account, asserts the same thing as 'The fangs of a wolf are advantageous' (i.e. to the class of fanged wolves). The ways in which the fangs are advantageous (what they do, their function) may also be specified, and in general an evolutionary account or an ecological report which merely stated that a trait was advantageous or adaptive would be regarded as incomplete. Yet this fact does not alter the logical character of the reformulation. Thus, 'The fangs of a wolf are advantageous, and they are employed in offense and defense'. As pointed out earlier, biologists sometimes conclude that a trait is adaptive, even though its use or function is unknown; moreover, not all adaptational sentences mention the function of a trait, even though it is known. The analysis presented here does not rule out such cases. In fact it makes it clear that the question about the use or function of a trait is different from the question about whether or not the trait is adaptive. Sentences about the use of an item provide supplemental information and are supported by different evidence. The sentence above asserts more than that fangs are adaptive for the class of individuals which have them, for it is equivalent to the conjunction of 'The fangs of a wolf are advantageous to it' and 'The fangs of a wolf are employed for (perform the function of) defense and offense'. It may well be true (and probably is) that what makes the fangs advantageous is just that they are employed for offense and defense. Yet as was stated in making the distinction between adaptation and function, it is possible to know the trait is adaptive without knowing its function – just as it is possible to know what the trait does (including its function) without knowing whether or not it is adaptive.

It should now be clear why sentences such as 'The fangs of a wolf are an adaptation for offense and defense' were earlier characterized as being doubly teleological. In accordance with the above analysis, such sentences are best interpreted as asserting both that a trait has a use or function and that the trait is adaptive for (advantageous to) its possessor. They are adaptational-functional compounds which can be expressed as the conjunction of an adaptational sentence and a functional sentence. Since the main interest here is in the analysis of adaptational sentences of the kind illustrated earlier, no further effort will be made to analyze adaptational-functional compounds. It is assumed, however, that the adaptational component can be treated in the same way as any sentence which claims that a trait is adaptive. An investigation of functional sentences and explanations goes beyond the scope of this paper.

A reformulation of adaptational sentences of the kind suggested above is clearly of little value unless some criterion for determining what traits are advantageous to the organism (and) or species is made explicit. Otherwise, it would be perfectly reasonable to object that, though the restatement contains no teleological terms, it is worthless, for the meaning of adaptational sentences still has not been specified.

An acceptable criterion is not difficult to find, for it is usually implicitly assumed in the majority of contemporary discussions of biological adaptation which are concerned with evolutionary relationships. It is ordinarily taken for granted that adaptation and selection are so related that selection determines those traits which are advantageous to the organisms which possess them.[8] More precisely, the advantage of a trait which makes it adaptive for those organisms which have it is Darwinian selective advantage, understood as differential mortality within the group, and distinguished from the more inclusive genetical selection. As a criterion, then, we may propose the following: A trait T is advantageous to its possessors if and only if, during a determinate period of time not greater than the length of a normal generation, and under the same or similar environmental conditions, the relative frequency of survival of those organisms which possess T is greater than the survival frequency of those which lack T, though similar in other respects.[9] In a darkly colored environment, for example, the color of individuals of the melanic variety of the British moth (*Biston betularia*) is advantageous to them, for their chance of survival is better than it is for individuals of the non-melanic variety.

The criterion which has been formulated is for the determination of the advantage which a trait bestows upon those organisms which possess it (i.e. the advantage for the class). The question of whether or not the trait is advantageous to the species is left open. Thus, in the kind of case considered, it makes no difference whether the trait is genotypically determined in a direct way (i.e. appears ontogenetically) or whether the trait is acquired (i.e. the genotype merely permits the acquisition), for since the concern is only with a class of organisms over a relatively short period of time, it makes no difference whether or not the trait is heritable. Stronger crocodiles, for example, probably have a better chance of living out their normal life span than do their weaker fellows, yet their superior strength is (presumably) not a heritable trait.

So far as determining whether or not a trait is advantageous to a species is concerned, a modification of the above criterion is required. Individuals of a species may possess a trait not possessed by other members of the species, and this trait, as determined by the original criterion, may be advantageous. Yet it may be that the possessors of the trait do not reproduce or reproduce less than those which lack this trait. Thus, though the trait may be advantageous to the individuals which possess it, it is not necessarily advantageous to the species. In order for a trait to be of advantage to the species, it must be heritable, but the advantage, once again, is selective advantage. The original criterion may be emended as follows: T is advantageous to species S in E if and only if T is heritable within S and during a determinate period of time, at least longer than one generation,[10] S continues to reproduce itself within E, and there is an increase in the relative number of organisms in S which possess T.[11] Selection here must be regarded as genetical selection, which includes Darwinian selection, for it is the genetic units which determine which traits are perpetuated by differential reproduction. The clause which demands there be a relative increase within the species of occurrences of the trait is a way of expressing the claim, which certainly must be made if the trait is called 'adaptive', that it is by virtue of the trait that the species has survived to reproduce. The clause does not, of course, claim that the trait is a sufficient condition (or even a necessary condition) for continuing reproductive success.

Another kind of case which makes this change in criterion necessary is worth mentioning. In general, traits which are advantageous to individual organisms are also advantageous to the species.[12] Thus, protective coloration enhances the survival chances of the individual compared with those which lack the trait and also makes it more likely that the species will endure, for over time individuals which possess the trait will make a greater contribution to the gene pool. In turn, this will increase the probability that a greater proportion of individuals in future generations will possess the trait. The converse of this relation does not always hold, however. As Dobzhansky has pointed out, some traits useful to the survival of the species may be judged harmful or frequently fatal to individuals, e.g. childbearing in mammals. ([4], p. 406) In any event, whether or not a given trait is advantageous to the individual and also to the species is a matter of factual inquiry, but since the two do not necessarily coincide, the criterion for the determination of advantage must be different in each case, if it is to represent correctly the distinction recognized in biological practice. Accordingly, when a trait is said to be adaptive, it must be made clear whether the claim is that it is adaptive for individual organisms or whether it is adaptive for the species. In general, this is done by biological writers on adaptation. (see [15], p. 86)

We are now enabled to provide an analysis of the meaning of some kinds of adaptational sentences. Those which assert that a trait is adaptive for organisms or for species can be interpreted as claiming that the trait is advantageous to the organism or species, and this, conjoined with the appropriate criterion for advantage, yields an equivalent reformulation of such sentences. The problem of providing an analysis of sentences which assert that an organism is adapted, that a species is adapted, and that one organism or species is better adapted than some other still remains. We wish then to show a way of providing logically equivalent sentences containing no adaptational or teleological terms for sentences of the following form:

(1) Organism O is adapted to environment E
(2) O_1 is better adapted to E than O_2
(3) Species S is adapted to E
(4) S_1 is better adapted to E than S_2

Sentences such as those above also seem best interpreted by reference to Darwinian and genetical selection. Darwinian selection, construed as differential mortality within a population, allows us to replace sentences like (1) with a sentence about the viability of a class of organisms under given environmental conditions. Sentences which make a comparative claim about organisms, like (2), require no treatment separate from (1), for those, like (1), involve an implicit comparison. That is, an organism may be said to be adapted to the environment only in comparison with some other organism which is not. As Medawar and others have observed ([16], p. 13), the other organism may be a member of the same species or a member of another species, for both will serve as parameters of comparison. Thus to say that an organism O_1 is adapted to E is the same as saying that it is better adapted than a certain O_2, and O_1 is better adapted than O_2 if and only if it has a greater chance of survival than O_2 during a specified time period and under the same or similar environmental conditions.[13] Accordingly, the evidence for such an adaptational claim is statistical, and though we are not able to say of some particular organism in either of the classes whether or not it will live out the normal life span (whatever this may be), we may still be in a position to conclude that the individuals which comprise one class have a greater chance of staying alive during a given stretch of time than do those which make up the comparative class.

When it comes to the analysis of the meaning of sentences which assert that a *species* is adapted or better adapted than some other, we must again turn to genetical selection. In general, sentences of the form 'species S is adapted to E' may be interpreted as equivalent to one asserting that during a definite time period, longer than at least one generation, S continues to reproduce itself within E. Thus, to say a species is adapted to a certain environment is to say no more and no less than that it can successfully reproduce under the environmental conditions during that time, and it is selection which determines whether or not a species continues to exist within the environment. It is genetical selection which must be understood as the criterion of adaptation, because the differential mortality of Darwinian selection is only one of the factors which is relevant to the differential reproduction which determines the success or failure of the species. Darwinian selection could be employed above, for there the concern was with the adaptation of individual organisms (a class) during a generational period, but sentences of kind (3) and (4) concern the continuation of the species over several generations.

The same considerations require the assumption of genetical selection in making adaptive comparisons of species. Sentences of the form 'Species S_1 is better adapted to E than S_2' can be regarded as asserting the same thing as a sentence which asserts that during a definite period of time, longer than one generation, S_1 increases numerically relative to S_2.[14] Comparisons other than this are possible, for clearly if the numerical ratio of S_1 to S_2 remains unaltered, the two are equally well adapted during that period; if the converse of the relation of S_1 to S_2 mentioned above is the case, then we should say that S_2 is better adapted than S_1. It need not be assumed, of course, that the increase or decrease in S_1 relative to S_2 is absolute. Simpson has pointed out that "abundance of one group relative to another may increase even though absolute numbers are small or decreasing." ([19], p. 161) The increase in relative number *is* genetical selection, and to say that one group is better adapted is only to say that it is selected and the other is not.

It might be mentioned in passing that this interpretation of comparative adaptational sentences also permits us to speak significantly about adaptational alternatives. Two species may be determined to be equally adapted to the same environment, even though they possess very different characteristics. Morphological and ecological studies, among others, might be made in order to determine in what ways each has solved the problems of the environment, but it is possible to determine that both are equally adapted without a knowledge of the mechanisms and environmental relations which render them such. (see [18], p. 400)

In summary: In all cases of the use of adaptational language which were distinguished initially, a sentence about Darwinian or about genetical selective advantage can be employed as an analysis of the meaning of adaptational sentences. (Adaptational-functional compounds are analyzed by breaking them into their components and treating the adaptational claim separately.) If this has been adequately shown, then it may be concluded that adaptational sentences can be provided with an empirical content and that they need not involve a covert reference to purpose in order to be meaningful.

III

One question still remains, namely, do adaptational sentences, when provided with an explicit empirical content, serve as explanations? It was initially observed that a sentence which mentions a trait as adaptive for a species, for example, might be construed as a teleological explanation. The explanandum phenomenon for which it might be taken to account would be the possession of a trait by a species, and the explanation might be supposed to consist in relating the trait to the 'end' or 'purpose' of species survival which it serves. If such a sentence is reformulated into one of the form 'Species *S* has *T* and *T* is advantageous to *S*' and this is conjoined with the independent criterion of species advantage, does this provide an explanation of why a species possesses the trait mentioned? No extended argument is necessary to show that it clearly does not. It need not be denied that the possession of a trait might well be a causal factor in the preservation of individual organisms and species. This is not at issue. To assert, however, that because a trait is adaptive in the relevant sense, then a species must have it is to commit the fallacy of affirming the consequent. It will be shown below that adaptational sentences, interpreted in the suggested fashion, are merely ways of expressing information about a species (or a class of its members) over an evolutionary period, information which must be accounted for by any adequate theory of evolution.

If, however, the explanandum question is changed from the form 'Why does *S* have *T*?' to 'Why does *S* with *T* survive in *E*?' or to 'Why do individuals of *S* which have *T* (the class) survive?', then the way is open for a statistical explanation. Taking the last and simplest kind of case, an explanation of why a mouse with a light coat color survives in an environment which has light-colored soil would be of the following form: "The probability that mice with a light coat color will survive in *E* is a certain degree *r*; mouse *a* has a light coat and is in *E*; therefore, it is probable to degree *r* that the mouse will survive." ([14], p. 18) Yet this is not what adaptational sentences of the kind being considered apparently purport to explain. What is more, the biologist generally wants an answer to the question of why it is that organisms with a certain trait have a better chance of survival than those which lack it. Such explanations as the ones illustrated are thus usually of little interest to him.

Adaptational sentences as they appear in contemporary biological theory are not explanatory principles which account for the possession of a trait by a class of organisms or species, as their implicit teleological character would suggest. To claim that they are would be to assume that there is some 'principle of adaptation' which is explanatory, analogous to a 'vital principle', rather than taking adaptive phenomena as being themselves proper subject matter for explanation. Yet if adaptational sentences do not provide explanations, then what role do they play, what position do they occupy, in evolutionary theorizing? The answer seems to be that, regarded in one way, sentences like 'The gray color of the British moth is adaptive' supply data for which a theory of evolution must account. Regarded in another, adaptational sentences are expressions of the *outcome* of selective processes, and it is only in relation to natural selection, Darwinian and genetical, that they can be given empirical significance. Presumably this is the way that evolutionists understand adaptational sentences when they speak of natural selection as the mechanism of adaptation and say that adaptation is the goal of evolution. ([20], p. 159; [7], p. 117)

In conclusion, let me briefly present this point in more detail and provide some illustration for it. As was mentioned earlier, a biologist is generally concerned with knowing more about a trait than whether or not it is adaptive. Having determined that it is adaptive, he ordinarily wishes to know also *why* it is adaptive. That is, even though it may be known that the trait bestows a selective advantage, this is a state of affairs which requires explanation. The biologist must investigate the relationship of the trait to environmental factors in order to account for the differential mortality or reproduction which constitutes the adaptive character of the trait. It is here, of course, that questions of function arise, for the biologist needs to learn whether the function served by the trait is connected with the differences in mortality or reproduction and how it is connected. Sometimes he must first find out what the function of a trait is and, more generally, what the trait does.

Empirical inquiry into the relation between the trait and environmental factors may reveal that a trait is adaptive for one or more of a variety of reasons. To name but a few typical ones, a trait may be adaptive because organisms which possess it are less subject to predation, are more viable physiologically, are more successful in securing food, are more resistant to certain diseases, and so on. Factors of this kind provide the basis for explaining why organisms which possess a certain trait are more likely to survive than ones which do not and why those which have the trait contribute differentially to the gene pool of the species. Thus, they figure in an explanation of why a trait is adaptive. Notice, putting the point another way, that such sentences as 'Organisms which have *T* are more resistant to diseases than ones which lack *T*' can be established independently of sentences like 'Organisms which have *T* have a greater probability of survival than ones which lack it', and the first sentence may enter into an explanation of the second.

Obviously, the factors which are explanatory in a given case must be discovered by inquiry. For illustration, consider the familiar case of the adaptive coloration of the British moth (mentioned several times above as an example) as investigated by Kettlewell. (see [13]; [14], p. 137ff.; [22], pp. 147ff.) By releasing a number of marked individuals of both the light and dark varieties of the moth and then recapturing a fair sample, Kettlewell discovered that there had been a decrease in the proportion of the light-colored forms during the intervening period. This established what had been suspected, namely that the survival chances of the dark moths were better than for the light ones in the sooty industrial environment of northern England. Thus melanism (the dark coloration) was recognized as an adaptive trait, but it was still necessary to explain why the melanic form was more likely to survive than the non-melanic. To be brief, Kettlewell established by observation that the moths are the prey of birds of several species and concluded that the differences in survival frequency was due to selective predation. The same data, plus information about changes in the environment due to industrialization, in conjunction with a theory of inheritance also serves to explain why there was a relative increase in the melanic forms in the moth population over several generations. Many other cases, differing in interesting ways, might be discussed, but this single example ought to be enough to make clear the original point: To show that a trait is adaptive is to present a phenomenon requiring explanation, and to provide the explanation is to display the success of the trait as the outcome of selection.[15]

University of Missouri, St. Louis

NOTES

[1] 'Synthetic theory', as commonly used in biology, refers to the theory of evolutionary phenomena which is a 'synthesis' of modifications of Darwin's theory and materials derived from other areas of biology, particularly genetics and population genetics. It is

the theory which is currently accepted by the vast majority of biologists and is sometimes called 'neo-Darwinian'. See [20], pp. 277–278.

[2] 'Adaptation' is a word used in a variety of ways in biology. The following are perhaps the most frequent: (1) To denote traits, both morphological and functional (The tail of a fish is an adaptation, as is the horse's ability to digest grass.); (2) To denote an organism or species which is in a certain state (The lion is a species which represents or is a case of an adaptation to conditions found in the savannas.); (3) To refer to that state itself (The lion, as a species, is in a state of adaptation.); (4) To refer to the process of which the state is an outcome.

This multiplicity of use is not as serious as it seems. Context alone is ordinarily enough to forestall confusions, but more importantly, there is no need to use the same term in every kind of case as above. At least in some instances there are unambiguous alternative adaptational terms which mark the differences. Stipulation in other cases can be made for the purposes of this paper. To avoid the indeterminacy, I shall usually, though not always, use the phrase 'adaptive trait', rather than 'an adaptation'. I shall never speak of an organism or species as being a case of adaptation. For this and for 'the adaptation' (i.e. the state), I shall speak only of an *adapted* organism or species. The only use of 'adaptation' as a verb form is (4). Since the process takes its name from its outcome, it only generates confusion to refer to both the state and the process by the same term. Accordingly, I here shall not refer to the changes and procedures by which an organism becomes adapted as 'adaptation'. Whatever those processes may be, they will simply be designated in some other way. In sum, 'adaptation' (except in straightforward context) will be used only in the phrase "an adaptation", and this is always to be read as referring to a trait which is asserted to be adaptive.

The uses in cases of kind (1) present a particular problem, for when a trait is not merely called 'adaptive' or characterized as 'an adaptation', but is said to be 'an adaptation for' something (e.g. swimming), the impression is given that adaptational sentences which mention traits are no different from functional sentences. This is a problem which will be dealt with later. (See [6] for a discussion of various other confusions which result from the use of adaptational language.)

[3] Marjorie Grene writes of the difficulty which the concept of adaptation presents to the synthetic theory: "It seems odd that the very theorists who most emphatically deny any shred of purposiveness to nature, should just as emphatically declare that all significant changes in nature are adaptive in character." ([8], p. 126)

[4] For numerous other examples see: [7], *passim*; [12], pp. 412–486; [1], *passim*.

[5] 'Trait' is another term which is notoriously vague in its biological usage. Warning colors, mimicry, functional changes under appropriate conditions, the physiological use of certain chemicals, structural characteristics, and a combination of these are all regarded as adaptive traits by biologists. They are generally roughly classified as morphological, physiological, biochemical, and behavioral. Yet the cases often intergrade, for physiological and behavioral characteristics are frequently connected with morphological ones. (See [20], p. 163). Thus the capacity to fulfill a function and the structures and mechanisms by which it is fulfilled are both traits of organisms.

[6] It is not necessary here to distinguish between the adaptation of a species and other groups. All that need to be noted is whether it is individuals (classes of individuals possessing a certain trait) or species which are asserted to be adapted or for which a trait is adaptive. I shall refer to any group or population which exists for more than one normal generation as a 'species'. This is not to be regarded as a definition of 'species', though, nor is it to be interpreted as asserting that it is species, as opposed to popula-

tions, which evolve. Moreover, 'organism' will be used to mean 'at least one individual of a species of the class of individuals possessing a given trait'. It is assumed that a biological species is not just a class of individuals living at a particular time.

[7] It is the contextual dependency of adaptation which permits biologists to speak of preadaptation, at least in one sense, for the term 'preadaptation' is sometimes used in sentences which are intended to assert the subjunctive conditional 'If *O* were in environment E_2 instead of E_1, then it would be adapted'. Thus, the prefix is not necessarily temporal, though it may be. See [12], p. 450.

[8] As Simpson ([9], p. 160) puts it, a trait is advantageous to an organism if it helps it 'remain alive'. Also see Canfield [2], who analyzes functional sentences in terms of survival chances.

[9] This formulation is not intended as a measure of adaptation, in the sense of adaptation of organisms by means of some function holding between the environment (statistically described) and the distribution of traits within a population. So far as I know, no satisfactory way has yet been found for doing this. This criterion and the ones which follow are relevant only to what may be called the qualitative concept of adaptation. Even then they just concern what in principle it is possible to discover from investigations (see note 11).

[10] During a period of time only slightly longer than one generation it may be impossible to determine whether or not a trait is of advantage to a species. Yet this is no objection to the criterion, for generally it is not possible to say that a trait is or is not adaptive for a species when only a relatively short time period is considered. The doubt is the same in either case, and ordinarily an investigation of at least several generations (experimentally or through the fossil record) is required.

[11] This criterion for advantage is similar to the one given long ago by E. J. Allen: "By an adaptation [i.e. adaptive trait] we mean nothing more than a character of an organism which has enabled a species to survive itself as such or survive until it is transformed into another species. It is survival that gives the measure of adaptation." Quoted in [3], p. 277 from *Proc. of the Linn. Soc. of London* (1928–1929), p. 119.

[12] Whether or not it is admissible to regard a single trait as adaptive for a species is a disputed point among evolutionists. In genetical selection, it is the genotype which is selected, and a genotype may determine a phenotype characterized by a variety of traits, so it would seem that only a phenotype can be regarded as advantageous to a species. In part this disagreement rests on the ambiguity of 'trait', for unit characteristics (e.g. color), complex characteristics (e.g. mechanisms which permit flight), and the very possession of a certain phenotype may be regarded as traits. This, in conjunction with the fact that only the genotype is selected, has sometimes led to the claim that all traits of reproducing organisms must be regarded as adaptive. Even granting that the genotype is the unit of selection, it does not follow that we cannot speak of phenotypic traits (of any kind) as adaptive, nor that because it is the genotype which is selected, we must regard all traits determined by it as adaptive. The comparison of a species *S* whose members have a certain trait *T* with another group (either another species or an ancestor species) whose members do not possess *T*, though resembling *S* in other respects, and which either did not survive or decreased in number relative to *S*, would allow us to conclude that *T* is advantageous to *S*. We need not conclude that all selected traits are adaptive traits, and the claim that *T* is an adaptive trait of *S* does not conflict with the claim that it is the genotype which is selected. It may be true that we are not able to determine whether or not every trait of the members of *S* is adaptive for *S*, but in principle it is possible to decide this. The decision is undeniably made in practice for

some traits, e.g. melanism vs. non-melanism. (For a discussion of traits and genotypic selection, see [5].)

[13] This is essentially Medawar's interpretation: "When we say that an organism *O* is adapted to ... *E* there is an implied comparison with an organism *O* that is not so well adapted" ([16] p. 13, note). Medawar also makes the environment a parameter.

[14] This formulation is a modification of Simpson's criterion for the success of a species. See [16], p. 161: cf. [24], p. 365.

[15] Work on this paper was supported during the summer of 1968 by a grant from the University of Missouri. For advice and criticism I am indebted to my colleagues Professors James Doyle and Henry Shapiro and to Professor Peter Kivy of Rutgers University-Newark.

BIBLIOGRAPHY

[1] Burnett, A. L. and Thomas Eisner, *Animal Adaptation*, N.Y., Holt, Rinehart, and Winston, 1964.
[2] Canfield, John, 'Teleological Explanation in Biology', *British Journal for the Philosophy of Science* **XIV** (1964), 285–295.
[3] Dawes, Ben, *A Hundred Years of Biology*, London, Duckworth, 1952.
[4] Dobzhansky, Th., 'Evolution and Environment', *Evolution after Darwin*, vol. I (ed. by Sol Tax), Chicago, University Chicago Press, 1960.
[5] Dobzhansky, Th., 'What is an Adaptive Trait?', *American Naturalist* **XC** (1956), 337–347.
[6] Ghiselin, Michael T., 'On Semantic Pitfalls of Biological Adaptation', *Philosophy of Science* **33** (1966), 147–154.
[7] Grant, Verne, *The Origins of Adaptations*, N.Y., Columbia University Press, 1963.
[8] Grene, Marjorie, 'Two Evolutionary Theories' (two parts), *British Journal for the Philosophy of Science* **9** (1958–1959), 110–127, 185–194.
[9] Hempel, Carl G., 'The Logic of Functional Analysis', *Aspects of Scientific Explanation*, N.Y., Free Press, 1965, pp. 297–331.
[10] Hempel, Carl G. and Oppenheim, Paul, 'The Logic of Explanation', *Readings in the Philosophy of Science* (ed. by Herbert Feigl and May Brodbeck), N. Y., Appleton, 1953, pp. 319–352.
[11] Hogben, Lancelot, *The Nature of Living Matter*, London, Kegan Paul, 1930.
[12] Huxley, Julian, *Evolution: The Modern Synthesis*, N.Y., Harper, 1942.
[13] Kettlewell, H. D., 'Selection Experiments on Industrial Melanism in the Lepidoptera', *Heredity* **9** (1955), 323ff.
[14] Maynard Smith, John, *The Theory of Evolution*, Baltimore, Penguin, 1966.
[15] Mayr, Ernst, *Systematics and the Origin of Species*, N.Y., Columbia University Press, 1942.
[16] Medawar, P. B., 'Problems of Adaptation', *New Biology* **2** (1951), 10–25.
[17] Nagel, Ernest, *The Structure of Science*, N.Y.: Harcourt, Brace, & World, 1961.
[18] Pittendrigh, C. S., 'Adaptation, Natural Selection, and Behavior', *Behavior and Evolution* (ed. by Anne Roe and George G. Simpson), New Haven, Yale University Press, 1958, pp. 390–417.
[19] Simpson, George G., *The Major Features of Evolution*, N.Y., Columbia University Press, 1953.
[20] Simpson, George G., *The Meaning of Evolution* (Rev.), New Haven, Yale University Press, 1967.

[21] Simpson, George G. and Beck, W. S., *Life: An Introduction to Biology* (2nd ed.), N.Y., Harcourt, Brace, & World, 1965.
[22] Tinbergen, Niko, *Curious Naturalists*, N.Y., Doubleday, 1968.
[23] Waddington, C. H., 'Evolutionary Adaptation', *Evolution After Darwin*, Vol. II (ed. by Sol Tax), Chicago, University Chicago Press, 1960, pp. 381–402.
[24] Wright, Sewall, 'Adaptation and Selection', *Genetics, Paleontology, and Evolution* (ed. by G. L. Jepsen, Ernst Mayr, and George G. Simpson), Princeton, Princeton University Press, 1949, pp. 365–390.

PART V

SPECIES PROBLEM

ERNST MAYR

SPECIES CONCEPTS AND DEFINITIONS

The importance of one fact of nature is being recognized to an ever increasing extent: that the living world is comprised of more or less distinct entities which we call species. Why are species so important? Not just because they exist in huge numbers, and because each species, when properly studied, turns out to be different from every other, morphologically and in many other respects. Species are important because they represent an important level of integration in living nature. This recognition is fundamental to pure biology, no less than to all subdivisions of applied biology. An inventory of the species of animals and plants of the world is the base line of further research in biology. Whether he realizes it or not, every biologist – even he who works on the molecular level – works with species or parts of species and his findings may be influenced decisively by the choice of a particular species. The communication of his results will depend on the correct identification of the species involved, and thus, on its taxonomy.

Yet, when I was first approached by the Chairman of the Division of Zoology of the American Association for the Advancement of Science to organize a symposium on the species problem I was, to put it mildly, hesitant. Much discussion of this subject in recent years suggested that there was perhaps no need for such a symposium. Ensuing correspondence, however, convinced me otherwise, and certain publications showed clearly that further thinking on this subject is welcome, if not necessary. The species problem continues to be one of the most disputed subjects in biology, in spite of the intense preoccupation with it during the past two hundred years. The recent publications by Spurway, Burma, and Arkell attest this. This symposium can be considered a success if it throws light on some of the disputed questions or even if it does nothing more than lead to a more precise phrasing of the basic points of disagreement.

One way of laying a foundation for such an investigation is to recall some of its history. Who was the first to realize that there is a species problem and what was his proposed solution? What were the subsequent developments? Time does not permit a thorough coverage of the field, but even a glance at the high lights is revealing. If we open a history of biology, the two names mentioned most prominently under the heading of 'Species' will be Linnaeus and Darwin. Linnaeus will be cited as the champion of two characteristics of the species, their constancy and their sharp delimitation (their 'objectivity'). One of the minor tragedies in the history of biology has been the assumption during the hundred and fifty years after Linnaeus that constancy and clear definition of species are strictly correlated and that one must make a choice of either believing in evolution (the 'inconstancy' of species) and then having to deny the existence of species except as purely subjective, arbitrary figments of the imagination, or, as most early naturalists have done, believing in the sharp delimitation of species but thinking that this necessitated denying evolution. We shall leave the conflict at this point and merely anticipate the finding made more than a hundred years after Linnaeus that there is no conflict between the fact of evolution and the fact of the clear delimitation of species in a local fauna or flora.

The insistence of Linnaeus on the reality, objectivity, and constancy of species is of great importance in the history of biology for three reasons. First, it meant the end of the belief in spontaneous generation as far as higher organisms are concerned, a belief which at that time was still widespread. Lord Bacon and nearly all leading writers of the pre-Linnaean period, except Ray, believed in the transmutation of species and the Linnaean conception "of the reality and fixity of species perhaps marks a necessary stage in the progress of scientific inquiry." (See Poulton, 1903, pp. lxxxiv–lxxxvii for further references on the subject). "Until about 1750 almost no one believed that species were stable. Linnaeus had to show that species were not erratic and ephemeral units before organic evolution as we know it could have any meaning." (Conway Zirkle in litt.) The idea that the seed of one plant could occasionally produce an individual of another species was so widespread that it died only slowly. We all know that it raised its ugly head once more during the past ten years. In spite of Redi's and Spallanzani's experiments spontaneous generation was still used in 1851 by the philosopher Schopenhauer as an explanation for the origin of higher categories. Linnaeus thus did for the higher organisms what Pasteur did one hundred years later for the lower.

A second reason why his emphasis was important is that it took the species out of the speculations of the philosophers who approached the species problem in the spirit of metaphysics and stated, for instance, that "only individuals exist. The species of a naturalist is nothing but an illusion." (Robinet, 1768) We shall return later to the point why species are more than merely an aggregate of individuals.

A third reason why the insistence on the sharp delimitation of species in the writings of Linnaeus is of historical importance is that it strengthened the viewpoint of the local naturalist and established the basis for an observational and experimental study of species in local faunas and floras, of which Darwin took full advantage.

Linnaeus was too experienced a botanist to be blind to the evidence of evolutionary change. Greene (1912) gathered numerous citations from his writings which clearly document Linnaeus' belief in the common descent of certain species, and Ramsbottom (1938) and Sirks (1952) have traced how Linnaeus expressed himself more and more freely on the subject, as his prestige grew. Paradoxically, Linnaeus did more, perhaps, to lay a solid foundation for subsequent evolutionary studies by emphasizing the constancy and objectivity of species than if he, like Darwin, had emphasized the opposite.

Darwin looked at the species from a viewpoint almost directly opposite to that of Linnaeus. As a traveler naturalist and particularly because of his studies of domesticated plants and animals he was impressed by the fluidity of the species border and the subjectivity of their delimitation. The views of both Linnaeus and Darwin underwent a change during the life of each. With Linnaeus the statements on the constancy of species became less and less dogmatic through the years. In Darwin, as the idea of evolution became firmly fixed in his mind, so grew his conviction that this should make it impossible to delimit species. He finally regarded species as something purely arbitrary and subjective. "I look at the term species as one arbitrarily given for the sake of convenience to a set of individuals closely resembling each other, and that it does not essentially differ from the term variety which is given to less distinct and more fluctuating forms.... The amount of difference is one very important criterion in settling whether two forms should be ranked as species or variety." And finally he came to the conclusion that "In determining whether a form should be ranked as a species or a variety, the opinion of naturalists having sound judgment and wide experience seems the only guide to follow." (Darwin, 1859) Having thus eliminated the species as a concrete unit of nature, Darwin had also neatly eliminated the problem of the multiplication of species. This explains why he made no effort in his classical work to solve the problem of speciation.

The seventy-five years following the publication of the *Origin of Species* (1859) saw biologists rather clearly divided into two camps, which we might call, in a somewhat oversimplified manner, the followers of Darwin and those of Linnaeus. The followers of Darwin, which included the plant breeders, geneticists, and other experimental biologists minimized the 'reality' or objectivity of species and considered individuals to be the essential units of evolution. Characteristic for this frame of mind is a symposium held in the early Mendelian days, which endorsed unanimously the supremacy of the individual and the nonexistence of species. Statements made at this symposium (Bessey, 1908) include the following: "Nature produces individuals and nothing more.... Species have no actual existence in nature. They are mental concepts and nothing more.... Species have been invented in order that we may refer to great numbers of individuals collectively." Taxonomists, one of the speakers claimed, did not merely name the species found in nature but actually 'made' them. "In making a species the guiding principle must be that it shall be recognizable from its diagnosis." A leftover from this period is the statement of a recent author: "Distinct species must be separable on the basis of ordinary preserved material."

It is a curious paradox in the history of biology that the rediscovery of the Mendelian laws resulted in an even more unrealistic species concept among the experimentalists than had existed previously. They either let species saltate merrily from one to another, as did Bateson and DeVries, defining species merely as morphologically different individuals, or they denied the existence of species altogether except as intergrading populations. Whether these early Mendelians considered species as continuous or discontinuous units, they all agreed in their arbitrariness and artificiality. There is an astonishing absence of any effort in this school to study species in nature, to study natural populations.

A study of natural populations had become the prevailing preoccupation in an entirely independent conceptual stream, that of the naturalists, which ultimately traces back to Linnaeus. The viewpoint of the naturalist was particularly well expressed by Jordan (1905), who stated "The units of which the fauna of a region is composed are separated from each other by gaps which, at a given place, are not bridged by anything. This is a fact which can be checked by any observer. Indeed, the activity of a local naturalist begins with the searching out of these units which with Linnaeus we call species." (For a more detailed discussion see Mayr, 1955.) Although this was the prevailing viewpoint among taxonomists, it was completely ignored by the general biologists by whom, as a result of Darwin's theory, "Species were mostly regarded merely as arbitrary divisions of the continuous and ever changing series of individuals found in nature... of course, active taxonomists did not overlook the existence of sharply and distinctly delimited species in nature – but as the existence of those distinct units disagreed with the prevailing theories, it was mentioned as little as possible." (Du Rietz, 1930) The two streams of thought are still recognizable today even though most geneticists, under the leadership of Dobzhansky, Huxley, Ford, and others, have swung into Jordan's camp. The principal opponents of the concept of objectively delimitable species are today found among philosophers and paleontologists. Publications maintaining this viewpoint are those of Gregg (1950), Burma (1949, 1954), Yapp (1951), and Arkell (1956). These are only the most recent titles in a vast literature, some of which is cited in the bibliography.

The point which is perhaps most impressive when one studies these voluminous publications is the amount of disagreement that has existed and still exists. The number of possible antitheses that have been established in this field may be characterized by such alternate views, to mention only a few, as follows:

Subjective *versus* objective;
Scientific *versus* purely practical;
Degree of difference *versus* degree of distinctness;
Consisting of individuals *versus* consisting of populations;
Only one kind of species *versus* many kinds of species;
To be defined morphologically *versus* to be defined biologically.

To give a well-documented history of the stated controversies would fill a book. As interesting as this chapter in the history of human thought is, the detailed presentation of the gropings and errors of former generations would add little to the task before us. Let us concentrate therefore on the gradual emergence of the ideas which we, today, consider as central and essential. Three aspects are stressed in most modern discussions of species, that (1) they are based on distinctness rather than on difference and are therefore to be defined biologically rather than morphologically, (2) they consist of populations, rather than of unconnected individuals, a point particularly important for the solution of the problem of speciation (3) they are more succinctly defined by isolation from non-conspecific populations than by the relation of conspecific individuals to each other. The crucial species criterion is thus not the fertility of individuals, but rather the reproductive isolation of populations. Let us try to trace the emergence of these and related concepts.

It is not surprising that species were considered merely 'categories of thought' by many writers in periods so strongly dominated by idealistic philosophy as were the eighteenth and nineteenth centuries. Thoughts as that expressed in the above quoted statement of Robinet were echoed by Agassiz, Mivart, and particularly among those paleontologists who considered their task merely the classification of 'objects' (=fossil specimens). In opposition to this, an increasingly strong school developed which considered species as 'definable', 'objective', 'real'. Linnaeus was, of course, the original standard bearer of this school to which also belonged Cuvier, De Candolle, and many taxonomists in the first half of the nineteenth century. They supported their case sometimes by purely morphological arguments such as Godron (1853) who stated: "c'est un fait incontestable que toutes les espèces animales et végétables se séparent les unes des autres par de caractères absolues et tranchées." Others used a more biological argument, as I will discuss below.

What is unexpected for this pre-Darwinian period, however, is the frequency with which 'common descent' is included in species definitions. When such an emphatically anti-evolutionary author as v. Baer (1828) defines the species as "the sum of the individuals that are united by common descent," it becomes evident that he does not refer to evolution. What is really meant is more apparent from Ray's species definition (1686) or a statement by the Swedish botanist Oeder (1764) that it characterizes species "dass sie aus ihres gleichen entsprungen seien und wieder ihres gleichen erzeugen." Expressions like 'community of origin' or 'individus descendants des parents communs' (Cuvier) are frequent in the literature. These are actually attempts at reconciling a typological species concept (with its stress of constancy) with the observed morphological variation.

Constancy was a property of species taken very seriously not only by Linnaeus and his followers but curiously enough also by Lamarck and by Darwin himself: "The power of remaining constant for a good long period I look at as the essence of a species" (letter to Hooker, Oct. 22, 1864). Such constancy in time was the strongest argument in favor of a morphological species concept, but it could be proved only by the comparison of individuals of different generations. Different morphological 'types' that are no more different than mother and daughter or father and son can safely be considered as conspecific. They are 'of the same blood'. It is obvious that this early stress of descent was essentially the consequence of a morphological species concept. Yet this consideration of descent eventually led to a genetic species definition.

Virtually all early species definitions regarded species only as aggregates of individuals, unconnected except by descent, as is evident not only from the writings of Robinet, Buffon, and Lamarck, but also of much more recent authors (e.g., Britton, 1908; Bessey, 1908). The realization that these individuals are held together by a supraindividualistic bond, that they form populations, came only slowly. Illiger (1800) spoke of species as a community of individuals which produce fertile offspring. Brauer (1885) spoke of the "natural tie of blood relationship" through which the "individuals of a species are held together," and which "is not a creation of the human mind... if species were not objective, it would be incomprehensible that even the most similar species mix only exceptionally and the more distant species never." Plate (1914) was apparently the first to state explicitly the nature of this bond: "The members of a species are tied together by the fact that *they recognize each other as belonging together* and reproduce only with each other. The systematic category of the species is therefore entirely independent of the existence of Man." Finally, in the language of current population genetics this community becomes the 'coadapted gene pool', again stressing the integration of the members of the population rather than the aggregation of individuals (a viewpoint which is of course valid only for sexually reproducing organisms).

The growth of thinking in terms of populations went hand in hand with a growing realization that species were less a matter of difference than of distinctness. 'Species' in its earlier typological version meant merely 'kind of'. This, as far as inanimate objects are concerned, is measured in terms of difference. But one cannot apply this same standard to 'kinds of' organisms, because there are various biological 'kinds'. Males and females may be two very different 'kinds' of animals. Jack may be a very different 'kind' of a person from Bill, yet neither 'kind' is a species. Realization of the special aspects of biological variation has led to a restriction in the application of the term species to a very particular 'kind', namely the kind that would interbreed with each other. The first three authors found[1] by me who state this clearly are Voigt (1817), "Man nennt Spezies... was sich fruchtbar mit einander gattet, fortpflanzt"; Oken (1830), "Was sich scharet und paaret, soll zu einer Art gerechnet werden"; and Gloger (1833), "What under natural conditions regularly pairs, always belongs to one species." (He stated that by stressing 'regularly' he wanted to eliminate the complications due to occasional hybridization.) Gloger later (1856) gave a different, but similar definition: "A species is what belongs together either by descent or for the sake of reproduction." It is interesting how completely all these definitions omit any reference to morphological criteria. They are obviously inapplicable to asexually reproducing organisms.

This is an exceedingly short outline of some of the trends in the development of a modern species concept. More extensive treatments can be found in the publications of Geoffroy St. Hilaire (1859), Besnard (1864), de Quatrefages (1892), Bachmann (1905), Plate (1914), Uhlmann (1923), Du Rietz (1930), Kuhn (1948), and other authors cited in the bibliography. Several conclusions are self-evident. One is that biological or so-called modern species criteria were already used by authors who published more than one hundred years ago, long before Darwin. Another is that a steady clarification is evident, yet that there is still much uncertainty and widespread divergence of opinion on many aspects of the species problem. It is rather surprising that not more agreement has been reached during the past two hundred years in which these questions have been tossed back and forth. This certainly cannot be due to lack of trying, for an immense amount of time and thought has been devoted to the subject during this period. One has a feeling that there is a hidden reason for so much disagreement. One has the impression that the students of species are like the three blind men who described the elephant respectively as a rope, a column, or a giant snake when touching its tail, its legs, and its trunk.

Perhaps the disagreement is due to the fact that there is more than one kind of species and that we need a different definition for each of these species. Many attempts have been made during the last hundred years to distinguish these several kinds of species, among the most recent being those of Valentine (1949) and Cain (1953). Camp and Gillis (1943) recognized no less than twelve different kinds of species. Yet, a given species in nature might fit into several of their categories, and in view of this overlap no one has adopted either this elaborate classification or any of the simpler schemes proposed before or afterwards.

SPECIES CONCEPTS

An entirely different approach to the species problem stresses the kaleidoscopic nature of any species and attempts to determine how many different aspects a species has. Depending on the choice of criteria, it leads to a variety of 'species concepts' or 'species definitions'. At one time I listed five species concepts, which I called the practical, morphological, genetic, sterility, and biological (Mayr, 1942). Meglitsch (1954) distinguishes three concepts, the phenotypic, genetic, and phylogenetic, a somewhat more natural arrangement. Two facts emerge from these and other classifications. One is that there is more than one species concept and that it is futile to search for *the* species concept. The second is that there are at least two levels of concepts. Such terms as 'practical', 'sterility', 'genetic' signify concrete aspects of species which lead to what one might call 'applied' species concepts. They specify criteria which can be applied readily to determine the status of discontinuities found in nature. Yet they are secondary, derived concepts, based on underlying philosophical concepts, which might also be called primary or theoretical concepts. I believe that the analysis of the species problem would be considerably advanced, if we could penetrate through such empirical terms as phenotypic, morphological, genetic, phylogenetic, or biological, to the underlying philosophical concepts. A deep, and perhaps widening gulf has existed in recent decades between philosophy and empirical biology. It seems that the species problem is a topic where productive collaboration between the two fields is possible.

An analysis of published species concepts and species definitions indicates that all of them are based on three theoretical concepts, neither more nor less. An understanding of these three philosophical concepts is a prerequisite for all attempts at a practical species definition. And all species criteria or species definitions used by the taxonomist in his practical work trace back ultimately to these basic concepts.

A. *The Typological Species Concept*

This is the simplest and most widely held species concept. Here it merely means 'kind of'. There are languages, as for instance German, where the term for 'kind' (*Art*) is also used for 'species'. A species in this concept is 'a different thing'. This concept is very useful in many branches of science and it is still used by the mineralogist who speaks of 'species of minerals' (Niggli, 1949) or the physicist who speaks of 'nuclear species'. This simple concept of everyday life was incorporated in a more sophisticated manner in the philosophy of Plato. Here, however, the word *eidos* (*species*, in its Latin translation) acquired a double meaning that survives in the two modern words 'species' and 'idea' both of which are derived from it. According to Plato's thinking objects are merely manifestations, 'shadows', of the eidos. By transfer, the individuals of a species, being merely shadows of the same type, do not stand in any special relation to each other, as far as a typologist is concerned. Naturalists of the 'idealistic' school endeavor to penetrate through all the modifications and variations of a species in order to find the 'typical' or 'essential' attributes. Typological thinking finds it easy to reconcile the observed variability of the individuals of a species with the dogma of the constancy of species because the variability does not affect the essence of the eidos, which is absolute and constant. Since the eidos is an abstraction derived from individual sense impressions, and a product of the human mind, according to this school, its members feel justified in regarding a species 'a figment of the imagination', an idea. Variation, under this concept, is merely an imperfect manifestation of the idea implicit in each species. If the degree of variation is too great to be ascribed to the imperfections of our sense organs, more than one eidos must be involved. Thus species status is determined by degrees of morphological difference. The two aspects of the typological species concept, subjectivity and definition by degree of difference, therefore depend on each other and are logical correlates.

The application of the typological species concept to practical taxonomy results in the morphologically defined species, 'degree of morphological difference' is the criterion of species status. Species are defined on the basis of their observable morphological differences. This concept has been carried to the extreme where mathematical formulas were proposed (Ginsburg, 1938) that would permit an unequivocal answer to the question whether or not a population is a different species.

Most systematists found this typological-morphological concept inadequate and have rejected it. Its defenders, however, claim that all taxonomists, when classifying the diversity of nature into species, follow the typological method and distinguish 'arche-types'. At first sight there seems an element of truth in this assertion. When assigning specimens either to one species or to another, the taxonomist bases his decision on a mental image of these species that is the result of past experience with the stated species. The utilization of morphological criteria is valuable and productive in the taxonomic practice. To assume, however, that this validates the typological species concept overlooks a number of important considerations. To begin with, the mental construct of the 'type' is subject to continuous revision under the impact of new information. If it is found that two archetypes represent nothing more than two 'kinds' within a biological species, they are merged into a single one. It was pointed out above that males and females are often exceedingly different 'kinds' of animal. Even more different are in many animals the larval stages, or in plants sporophyte and gametophyte, or in polymorph populations the various genotypes. A strictly morphological-typological concept is inadequate to cope with such intraspecific variation. It is equally incapable of coping with another difficulty, namely an absence of visible morphological differences between natural populations which are nevertheless distinct and reproductively isolated, and therefore to be considered species. The frequent occurrence of such 'cryptic species' or 'sibling species' in nature has been substantiated by various genetic, physiological, or ecological methods. They form another decisive argument against defining species on a primarily morphological basis. Any attempt in these two situations to define species 'by degree of difference' is doomed to failure. Degree of difference can be specified only by a purely arbitrary decision.

More profound than these two essentially practical considerations is the fact that the typological species concept treats species merely as random aggregates of individuals which have the 'essential properties' of the 'type' of the species and 'agree with the diagnosis'. This static concept ignores the fact that species are not merely classes of objects but are composed of natural populations which are integrated by an internal organization and that this organization (based on genetic, ethological, and ecological properties) gives the populations a structure which goes far beyond that of mere aggregates of individuals. Even a house is more than a mere aggregate of bricks or a forest an aggregate of trees. In a species an even greater supraindividualistic cohesion and organization is produced by a number of factors. Species are a reproductive community. The individuals of a species of higher animals recognize each other as potential mates and seek each other for the purpose of reproduction. A multitude of devices insures intraspecific reproduction in all organisms. The species is an ecological unit which, regardless of the individuals of which it is composed, interacts as a unit with other species in the same environment. The species, finally, is a genetic unit consisting of a large, intercommunicating gene pool whereas each individual is only a temporary vessel holding a small portion of this gene pool for a short period of time. These three properties make the species transcend a purely typological interpretation or the concept of a 'class of objects'.

The very fact that a species is a gene pool, with numerous devices facilitating genic intercommunication within and genic separation from without, is responsible for the morphological distinctness of species as a byproduct of their biological uniqueness. The empirical observation that a certain amount of morphological difference between two populations is normally correlated with a given amount of genetic difference is undoubtedly correct. Yet, it must be kept in mind at all times that the biological distinctness is primary and the morphological difference secondary. As long as this is clearly understood, it is legitimate and indeed very helpful to utilize morphological criteria. This caution has been exercised, consciously or unconsciously, by nearly all proponents of the morphological species concept. As pointed out by Simpson (1951) and Meglitsch (1954), they invariably abandon the morphological concept when it comes in conflict with biological data. This was true for Linnaeus himself and for his followers to the present day.

The typological species concept has a certain amount of operational usefulness when applied to inanimate objects. Ignoring the population structure of species, however, and incapable of coping with the facts of biological variation, it has proved singularly inadequate as a conceptual basis in taxonomy. Much of the criticism directed against the taxonomic method was provoked by the application of the typological concept by taxonomists themselves or by other biologists who mistakenly considered it the basis of taxonomy.

B. *The Second Species Concept*

This is sometimes called the nondimensional species concept and has no generally accepted designation. The essence of this concept is the relationship of two coexisting natural populations in a nondimensional system, that is, at a single locality at the same time (sympatric and synchronous). This is the species concept of the local naturalist. It was introduced into the biological literature by the English naturalist John Ray and confirmed by the Swedish naturalist Linnaeus. It is based not on difference but on distinction, and this distinction in turn is characterized by a definite mutual relationship, namely that of reproductive isolation. The word 'species' is here best defined in combination with the word 'different'. The relationship of two 'different species' can be objectively defined as reproductive isolation. We have, thus, an objective yardstick for this species concept, something that is absent in all others. Philosophers have objected to the use of the terms 'objective' or 'real' for species, and it may be more neutral to use the terms arbitrary or nonarbitrary (Simpson, 1951). Presence or absence of interbreeding of two populations in a nondimensional system is a completely nonarbitrary criterion. Since the nondimensional species concept is based on a relationship, the word species is here equivalent to words like, let us say, the word brother, which also has a meaning only with respect to a second phenomenon. An individual is a brother only with respect to someone else. Being a brother is not an inherent property as hardness is a property of a stone. Describing a presence or absence relationship makes this species concept nonarbitrary.

This species concept seems so self-evident to every naturalist that it is only rarely put in words. That the species is more than an aggregate of individuals, held together by a biological bond, has long been realized, as was pointed out in the historical survey above. The interbreeding within the species is more conspicuous, and it was thus more often emphasized than is the reproductive isolation against other species. Eimer, as early as 1889 (p. 16) defined species as "groups of individuals which are so modified that successful interbreeding [with other groups] is no longer possible." The first author, however, who stated the nondimensional species concept in its full extent and implication was Jordan (1905).

In spite of its theoretical superiority, the nondimensional species has a number of serious drawbacks (which will be discussed later), particularly its limitation to sexually reproducing species and to such without the dimensions of space and time. Yet, as a basic, nonarbitrary yardstick, this is the species concept on which we have to fall back whenever we encounter a borderline situation.

C. *The Third Species Concept*

This is a concept of an entirely different kind, it is the concept of the polytypic or multidimensional species. In contradistinction to the other two concepts, of which one is based on a degree of difference, the second one on the completeness of a discontinuity, this concept is a collective one. It considers species as groups of populations, namely such groups as interbreed with each other, actually or potentially. Thus this species concept is a concept of the same sort as the higher categories, genus, family, or order. Like all collective categories it faces the difficulty, if not impossibility, of clear demarcation against other similar groupings. What this species gains in actuality by the extension of the nondimensional situations in space and time, it loses in objectivity. As unfortunate as this is, it is inevitable since the natural populations, encountered by the biologist, are distributed in space and time and cannot be divorced from these dimensions. Thus, this species concept likewise has its good and its bad points.

SPECIES DEFINITIONS

All our reasoning in discussions of 'the species' can be traced back to the stated three primary concepts. As concepts, of course, they cannot be observed directly, and we refer to certain observed phenomena in nature as 'species', because they conform in their attributes to one of these concepts or to a mixture of several concepts. From these primary concepts, just discussed, we come thus to secondary concepts, based on particular aspects of species. We have already mentioned the so-called morphological species concept, which, in most cases, is merely an applied typological concept, using morphological criteria. The case of the so-called genetic species concept shows that all three of the basic concepts can be expressed, on this level, in genetic terms. Some geneticists, for instance, subscribed to the typological concept and defined species by the degree of genetic difference as did Lotsy or DeVries; others stressed the genetic basis of the isolating mechanisms between species thereby endorsing the nondimensional species concept; still others finally emphasized the gene flow among interbreeding populations in a multidimensional system, thus adopting the multidimensional collective species concept. All three groups of geneticists thought they were dealing with a uniquely 'genetic species concept', yet they were merely observing secondary manifestations of the primary concepts.

It is evident from the analysis of the morphological and genetic species concepts, that such derived concepts are attempts to deal directly with the discontinuities in nature. In the past, almost every taxonomist worked with his own personal yardstick based on a highly individual mixture of elements from the three basic concepts. As a consequence one taxonomist might call species every polymorph variant, a second one every morphologically different population, and a third one every geographically isolated population. Such lack of standards, which is still largely characteristic for the taxonomic literature, has been utterly confusing to taxonomists and other biologists alike. It has therefore been the endeavor of many specialists within recent decades to find a standard yardstick, on which there could be general agreement. A historical study of species definitions indicates clearly a trend toward acceptance of a synthetic species definition, often referred to as 'biological species' definition. It is essentially based on the nondimensional ('reproductive gap') and the multidimensional ('gene flow') species concepts. Nearly all species definitions proposed within the last fifty years incorporate some elements of these two concepts. This is evident from the species definitions of Jordan (Mayr, 1955), Stresemann (1919), and Rensch (1929). Du Rietz (1930) called the species "a syngameon... separated from all others by... sexual isolation." Dobzhansky (1935) was apparently the first geneticist to define species in the terms customary among naturalists and taxonomists, namely interbreeding and reproductive isolation; other recent definitions are variants of the same theme. Mayr (1940) defined species as "groups of actually or potentially interbreeding natural populations which are reproductively isolated from other such groups." Simpson (1943) gave the definition "a genetic species is a group of organisms so constituted and so situated in nature that a hereditary character of any one of these organisms may be transmitted to a descendent of any other," and Dobzhansky (1950) defined the species as "the largest and most inclusive... reproductive community of sexual and cross-fertilizing individuals which share in a common gene pool."

It might be useful to mention some qualifications which are often included in species definitions but needlessly so. Anything that is equally true for categories above and below species rank should be omitted, since there is no sense burdening a species definition with features which do not help discrimination between species and infraspecific populations.

(1) Species characters are adaptive. This component of Wallace's (1889) species definition was correctly rejected by Jordan (1896). Adaptiveness is not diagnostic for species characters and not even necessarily true. Not every detail of the phenotype needs to be adaptive as long as the phenotype as a whole is adaptive and the genotype itself is the result of selection.

(2) Species are evolved and evolving. Again this is true for the entire organic world from the individual to the highest categories and adds nothing to the species definition.

(3) Species differ genetically. This is only the morphological species concept expressed in genetic terms. It does not permit discriminating species from infraspecific populations or from individuals.

(4) Species differ ecologically. This qualification is unnecessary and misleading for the same reasons as the genetic one. Ecological differences exist for all ecotypes within species and in general for all geographical isolates. Conspecific populations are sometimes more different ecologically than are good species.

A yardstick, such as the biological species concept, is not automatic. To apply it properly requires skill and experience. This is particularly true in the recognition of situations where it cannot be applied, for one reason or another, and where the worker has to fall back on the criterion of 'degree of difference'. It will be one of the tasks of this symposium to investigate to what extent the diversity of animal and plant life permits application of the standard yardstick of the biological species concept. When can it not be applied and for what reasons? What types of difficulties are there? And finally, are there perhaps other basic concepts in addition to the stated ones? By approaching the species problem with new questions and new material, perhaps this symposium can make a contribution to its solution.

Harvard University

NOTE

[1] Still earlier statements can no doubt be found in the extensive literature, particularly on hybridization.

BIBLIOGRAPHY

Arkell, W. J.: 1956, 'The Species Concept in Paleontology', *Systematics Assoc. Publ. No.* 2, pp. 97–99.

Bachmann, H.: 1905, 'Der Speziesbegriff', *Verhandl. schweiz. naturforsch. Ges.* **87**, 161–208.

Baer, K. E. von: 1828, *Entwickelungs–Geschichte der Thiere, Königsberg.*

Besnard, A. F.: 1864, 'Altes und Neues zur Lehre über die organische Art (Spezies), *Abhandl. zool. mineral. Ver. Regensburg* **9**, 1–72.

Bessey, C. E.: 1908, 'The Taxonomic Aspect of the Species Question', *Am. Naturalist* **42**, 218–24.

Brauer, F.: 1885, 'Systematisch-zoologische Studien', *Sitzber. Akad. Wiss. Wien* **91** (Abt. 1), 237–413.

Britton, N. L.: 1908, 'The Taxonomic Aspect of the Species Question', *Am. Naturalist*, **42**, 225–42.

Burma, B. H.: 1949a, 'The Species Concept: A Semantic Review', *Evolution* **3**, 369–70.

Burma, B. H.: 1949b, 'The Species Concept: Postscriptum', *Evolution* **3**, 372–73.

Burma, B. H.: 1954, 'Reality, Existence, and Classification: A Discussion of the Species Problem', *Madroño* **7**, 193–209.

Cain, A. J.: 1953, 'Geography, Ecology and Coexistence in Relation to the Biological Definition of the Species', *Evolution* **7**, 76–83.

Camp, W. H. and Gillis, C. L.: 1943, 'The Structure and Origin of Species', *Brittonia* **4**, 323–85.

Darwin, C.: 1859, *On the Origin of the Species by Means of Natural Selection*, London.

Dobzhansky, T.: 1935, 'A Critique of the Species Concept in Biology', *Phil. Sci.* **2**, 344–55.

Dobzhansky, T.: 1950, 'Mendelian Populations and Their Evolution', *Am. Naturalist* **84**, 401–18.

Doederlein, L.: 1902, 'Über die Beziehungen nahe verwandter "Thierformen" zu einander', *Z. Morphol. Anthropol.* **26**, 23–51.

Dougherty, E. C.: 1955, 'Comparative Evolution and the Origin of Sexuality', *Systematic Zool.* **4**, 145–69.

Du Rietz, G. E.: 1930, 'The Fundamental Units of Botanical Taxonomy', *Svensk. Bot. Tidsskr.* **24**, 333–428.

Eimer, G. H. T.: 1889, *Artbildung und Verwandtschaft bei Schmetterlingen*, Jena, Vol. II, p. 16.

Geoffroy Saint Hilaire, I.: 1859, *Histoire naturelle générale des règnes organiques* **2**, 437, Paris.

Ginsburg, I.: 1938, 'Arithmetical Definition of the Species, Subspecies and Race Concept, with a Proposal for a Modified Nomenclature', *Zoologica* **23**, 253-86.

Gloger, C. L.: 1833, *Das Abändern der Vögel durch Einfluss des Klimas*, Breslau.

Gloger, C. L.: 1856, 'Ueber den Begriff von "Art" ("Species") und was in dieselbe hinein gehört', *J. Ornithol.* **4**, 260–70.

Godron, D. A.: 1853, *De l'espèce et des races dans les êtres organisés et spécialement de l'unité de l'espèce humaine*, Paris, 2 vols.

Greene, E. L.: 1912, 'Linnaeus as an Evolutionist', in *Carolus Linnaeus*, pp. 73–91, C. Sower & Co., Philadelphia, Pa.
Gregg, J. R.: 1950, 'Taxonomy, Language and Reality', *Am. Naturalist* **84**, 419–35.
Huxley, J.: 1942, *Evolution, the Modern Synthesis*, Allen and Unwin, London.
Illiger, J. C. W.: 1800, *Versuch einer systematischen vollständigen Terminologie für das Thierreich und Pflanzenreich*, Helmstedt.
Jordan, K.: 1896, 'On Mechanical Selection and Other Problems', *Novit. Zool.* **3** 426–525.
Jordan, K.: 1905, 'Der Gegensatz zwischen geographischer und nichtgeographischer Variation', *Z. wiss. Zool.* **83**, 151–210.
Kuhn, E.: 1948, 'Der Artbegriff in der Paläontologie', *Eclogae Geolog. Helv.* **41**, 389–421.
Lorkovicz, Z.: 1953, 'Spezifische, semispezifische und rassische Differenzierung bei *Erebia tyndarus* Esp.', *Rad. Acad. Yougoslave* **294**, 315–58.
Mayr, E.: 1940, 'Speciation Phenomena in Birds', *Am. Naturalist* **74**, 249–78.
Mayr, E.: 1942, *Systematics and the Origin of Species*, Columbia University Press, New York, N.Y.
Mayr, E.: 1949, 'The Species Concept: Semantics versus Semantics', *Evolution* **3**, 371–72.
Mayr, E.: 1951, 'Concepts of Classification and Nomenclature in Higher Organisms and Microorganisms', *Ann. N.Y. Acad. Sci.* **56**, 391–97.
Mayr, E.: 1955, 'Karl Jordan's Contribution to Current Concepts in Systematics and Evolution', *Trans. Roy. Entomol. Soc. London*, 45–66.
Mayr, E.: 1956, 'Geographical Character Gradients and Climatic Adaptation' *Evolution* **10**, 105–8.
Mayr, E. and Rosen, C.: 1956, 'Geographic Variation and Hybridization in Populations of Bahama Snails (*Cerion*)', *Am. Museum Novit.* **1806**, 1–48.
Mayr, E., Linsley, E. G., and Usinger, R. L.: 1953, *Methods and Principles of Systematic Zoology*, McGraw-Hill Book Co., New York, N.Y.
Meglitsch, P. A.: 1954, 'On the Nature of the Species', *Systematic Zool.* **3**, 49–65.
Niggli, P.: 1949, *Probleme der Naturwissenschaften* (Der Begriff der Art in der Mineralogie), Basel.
Plate, L.: 1914, 'Prinzipien der Systematik mit besonderer Berücksichtigung des Systems der Tiere', in *Die Kultur der Gegenwart*, III (iv, 4), pp. 92–164.
Poulton, E. B.: 1903, 'What is a Species?', *Proc. Entomol. Soc. London* for 1903, lxxvii–cxvi.
Quatrefages, A. de: 1892, *Darwin et les précurseurs français*, Paris.
Ramsbottom, J.: 1938, 'Linnaeus and the Species Concept', *Proc. Linnean Soc. London* (150 session), 192–219.
Ray, J.: 1686, *Historia Plantarum*, p. 40.
Rensch, B.: 1929, *Das Prinzip geographischer Rassenkreise und das Problem der Artbildung*, Bornträger Verl., Berlin.
Schopenhauer, A.: 1851, *Parerga und Paralipomena: kleine philosophische Schriften*, Vol. 2, pp. 121–22, Berlin.
Simpson, G. G.: 1943, 'Criteria for Genera, Species, and Subspecies in Zoology and Paleozoology', *Ann. N.Y. Acad. Sci.* **44**, 145–78.
Simpson, G. G.: 1951, 'The Species Concept', *Evolution* **5**, 285–98.
Sirks, M. J.: 1952, 'Variability in the Concept of Species', *Acta Biotheoretica* **10**, 11–22.
Spring, A. F.: 1838, *Ueber die naturhistorischen Begriffe von Gattung, Art und Abart*

und über die Ursachen der Abartungen in den organischen Reichen, Leipzig.
Spurway, H.: 1955, 'The Sub-Human Capacities for Species Recognition and Their Correlation with Reproductive Isolation', *Acta XI Congr. Intern. Orn.*, Basel, 1954, pp. 340–49.
Stresemann, E.: 1919, 'Über die europäischen Baumläufer', *Verhandl. Orn. Ges. Bayern* **14**, 39–74.
Sylvester-Bradley, P. C.: 1956, 'The Species Concept in Paleontology. Introduction', *Systematics Assoc. Publ.*, No. 2.
Thomas, G.: 1956, 'The Species Concept in Paleontology', *Systematics Assoc. Publ.*, No. 2, pp. 17–31.
Uhlmann, E.: 1923, 'Entwicklungsgedanke und Artbegriff in ihrer geschichtlichen Entstehung und sachlichen Beziehung', *Jena. Z. Naturw.* **59**, 1–114.
Valentine, D. H.: 1949, 'The Units of Experimental Taxonomy', *Acta Biotheoretica* **9**, 75–88.
Voigt, F. S.: 1817, *Grundzüge einer Naturgeschichte als Geschichte der Entstehung und weiteren Ausbildung der Naturkörper*, Frankfurt a.M.
Wallace, A. R.: 1889, *Darwinism: An Exposition of the Theory of Natural Selection, with Some of Its Applications*, London.
Yapp, W. B.: 1951, 'Definitions in Biology', *Nature* **167**, 160.

VERNON PRATT

BIOLOGICAL CLASSIFICATION*

INTRODUCTION

From time to time biologists profess themselves to be searching for what they call a 'Natural Classification', and in what follows I want to propose an answer to the question of what it is they are looking for.

It may be helpful at the outset to indicate intuitively the sort of notion of which we shall be seeking a rigorous or philosophical understanding.

If we take a group of objects, say the plants growing in a garden, we can divide them into groups in innumerable different ways. We could, for example, classify them according to whether or not their stems were over a foot in length. Or we could group them according to their smell, putting plants with attractive smells in one group and those with unpleasant smells or none at all in a second. Or we might set up just twelve categories, and ascribe individuals to them on the basis of their month of flowering.

It might be said, however, that in following any of these procedures we would often be involved in separating individuals which on the basis of untutored observation seemed very similar. On the first criterion, for example, what according to orthodox modern taxonomy we should call two individuals belonging to the same species (say, the White Dead-Nettle) might be allotted to different groups since only their height is taken into consideration; and observation, even unprejudiced by orthodox taxonomy, would suggest strongly that two such individuals were very like each other, in spite of the difference in height. When the same unprejudiced eye is cast over the rest of the individuals belonging to the two groups thus created, it is likely to report that not only are individuals apparently very similar to each other being separated, but some individuals which seem on every count except height to differ radically from each other are being grouped together: the apple tree goes along with the rambling rose, the Shepherd's Purse with the Sphagnum moss (using for the sake of reference the categories of orthodox taxonomy).

Both of the other two criteria we suggested would give rise to the same feeling: the feeling perhaps we might say of *arbitrariness*. In using these criteria we would seem to be grouping the individuals without any real regard for the resemblances, relations, and differences which exist among the plants themselves: we are, it could be said, classifying them *artificially*. Such an expression holds out the hope that it might be possible, as an alternative, to classify these individuals (and organisms in general) non-artificially or *naturally*. Taxonomists (botanical, in the following quotation) would then be looking, in the words of William Whewell, for something 'not of their own creation; – not anything merely conventional or systematic; but something which they conceive to exist in the relations of the plants themselves; – something which is without the mind, not within; – in nature, not in art; – in short a natural order' (Whewell, 1840, volume 1, p. 474). Pierre Duhem expresses the same intuition, this time as regards the zoologist. In looking for a natural classification he would be looking, Duhem says, for a system whose relations 'correspond to *real* relations among the associated creatures brought together and embodied in his abstractions.' (Duhem, 1906, p. 25, my italics).[1]

So much, then, by way of indicating (simply by an appeal to intuition) that there is a distinction between artificial and natural classification. But is our intuition correct? Is there such a thing as a natural as opposed to an artificial way of classifying organisms? If there is, what are the grounds upon which the distinction is made?

I. SOME QUESTIONS DISTINGUISHED

Several questions arise, connected though distinct; and an understanding of the theorising about classification that has gone on (largely in prolegomena to works devoted to substantive taxonomy) depends in large part on distinguishing between them. Most classifications, and all the biological classifications that have seriously been canvassed, are based on resemblances and differences between things. Things that resemble each other are placed together, and separated from other things from which they differ. But this is an almost completely formal statement, because the notion of two things *resembling* each other remains empty until the respect or respects of resemblance are indicated. With dubious exceptions the members of any pair of things resemble, and differ from, each other in an indefinite number of ways.[2] "Resemblance" as Renford Bambrough says, "is unintelligible except as resemblance *in a respect.*" (Bambrough, 1961, p. 204).[3] In defining a scheme of classification therefore, whether a new one is being proposed or an attempt made to provide a rationale for an existing one, we must have an answer to the question: What are the respects (or what is the respect) in which we are to compare things in order to apply to them the classification in question? We might express this as the question, What is the classification's basis? – or, What is the principle of unity embodied in its groupings?

One recommendation that has often been made is that we should look to the evolutionary relationships that hold between organisms and classify them on that basis. Thus if two species are alike in springing from a certain ancestor-group, and a third originates from a different source, that might be a reason for classifying the first two together and separately from the third. This would be, however, just one possibility, for there is more than one way of classifying organisms phylogenetically and different people have recommended different ways (Section IIB). For a full statement of the respects in which organisms were to be compared we should have to go into each of the varieties. Nevertheless, 'evolutionary relationships' constitutes a distinctive outline answer to the question (let us call it the first) we have just distinguished. 'In respect of their reproductive apparatus' is a second, which has also been widely influential in the past; and 'In respect of their genotypes' a third, for obvious reasons of more recent importance.

I stress that all these answers give an outline indication only, and we shall explore two of them in a little more detail later (Section II). But at present we are simply concerned to distinguish as sharply as possible between the different questions which a complete rationale of a classification must answer.

A second question asks what job or jobs a classification of living things is expected to help with. Some have thought important the rather humble job of aiding the memory (e.g. E. Pigeon on his preface to Cuvier's *Le Régne Animal* (Cuvier, 1817)), and something of the same idea may be at the bottom of the more modern-sounding suggestion that classification should above all provide an instrument for storing biological information in such a way as to make it easily accessible (Mayr, 1969, p. 78; Heslop-Harrison, 1962, p. 14.) Another closely related idea is that it should enable us to express our biological knowledge in the most economical way possible. Cuvier says, for example, that a soundly based system of classification will provide the means of expressing the properties of the individuals so classified in the fewest, most comprehensive, most significant terms. Mill thought a classification ought to serve the heuristic function of generating hypotheses (Mill, 1843, book 4, chapter 7); Mayr is one who defends the further idea that the hypotheses generated should have a good likelihood of being true, so that one of the functions of a classification on his view is that of generating reliable predictions (Mayr, 1969, pp. 79–80. See also Hempel, 1965, p. 146.)

There is no *a priori* reason why the performance of one of these jobs should preclude the performance of any of the others; and one might easily think that the more jobs a classification does, the better it is. Indeed, the philosopher J. S. L. Gilmour, for example (1951, p. 401), has suggested that it is the capacity of a classification to serve a number of different functions which makes it 'natural' as opposed to artificial – although his suggestion is I think misconceived since it implies that functions can be counted independently of being specified (cf. the parallel point about the notion of a character, Section IIC)[4]; as I hope to make plain as we proceed we are likely to achieve a better understanding of the natural/artificial distinction if we wonder *why* some classifications are more generally useful than others.

The two questions so far distinguished clearly bear on each other. It might well seem that the logical procedure (if one were to begin classification *ab initio*) would be to decide what functions the proposed classification would be required to fulfil, and then to choose the principle of unity likely to create the classification best suited to those ends. But such a possibility immediately points to a third question: How is it that employing a particular principle of unity leads to a classification which serves its intended functions? For example, Why should it be that a phylogenetic basis for one's classification produced the most efficient device possible for storing retrievable biological information? (Mayr, 1969, p. 78.) Or, what is the mechanism which secures that groups set up on a genotypical basis are helpful in deriving reliable predictions?

Depending on the answers given to the other questions this third one may or may not arise. We do not, for example, need to have an explanation of how a classification based on phylogeny provides us with a means of identification: any classification worth the name would surely do that.

But in other cases it is not obvious why the suggested basis should yield a classification capable of doing the job assigned to it, as in the two examples just mentioned.

Also, the kind of answer required by our third question may differ from case to case. Sometimes, what needs to be shown is a logical connection between the principle of unity and the capacity of a classification to perform a certain function. Some theorists for example have urged that an important task of classification is the provision of a framework which will allow biological knowledge to be expressed 'economically' – and the greater the economy, the better the classification. When some of these theorists also suggest that 'overall resemblance' should be the basis of classification (see Section IIC) it may be that they have in mind a connection of this logical kind between economy of expression and the basis they suggest. On the other hand, our third question often demands factual information. In relating a phylogenetically based classification to its ability to perform the function of generating reliable predictions, for example, reference will very likely be made to our knowledge of evolution and genetics.

The practising taxonomist will however be interested primarily in a question that has still to be raised: How are taxonomically important characters to be picked out from those that are unimportant? If we are trying to group organisms phylogenetically, for example, which of their characteristics can we regard as indicative of evolutionary relationships, and which can we ignore as unhelpful? For the practical task of classifying we shall need to know some principles whereby this distinction may be made.

Once again, it seems (at first sight anyway) that for some principles of unity this fourth question does not arise. It has been held, for example, that in grouping things according to 'overall similarity' it is possible, and greatly desirable, not only to take every character into account, but to regard each as having the same taxonomic importance as any of the others. If such an idea were a coherent one (though I shall argue that it is not) it clearly rules any criteria of taxonomic importance out of court. But equally clearly for some principles of unity, our fourth question is legitimate and must be answered.

II. SOME THEORIES OF BIOLOGICAL CLASSIFICATION

We turn now to consider three (and a half) 'theories' of biological classification as a preliminary to presenting our own. None of these theories is silly, and they all – even the theory of archetypes which we are to discuss first and which is often represented as a pre-scientific phantasy – contribute substantially to the fuller understanding that now, I believe, is possible. In many ways that fuller understanding comes of seeing how these several insights can be synthesised into a single account.

A. *The Archetype Theory*

It is assumed by the theory of archetypes that we have to regard the living organism as constructed, or derived from ancestors which were constructed, according to a 'blueprint' or plan – or, to use its own word, an 'archetype'; and it holds accordingly, that we ought, in classifying organisms, to form our groups according to the similarities the blueprints associated with the organisms bear to each other.

To begin with, the taxonomist will form a series of groups each consisting of organisms 'built' according to the very same plan as each other: we may think of this as constituting classification at the 'species' level. But he will find that some plans, while not being identical in detail with certain others, are yet identical with them at a certain level of generality. So arises the possibility of bringing together the first order groups into higher level categories, where the plans associated with each member group share a common general outline. Parallel considerations lead us to admit the possibility of forming third order groups, and fourth, and so on: the groups of a given order uniting some of the next lowest rank in virtue of the fact that their plans, at a certain level of generality, possess a common form. Thus would be generated the hierarchy of categories we are familiar with in orthodox classification – individuals being united into species, species into genera, genera into families, and so on.

How, on the archetype approach, is the practising taxonomist to know which characters to regard as taxonomically significant, as reliable indicators of archetype relationships? The answer of one such theorist, Cuvier, is instructive. First, he suggests, we should regard as important the characters which heavily influence the rest of the organism – those which are 'most essential' to its life-form. As Cuvier himself puts it: 'The parts, properties, or traits of conformation which ... exert the most marked influence over the ensemble of the entity, are those which are called the *important characters*, the *dominant characters*' (cf. Cain, 1959, p. 188).

This suggestion clearly stems from the emphasis Cuvier places throughout his zoological work on regarding the organism as a 'whole' made up of complexly interacting parts. You knew, Cuvier thought, that if an organism possessed a certain kind of respiratory system, for example, the rest of its internal organisation must fit in with this: the various component parts which together made up the whole system which is the organism must be correlated with each other. Some features would clearly have wider implications than others: the colour of an organism would not have the profound effect on its total organisation as would, say, its circulatory system. It is those with the larger implications which Cuvier, reasonably enough, is suggesting are the best guide to what perhaps we may call, rather tendentiously, an organism's 'fundamental organisation' – its archetype.

Cuvier's second criterion is less interesting. We should consider those characters as most important, he says, which are most constant throughout the group, so that the circulatory system must be regarded as significant when we are considering mammals, since although other features vary from mammal to mammal, the circulatory system remains the same. But as Cain argues (1959), when directed at taxonomists attempting to set up groups of organisms, this second suggestion is patently circular. If our project is the selection of characters to use in establishing groups, we cannot appeal to *constancy* – making reference as it does to groups which *ex hypothesi* have not yet been set up.

Cain argues that Cuvier's first criterion is circular, too. Most of the physiological theory which provided the means of assessing the influence a particular feature exercised on the functioning of the whole was itself based, Cain tells us, on observing which features were constant within the already recognised groups. Be that as it may, modern physiology is differently founded, and I believe an adequate theory of classification must take Cuvier's first criterion very seriously. The insight on which it is based had been vindicated by subsequent discoveries, particularly in genetics, and forms an essential element of the view we shall shortly develop.

B. *The Phylogenetic Theory*

Those who take the evolutionary or phylogenetic view argue that our grouping of organisms should reflect the attempt to set forth the groups and relationships between groups which have been produced by the process of evolution. This general programme can however be realised in a number of different ways: it is in fact inadequate to speak of *the* groups produced by evolution, for at least at levels higher than the species the workings of evolution can be thought of as producing different groupings depending on which aspect of those workings is stressed.

If you have, for example, three groups, *A*, *B*, and *C*, two different questions about their evolutionary relationships arise. The first concerns how long ago and by what route they each arose from their common ancestor-group. If we classify simply on the basis of this consideration we might put *A* and *B* together and separated from *C* on the grounds that the ancestral lineage gave off *C* a long time ago, and then subsequently split, to form *A* and *B*.

A second question arises, however. Because the rate of evolutionary change can vary widely in different lines, two groups with a common origin may be genetically and phenetically either closely similar or widely dissimilar, depending on the degree of divergence that has occurred. If we suppose *B* in our previous example to have diverged rapidly from *A*, which, let us suppose, has evolved in parallel with *C*, there would be a case for classing *A* and *C* together and apart from *B*.

The first scheme, which only takes into account the order in which the groups in question diverged from a common ancestor-group, is most conveniently called 'cladism' (Mayr, 1969, pp. 70ff.); and its results will be quite different from those which ensue if you take the relative divergence of groups into the reckoning.

The important question thus arises: What criterion or criteria can be appealed to in order to choose between these alternative schemes? Our own remarks would suggest that we should choose in the light of the job or jobs we want our classification to assist us with. Ernst Mayr, as a defender of the second alternative against the 'cladism' of W. Hennig and his followers, seems to adopt this approach. He argues that the ends of classification will be better served if the groups one sets up are a true reflection of the relationships that exist among the genetic programmes, or genotypes, of the organisms concerned: "When we classify organisms, classification by phenotype is only the first step. As the second step we attempt to infer the genotype, the evolved genetic programme, which has a far greater explanatory and predictive value than the phenotype" (Mayr, 1969, p. 77).

But this crucial reference to the genotype – and it is made throughout his book – makes one ask whether Mayr is a true 'phylogeneticist' in classificatory theory at all. It shows that it is not fully revealing to say that for Mayr the principle of unity in classification is simply evolutionary relationships. He wishes rather to erect groupings on the basis of evolutionary relationships *in so far as these are correlated with resemblances and differences between genotypes.* As it happens, evolution has produced groupings of genetically similar organisms; but if it hadn't, we would in classifying ignore it and concentrate on genetic relationships directly. As Mayr himself puts it: "Evolutionism, as a philosophical basis for classification, is a valid approach only if and when natural groups of organisms are the result of divergent evolution. If reticulate evolution were common owing to the frequent fusion of previously separated evolutionary lines, or if convergence were frequently so complete as to lead to groupings that could not be unmasked as having a polyphyletic origin, then the claims of evolutionism as a proper theoretical basis for classification would indeed be questionable" (Mayr, 1969, pp. 78–79).

We could put this equally well, I suggest, by saying that evolutionary relationships are not, for Mayr, the ultimate principle of unity which biological classification should employ: rather he is suggesting it should classify according to genetic similarities, using evolutionary relationships as simply an extremely useful guide. It is when we begin to answer our fourth question, about the characters we are to take as taxonomically significant if our object is to classify according to genetic similarity, that evolution comes in. Because by and large evolution has produced groups of genetically similar individuals, knowing in some degree how evolution works will assist us in discerning the aspects of the phenotype – that is to say, roughly, the observable characters of an organism – which serve as the most reliable indicators of the genotype.

The 'phylogenetic' view therefore as we see it developed in one of its most powerful advocates is not quite what the term might suggest on first acquaintance. And in going on now to consider a different kind of approach to classification theory that has been canvassed in recent years we shall find there, most interestingly, an exactly parallel state of affairs: a theory which despite every first appearance makes an appeal to genotypical considerations for its ultimate rationale.

C. *'Overall Resemblance' Theories*

Critics of the idea that evolutionary relationships should form the basis of classification have usually urged, as their principal objection, that we possess too little information to enable us to classify organisms in this way. Partly because our factual knowledge is at present so incomplete, but partly also because of the phenomena of homoplasy (convergent evolution, parallelism, mimicry *etc.* – see Simpson, 1961, p. 78) it is all too frequently impossible to discern the precise evolutionary relationship that obtains between organisms. To recommend basing classification on phylogeny is therefore, it is claimed, to suggest founding it upon ignorance. (Sokal and Sneath, 1963, §2.3; Cain and Harrison, 1960; Thompson, 1952, *etc.*) So the critics argue. Instead it is proposed that classification should be based on 'the resemblances existing *now* in the material at hand' (Sokal and Sneath, 1963, p. 55), that is, on 'a general comparison' between organisms which takes into account 'every phenotypical character'. We are to compare organisms as a whole with each other and classify them on the basis of the 'overall similarity' we will thus discern.

Some theorists who recommend this view see it as contrasting favourably with the principles of some classifications in the past which, based on comparisons involving just one or a few characters only, were for that reason to be regarded as 'artificial'. A. J. Cain declares, in a chapter not entirely consistent with itself, that the only difference between a natural and an artificial system is that more characters are taken into account by the former than by the latter: "Every possible intermediate between an obviously artificial and an obviously natural classification can exist. All depends on the number of characters taken into account" (Cain, 1954, p. 18).

We may notice the strangeness of Cain's view here. His notion of 'natural' does not, I think, square with other people's idea of it, either historically or in our own time; nor does it mark a distinction which promises to be of much use. Surely much more important for the nature and value of a classification than the number of characters that have to be taken into consideration in applying it is the criterion or criteria by reference to which the 'consideration' is conducted. I might, to take an instance, try to classify plants into those that would be suitable to give as gifts for Christmas and those which would be inappropriate for this purpose; and in asking my question Is this plant suitable? I might try and take every feature into consideration. Cain would have to call this a natural classification; but, according to *my* reading of biologists' discussions (I must simply leave this for others to judge – a list of references would have to be too long to prove my point) I don't think many people would agree with him. And would there be much point in pressing the word 'natural' to this unaccustomed use? I cannot myself see the value of making the distinction Cain wants the word to mark.

Cain's view of 'naturalness' (as expressed in the quoted sentences) is not alone in feeling the prick of the point I raise here however: it is seriously damaging to the concept of classification on the basis of 'overall similarity' in its entirety.

Let us try to imagine a taxonomist attempting to classify according to 'overall similarity'. He will begin, we may think, by comparing each individual with each other individual in turn and drawing up a table showing the characters each such pair of individuals shares. But once this initial task is completed (our imagination has to transcend the limitations of logic here, as I shall argue later) he must go on to interpret his findings. How are his data to be processed in order to result in the required classification?

Two possibilities seem open. The first, and one might think the most simple-minded, is to add up the total number of similarities between the members of the pairings and then to classify on the basis of numbers of shared characters. Individuals with many characteristics in common would thus be classified together and away from individuals with which they shared few common characters. We may call this the numerical approach.[5]

Contrasted with it is the probably more intuitively appealing alternative of regarding some characters as more important, taxonomically, than others. Individuals would then be grouped together if they resembled each other in *important* ways, and away from those with which they had little *important* in common. Those who take this second alternative, however, must have an answer to the question: By what criterion or criteria is the importance of a character to be judged? For in rejecting the notion that all characters are to count equally they must be appealing, tacitly or explicitly, to some principle whereby differential weight is to be attached to them.

At this point, this second alternative of the proposal to classify on the basis of 'overall similarity' collapses into a type of proposal we have already considered. Cain, to revert to the chapter we referred to above, suggests that characters are taxonomically important insofar as they reflect the 'basic theme' of the organism, or its 'general plan' (Cain, 1954, p. 19), which makes him (at least in the chapter in question), a contemporary exponent of an archetype theory.[6] Others have held that the taxonomic importance of a character hangs on whether it offers a reliable indication of evolutionary relationships: clearly this is the phylogenetic view already discussed. And still others have proposed that it is its power to indicate the genotype of the organism that confers taxonomic significance upon a character.

Thus the non-numerical version of the 'overall similarity' view is not to be distinguished as an independent theory at all. It quite fails to provide a distinctive answer to the first of our analytic questions (What is the classification's basis?): the true basis of a proposed classification, be it phylogenetic, or genotypical, or archetypal, or something else, will only be indicated by citing the principle of weighting to be employed.

Is there anything better to be said for the numerical version of the 'overall resemblance' approach? Although it has enjoyed a certain amount of intelligent and enthusiastic support in recent years it seems to me that this doctrine too suffers from a fundamental weakness of principle. The taxonomist is recommended to classify on the basis of judgments of the form 'Individual *A* has more characters in common with *B* than it has with *C*' and the possibility of judgments of this form entails that an organism can be regarded as having a fixed and definite number of characters. *But there is no such number as the number of characters that an organism has.* Any individual living thing can be described in a variety of ways, and the number of characters it is thought of as possessing will depend on the description being employed. Of course, once a description has been given, characters can be counted: but the number is 'description-relative'. To speak as though a given individual had a fixed and definite number of characters is thus like speaking of a bag of flour as though it contained a fixed and definite number of heaps: in fact, you can only count heaps once the flour has been poured out, just as you can only count characters once the individual has been described in a particular way.[7]

Numerical taxonomy therefore does not in my view present a coherent theory of classification (and I have provided elsewhere the very necessary expansion of the argument I sketch here – see my [1972]): but its interest does not stop at this point. For if the writings of its proponents – we consider specifically here *The Principles of Numerical Taxonomy*, by R. R. Sokal and P. H. A. Sneath (1963) – are examined carefully, some curious elements are to be discerned, quite alien to their surroundings, which have somehow wormed their way into the numerical taxonomist's arguments.

In the particular case of the authors we have mentioned the plain fact is that while they declare in some places (as we have seen) that their proposal is to banish all differential weighting, in others they explicitly embrace a principle of weighting: in some places they urge all characters are to be considered, and all equally, but in others they discuss how taxonomically significant characters are to be picked out from those that may be safely ignored. Here we are concerned not so much with the contradictions which these writers get into but with the principle of weighting which in spite of themselves, it seems, they feel it right to employ: for, very interestingly, it comes close to the genotypical principle we found defended by Mayr.

To be used for taxonomic purposes, think Sokal and Sneath, characters should reflect the genotype of the organism concerned: "It is generally considered that only genetically determined characters should be used in orthodox taxonomy, and with this we concur" (Sokal and Sneath, 1963, p. 92). Again: "It is undesirable to use attributes which are not a reflection of the genotypes of the organisms themselves" (Sokal and Sneath, 1963, p. 66). I think it emerges clearly from their discussion that despite their own self-interpretation they in fact look to genotypical similarities for their principle of unity.

The importance of the hints we find in Mayr, and then echoed in the ostensibly opposed writings of Sokal and Sneath, will become clear as we pass on now to make a more positive approach to the problems we have seen others trying to solve.

III. TOWARDS AN IMPROVED THEORY

I begin with the conviction that the significance of the job which a classification is required to perform has been underestimated by many taxonomic theorists and grossly underestimated by the rest.

Unless the biologist is able to think of the individual organism he is studying as a representative of a group his results cannot be anything other than particular statements, asserting truths about individuals. The limitation of such a predicament will be obvious. For science to be possible at all, individuals must be referrable to categories, and the categories must be such that a scientist must be able to think that when certain circumstances obtain in establishing facts about particular individuals he is thereby establishing facts about members of the category in general.[8]

Inferences of this general kind must have a place in every science[9] – their justification of course constitutes the celebrated problem of induction – and to recognise the central role of induction in biology in particular – is to recognise the central importance of assigning organisms to groups such that induction is, in certain circumstances at least, reliable within them.

While it may be right therefore to say that there is no limit to the number of possible classifications of biological objects (Bambrough, 1961, p. 203, to take a recent example) it seems to me to be wrong to deny (as Bambrough does) that any may be more fundamental than others. A classification of biological individuals into groups within which induction works seems more fundamental than any other, in the sense that without it no scientific biology could go on. This suggests, I believe, the most satisfactory way of construing the dog-eared question of whether there are 'natural kinds' of organisms: it is best understood as the question of whether organisms *can* be sorted into groups of this fundamental kind. And a classification into such groups has best title, I suggest, to the accolade of 'natural'.

If this approach is right, our object in biological classification is to set up what we may call for short 'inductive groups'. There is of course no guarantee *a priori* that such an object is achievable. One can easily imagine groups of things within which inductive inferences are not in any way reliable, and there would be no way of telling, in advance of acquiring relevant factual information, whether biological organisms could be grouped more 'naturally' than this or not. The success of biology as a science, on the other hand, however, shows that as a matter of fact its inductive inferences have often been sound – the literature is full of results which later experience has substantiated – and so we have it confirmed that much, at least, of the classification actually employed in biology is natural.[10] Largely by intuition – by the method of *tâtonnement* or 'groping in the dark' of which de Candolle speaks (see Cain, 1959, pp. 202–3) – biologists have been able to set up groups within which some measure of reliability in inductive inference is achieved.[11]

Our problem is therefore that of explaining how this has been possible. Why have biologists been able to set up natural groups? What facts about living things gives the search for a natural classification the hope of success? In applying ourselves to this question we shall hope for illumination on another score, too. The practising taxonomist will need to know how to translate the very abstract injunction to set up inductive groups into practical terms. How are such groups to be erected? By what empirical criteria shall we be able to tell whether *A* and *B* are to belong to the same group but *C* and *D* to a different one?

Our strategy in what follows will be to show how a sophisticated phylogenetic-genotypical theory such as Mayr's comprises a set of principles which, generally, have the effect of setting up the type of groups we have specified. In setting up groups which reflect (in a certain way) evolutionary relationships, we are, it will be argued, setting up inductive groups. For an answer to the practical question mentioned above we can therefore refer to such texts as Mayr's and Simpson's where the empirical criteria for setting up phylogenetically based groupings are discussed.

Why is it then that the appropriate kind of phylogenetically based classification sorts organisms into inductive groups? In outline the answer is clear, and we have already noticed Mayr indicating it. The sort of phylogenetic grouping he recommends sorts individuals into groups whose members have similar genetic endowments; and because the form an organism takes in a given environment is determined by its genotype, similar organisms (in similar environments) will result from similar genotypes. If a fact about an individual organism is established, therefore, it seems that we can reliably infer that similar things will be true of all individuals (in the same environment) which have similar genotypes. In other words, groups of organisms with similar genotypes which will result from the appropriate kind of phylogenetic classification, will be inductive groups: and thus the required explanation is provided.

Yet it was right to say (above) that we have here an *outline* explanation only, for the import of the convenient word 'similar', very busy in the account just presented, has not been sufficiently indicated. As we have already pointed out, with a few dubious exceptions any one thing is similar in some respects to any other, and in particular any two genotypes resemble each other in some respects as well as differing from each other in others.

If, therefore, one wishes to claim that the species (say) is a natural category in virtue of the fact that individuals belonging to the same species have similar genotypes, one must go on to specify the similarity. It is to this task that we now address ourselves.

It is important to realise at the outset that not all taxonomic categories in orthodox use need have the same rationale. The facts that make a category natural need not be the same in every case. Actually we shall see that a species – 'biologically' conceived as an interbreeding natural population reproductively isolated from other such groups (Mayr, 1969, p. 26) – is a natural group for reasons other than those which make higher groupings natural. Accordingly, our exposition treats of the two types of group separately.

How then are we to specify the similarity that holds between all genotypes belonging to the same species? In the following way: two genotypes from the same species resemble each other to the extent that in general any gene of one could replace the corresponding gene of the other and form a genotype which would produce a viable organism. Corresponding genes (alleles) within the species can thus generally be said to be 'interchangeable': the system to which a genotype gives rise (i.e. the organism as a whole) is able to function successfully irrespective of the particular allele which occupies any given locus.[12]

What we assert here is the consequence of certain genetic facts, which may be briefly rehearsed: an organism's genetic complement comprises a set of chromosomes derived from one parent, together with a set of chromosomes derived from the other parent, each member of which corresponds with just one chromosome (its homologue) in the other set. During the generation of gametes homologous chromosomes pair and genetic material is interchanged between them in a process called 'crossing-over'. Generally, crossing-over can occur at any point along the paired chromosomes, and so *any* element of one can be exchanged for its corresponding element in the other. In time, therefore, it is possible for any gene to be replaced by any of its alleles: and this means that if a high proportion of viable genotypes (i.e. genotypes capable of generating viable offspring) are to be produced corresponding genes must be in general intersubstitutable: a viable genotype must result no matter which of a gene's alleles occupies a given locus.

A species is in this respect like a fleet of cars, the corresponding parts of which are interchangeable. A manufacturer, briefed to hand-build such a fleet, would have to build each so that it is closely similar to each of the others: and the exact sense of 'closely similar' here is given by the indicated requirement, viz. that each car must be such that its functioning would not be seriously impaired by the substitution for any of its parts of a corresponding part belonging to any of the other cars of the fleet. Minor variations among the cars would of course be possible. But they would have to be so minor that they had no bearing on the working of the wholes to which they belonged.

It is the fact that the genotypes belonging to a single species are composed of 'interchangeable (corresponding) parts' that makes the species what we have called an inductive group. For it implies that there is a level of discussion at which in talking about an individual one is also talking about all the individuals of the species. There is also of course a level at which variation is possible within the species – a gene's different alleles each have different effects. But very few of those effects must be of such a kind as to prevent the successful functioning (in any of its possible combinations with other genes at the other loci) of the resultant organism; for otherwise too many non-viable offspring would be produced for the species as a whole to compete successfully with its rivals.

Suppose then that one discovered a certain fact about an individual member of a species. It might be a fact which we could reliably infer to hold of all the other members of the species; but it might be a fact peculiar to the particular individual studied. How are we to decide between these two possibilities?

The answer suggested by our remarks above is that we can tell by bringing to bear, after first acquiring, an understanding of the workings of the functioning system which is the organism in question. If the discovered fact is of central importance to the functioning of the system as a whole then it can reliably be regarded as holding generally; but if it has little bearing on the system's working to regard it as generally true might easily be misleading.[13]

It might be helpful to take up our analogy once again.[14] Suppose our imaginary manufacturer has built all but one of the cars required for the fleet when an assistant takes him on one side to point out that a machine very similar to the type they are now building was constructed by the firm some time ago as part of a cancelled order, and that it is now sitting uselessly in the showroom. Wouldn't that machine do as the final member of the fleet?

The manufacturer knows that it can be used in this way only if it meets the 'interchangeable parts' criterion, and he is faced with deciding whether the apparent differences between the left-over machine and the ones he has just built are superficial or substantial from this point of view. All I wish to remark is that in making his decision it will be his engineering knowledge that he will be drawing on. Knowing how cars work, it will be a relatively easy job for him to inspect the odd man out and determine whether its parts could replace and be replaced by those belonging to the fleet proper. Similarly, it is by his engineering knowledge – his knowledge of how the organism in question functions – that the biologist, I am suggesting, distinguishes between characteristics likely to be true of the species as a whole and those that may well be idiosyncratic.

This then accounts for the reliability of induction in certain circumstances within the species. But can species themselves be arranged into inductive groups – in other words, can higher *taxa* be natural? We now try to show that this is indeed a possibility. Once again the general thesis that natural higher *taxa* unite individuals (and thus the lower order groups they belong to) with similar genotypes holds true; and once again we must develop this hint and specify the nature and limits of the similarity in question.

There are two elements to my solution of this problem, and it will probably be clearest to separate them. The first represents an attempt to describe in the abstract one possible set of circumstances which would allow natural higher taxa to be set up; and as a second step I try to show that the mechanism of evolution has produced such a set of circumstances, so that in fact organisms can be so classified. One might say that I first set up a model, and then attempt to show that the model is based on axioms that reflect the facts so that we can accept it as a possible explanation of how natural higher taxa are possible.

We develop the model by considering a system of which the parts or components are themselves systems: i.e. a whole which functions in virtue of the ordered interaction of sets of components organised into sub-systems. If one thinks of a heating furnace as a system, the notion of a sub-system would be illustrated by the apparatus which supplies the fuel, and also by the mechanism by which the furnace is turned off and on as the situation demands. For the functioning of the system as a whole, the mechanism of each of its sub-systems is a matter of indifference. It is only necessary that each sub-system perform its allotted role, and if there are alternative ways in which that role may be performed it will not make any difference to the properties of the system as a whole which alternative is actually operative. In terms of our example, it doesn't matter whether the control is mediated mechanically or electrically, so long as each is as effective as the other in reflecting the heating demands of the environment for the temperature of which the furnace is responsible.

There would be a level of discussion, therefore, at which the properties and functioning of the system as a whole might be spoken of without any implications being made regarding the mechanisms by which the sub-systems performed their role. Moreover, it would be possible to have a group of systems which were identical as far as their overall organisation went, but which differed from each other in respect of the internal organisation of their sub-systems. The vital point for our present purpose is that such a group of systems would be 'inductive': if one knew that a certain individual system belonged to such a class as we are imagining, and one established some facts about this individual as a result of studying it, it would under certain circumstances be possible to infer that these facts held of the other members of the group. And the 'certain circumstances' would be these: when the facts established were facts about the properties of the system as a whole – at a level, that is to say, at which the internal organisation of the sub-systems was irrelevant.

It is easy to see, further, that what is true of systems with two levels of organisation – the system and the sub-systems that make it up – can be generalised to apply to multi-level systems, that is, to systems complicated by possessing one or more intermediate grades of organisation between the level of the system as a whole and the sub-systems of the case just discussed. For any given level one could imagine a group made up of systems identical from that level upwards, but otherwise showing variation. And such groups would be inductive: for in each case facts established in connection with a single member of the group, provided it concerned the right level of organisation, could be inferred to hold for the other members too. One can therefore imagine a hierarchy of inductive groups, new groupings being possible at each level of organisation. If we call the level of organisation at and above which members of the first order groups are identical the *first* level of organisation, then we shall say that second order groups will unite first order groups sharing a common organisation at the second level and above; third order groups will unite second order groups which have the same organisation at the third level and above; and so on. For each level of organisation there will be the possibility of inductive groupings, and as the level of organisation gets higher, so the groupings become more inclusive and so fewer. In brief, then: corresponding to the hierarchical organisation of the individuals concerned one night erect a natural hierarchical classification of them. So much for the model. How does it relate to what we know of organisms and their relationships?

Perhaps the picture of the organism as a system built out of a hierarchy of sub-systems according to the rigorously tidy pattern envisaged in the model is something of a simplification, yet I think the general approach is valid enough. An organisation *can* generally speaking be regarded as a hierarchy of systems, and we appeal to the work of J. Z. Young, for example, as vindicating this conception (see particularly the first chapter of his [1959]). And insofar as the relationship between the levels of organisation departs from the tidy hierarchical pattern, so far, on my own view, will the tidy hierarchical classification of the organisms in question run into anomalies.

But even if it is legitimate to regard organisms as hierarchies of systems we have not yet shown that they can be classified in the way indicated: for one can easily imagine a collection of individuals all with hierarchical organisation but all with quite *different* hierarchical organisations. At this point we must turn to the mechanism whereby the different species of living things have arisen: that is, the process of evolution. We shall see that it is this process which has resulted in the organisations of organisms bearing to each other the resemblances represented in the model. In other words, we are to show how evolution has generated a hierarchy of inductive groups of organisms which reflects the hierarchy of levels in the organisation of the individuals that compose them.

The central observation to make is that the huge variety of different species of organisms that we know of have evolved, over the millennia, from relatively very few, or even a single, progenitor-groups or group; and by a process, speciation, which sets very close limits on the size of the steps by which these ultimately extremely wide divergences can be achieved. This is because new species arise, typically, not as a result of a single individual developing a large variation and begetting a host of offspring but by partition of the parent species. (See e.g. Cain [1954], chapter 8.) A section of the latter typically becomes spatially isolated so that interbreeding with the remainder of the parent species fails to occur. The newly delimited gene pool then responds to selection pressures now peculiar to it and so the isolated population diverges from its parent group. Here is the limitation on the degree of difference that can exist between parent and offspring gene pools: not only do the genotypes of the offspring species have to be mutually compatible in the way already specified (*above*, pp. 388 ff.) but at the point at which the spatial barrier becomes operative they must also be compatible with the genotypes of the parent population. For at that point parent and offspring populations are capable of interbreeding – only the spatial separation prevents it. In consequence fundamental changes in the structure and function of the evolving organism – that is, modifications to the higher levels of organisation – have to wait on the accumulation of modifications at less fundamental levels; as evolution proceeds, different modifications affecting the same level of organisation (which will occur in so far as the appropriate environments present themselves) generate a number of *taxa* of the same rank. Then, with the accumulation of a coherent set of modifications at one level, a modification of the level next above, and so a higher rank of *taxa*, becomes possible. Each higher category in a natural scheme of classification will thus correspond to a level of organisation in the hierarchy of systems of which the organisms to be classified consist, and the number of higher ranks it is legitimate to recognise will depend on the number of levels of organisation it is legitimate to recognise in the organisms concerned.

IV. CONCLUSION

In outline our 'theory' of classification is then this. A natural classification will be one that classifies on the basis of genotypical similarity. The job we need a classification to help us with is the job of studying organisms scientifically. In particular we have pointed out that a basic division of individuals into inductive groups is an absolute precondition of investigating them scientifically. The secondary organisation of those basic groups into a hierarchy of wider groupings is not necessary in the same way; yet clearly it has the effect of making the scientific method more powerful in as much as it enables certain scientific findings (those concerning the appropriate level of organisation) to be applied beyond the confines of the basic group within which they were established. Why is it that classifying according to genotype should produce inductive groups? The answer is given by the elementary facts of genetics: a similar genotype, other things being equal, will produce a similar organism – and we took pains to define the similarity in question here. How do we set up groups of genotypical similarity? Of course there is no unproblematic method. But because the process of evolution has generated groups of the kind in question we shall move towards our goal by attempting to set up the appropriate kind of phylogenetic classification.

The substance of our discussion therefore amounts in a sense to a vindication of the phylogenetic approach characteristic of those who perhaps constitute today's mainstream taxonomic theorists. It is the rationale of such an approach that we have chiefly been attempting to clarify. An evolutionary classification is not self-justifying, as some writers seem to suppose: a taxonomist is trying to do more than produce a tool for the specialist in evolution. Those who go on to justify evolutionary groupings in terms of the fact that they constitute at the same time groups of genetic similarity present a much more satisfactory account, it seems to me. I simply take two philosophical paces beyond this and attempt to say, first, what precisely must be meant by 'similarity' in this context, and second, why it is that groups of genetic similarity in this sense are the sort of groups which are all-important for biological science.

University College, Cardiff

NOTES

* I wish to thank Professor Peter Alexander, Professor Dorothy Emmet, Professor D. Hull, Dr Humphrey Palmer, Dr Roger Woolhouse, the Editor and the members of the L.S.E. Philosophy Seminar for their comments on an earlier draft of this paper, and to acknowledge the hospitality of the University of Ife, Nigeria, where it was written.

[1] Duhem is interested in using the notion of a natural classification to illuminate that of a scientific theory, rather than in naturalness itself. His account of the latter is completely undeveloped: to say that a natural classification is one that 'corresponds to reality' is of course to *pose* the philosophical problem, not to answer it.

[2] For this reason the following definition of a natural kind seems less than illuminating: "A Natural Kind is a group of objects which have *many* (perhaps indefinitely many) features in common" (Price, 1953, p. 7).

[3] Or Popper: "Two things which are similar are always similar *in certain respects*" (Popper, 1955, p. 420; his italics).

[4] See Simpson's comment in his (1961), p. 25, and Heslop-Harrison's (1962), pp. 21–2. Gilmour makes a different suggestion also, that "there will be one, as it were, general-purpose classification, more natural than others, in the sense that more inductive generalisations can be made about its classes" (Gilmour, 1951, p. 402). This and other remarks in that paper are a good deal closer to the view I try to defend in the present paper.

[5] Sophisticated statistical techniques are applied in order to erect a hierarchy of categories, but this leaves the principle of the method untouched.

[6] Cain's remarks on p. 188 of his [1959] are consonant with this position, but in later papers (e.g. his (1962)) he seems to be advocating a 'numerical' position. "An analysis of the taxonomist's judgement of affinity," over which he collaborated with G. A. Harrison, did much indeed to initiate this approach (Cain and Harrison, 1958).

[7] Compare Gilmour (1961), p. 100.

[8] This should not be taken to imply that I believe his findings will be best represented by expressions of the form 'All *A*'s are *B*'s: it may be true that the Herring-gull has a red spot on the beak and yet false that all Herring-gulls have red spots on their beaks (some might have lost their beaks through disease, for example). But in inferring truths about 'the' Herring-gull from truths about the particular Herring-gull or gulls actually studied the scientist is nonetheless involved in making claims about individuals other than those actually studied.

[9] But the details differ. In chemistry, for example, a study not of things but of the substances of which things are made, the inferences are from a truth about a specimen of the substance to a truth about the substance in general. This brings out one reason why it is misleading to speak of 'the' problem of induction.

[10] "That there are or have been regularities, for whatever reason, is an established fact of science" (Quine 1969, p. 13).

[11] I say 'groping in the dark' because inductive groups were being set up before the genetic facts which provide their rationale were known. Hull reads something more than this into '*tâtonnement*' (Hull, 1967, p. 180). But this is a point of interpretation. On the point of substance see Quine (1969), p. 15.

[12] It is true that the gene pool of a species can contain *some* genes which in certain combinations produce non-viable offspring – e.g. the homozygous occurrence of the gene responsible for sickle-celled anaemia. But these, I think, must be exceptions to the general rule.

[13] The radical differences between the reproductive and related organs of male and female within the same species presents a complication: it would be easy, on the suggestion we are proposing, to regard the rabbit's ovaries for example as playing a key role in the functioning of the animal's system as a whole, and thus conclude that it would be safe to think of all rabbits as having ovaries: but this kind of error would not be likely to survive a more complete understanding of how reproduction in the organism in question was achieved.

[14] I wish nothing of substance to hang on my analogies of course: they are meant as expository devices only.

BIBLIOGRAPHY

Bambrough, R.: 1961, 'Universals and Family Resemblances', *Proceedings of the Aristotelean Society* **61**, 207–22. (Page references are to reprint in G. Pitcher (ed.): *Wittgenstein, The Philosophical Investigations*, 1968, pp. 186–204.)

Cain, A. J.: 1954, *Animal Species and Their Evolution*, London, Hutchinson.

Cain, A. J.: 1959, 'Deductive and Inductive Methods in Post-Linnean Taxonomy', *Proceedings of the Linnean Society of London* **170**, 185–217.

Cain, A. J.: 1962, 'The Evolution of Taxonomic Principles', in G. C. Ainsworth and P. H. A. Sneath (eds.): *Microbial Classification*, Cambridge, Univ. Press, pp. 1–13.

Cain, A. J. and Harrison, G. A.: 1958, 'An Analysis of Taxonomist's Judgement of Affinity', *Proceedings of the Zoological Society of London* **131**, 85–98.

Cain, A. J. and Harrison, G. A.: 1958, 'Phyletic Weighting', *Proceedings of the Zoological Society of London* **135**, 1–31.

Cuvier, G. L. C. F.: 1817, *Le Règne Animal*, Paris.

Duhem, P.: 1906, *La Théorie physique, son objet et sa structure*. (Page reference is to the English translation of the second, 1914, edition: *The Aim and Structure of Physical Theory*, Princeton, Univ. Press 1954.)

Gilmour, J. S. L.: 1951, 'The Development of Taxonomic Theory since 1851', *Nature* **168**, pp. 400–2.

Gilmour, J. S. L.: 1961, 'The Mathematical Assessment of Taxonomic Similarity, Including The Use of Computers', *Taxon* **10**, 97–101.

Hempel, C. G.: 1965, *Aspects of Scientific Explanation*, New York, Free Press.

Hennig, W.: 1966, *Phylogenetic Systematics*, Urbana, Univ. Ill. Press.

Heslop-Harrison, J.: 1962, 'Purposes and Procedures in the Taxonomic Treatment of Higher Organisms', in G. C. Ainsworth and P. H. A. Sneath (eds.): *Microbial Classification*, pp. 14–36.

Hull, D. L.: 1967, 'Certainty and Circularity in Evolutionary Taxonomy', *Evolution* **21**, 174–189.

Huxley, J. S.: 1940, *The New Systematics*, London, Oxford Univ. Press.

Mayr, E.: 1968, *Principles of Systematic Zoology*, New York, McGraw Hill.

Mill, J. S.: 1843, *A System of Logic*, 8th ed., New York, Harper, 1882.

Popper, K. R.: 1935, *Die Logik der Forschung*. (Page reference is to the expanded English edition: *The Logic of Scientific Discovery*, London, Hutchinson, 1959.)

Pratt, V.: 1972, 'Numerical Taxonomy – A Critique', *Journal of Theoretical Biology* **36**, 581–92.

Price, H. H.: 1953, *Thinking and Experience*, Cambridge, Harvard Univ. Press.

Quine, W. V. O.: 1969, 'Natural Kinds', in N. Rescher (ed.): *Essays in Honour of Carl G. Hempel*, pp. 5–23.

Simpson, G. G.: 1961, *Principles of Animal Taxonomy*, New York, Columbia Univ. Press.

Sokal, R. R. and Sneath, P. H. A.: 1963, *Principles of Numerical Taxonomy*, San Francisco, Freeman.

Thompson, W. R.: 1952, 'The Philosophical Foundations of Systematics', *Canadian Entomologist* **84**, 1–16.

Whewell, W.: 1840, *The Philosophy of the Inductive Sciences*, London.

Young, J. Z.: 1959, *The Life of Mammals*, Oxford, Clarendon Press.

DAVID L. HULL

CONTEMPORARY SYSTEMATIC PHILOSOPHIES

During the past decade, taxonomists have been engaged in a controversy over the proper methods and foundations of biological classification. Although methodologically inclined taxonomists had been discussing these issues for years, the emergence of an energetic and vocal school of taxonomists, headed by Sokal and Sneath, increased the urgency of the dispute. This phenetic school of taxonomy had its origins in a series of papers in which several workers attempted to quantify the processes and procedures used by taxonomists to classify organisms. Of special interest was the process of weighting. These early papers give the impression that the primary motivation for the movement was the desire to make taxonomy sufficiently explicit and precise to permit quantification and, hence, the utilization of computers as aids in classification [22, 23, 41, 91, 106, 107, 111, 112]. The initial conclusion that these authors seemed to come to was that taxonomy, as it was then being practiced, was too vague, intuitive, and diffuse to permit quantification. Hence, the procedures and foundations of biological classification had to be changed.

The central issue in this dispute, however, has not been quantification but the extremely empirical philosophy of taxonomy which the founders of phenetic taxonomy seemed to be propounding [54, 79]. The pheneticists' position on these issues is not easy to characterize because it has undergone extensive development in the last few years.[1] The words have remained the same. Pheneticists still maintain that organisms should be classified according to overall similarity without any a priori weighting. But the intent of these words has changed. However, one thing seems fairly certain. Pheneticists believed that there was something fundamentally wrong with taxonomy as it was being practiced, especially as set out by such evolutionists as Dobzhansky, Mayr, and Simpson.[2] Later, a third group of taxonomists, led by Hennig, Brundin, and Kiriakoff, entered the dispute, appropriating the name phylogenetic school for themselves. The evolutionists and the phylogeneticists agree that evolutionary theory must play a central role in taxonomy and that biological classification must have a systematic relation to phylogeny. They disagree only over the precise nature of this relation. For the purpose of this paper *evolutionary taxonomy* will refer to the views of the Dobzhansky-Mayr-Simpson school and *phylogenetic taxonomy* will refer to the views of Hennig, Brundin, and Kiriakoff. Together, these two schools wil be referred to as *phyleticists* in contrast to the pheneticists.

Although the emphasis of this paper will be on contemporary systematic philosophies and not on the role of quantification in taxonomy, some of the resistance which phenetic taxonomy met was due to a blanket distaste on the part of some taxonomists for mathematical techniques as such and, in particular, for the pheneticists' attempt to quantify taxonomic judgment [104, 105]. When Huxley called for 'more measurement' in the *New Systematics* [68], he did not have in mind the processes by which taxonomists judge affinity. It is easy to sympathize with both sides, with the biologists who were less than elated over the prospect of learning all the new, high-powered notations and techniques that were beginning to flood the literature and with the pheneticists whose work was rejected on occasion, not because the particular mathematical techniques suggested were inadequate, but because they were mathematical. Happily, this aspect of the conflict has largely abated, although pockets of resistance still remain. The question is no longer whether or not to quantify but which are the best methods for quantifying.[3]

Recognition should also be made of the majority of taxonomists who, though they consider themselves mildly evolutionary in outlook, feel that all such disputes over foundations and methodology are idle chatter. Taxonomy is not the kind of thing one has to talk about. One just does it. The closest approximation to a spokesman for this group is R. E. Blackwelder, but he is atypical of the majority for which he speaks since he still advocates essentialism in almost its pristine, Aristotelian form [7–13, 15, 121–123]. The inadequacy of essentialism as a philosophical foundation for biological classification has been discussed so extensively that nothing more needs to be said here [63, 65, 83, 86].

Not only will this paper be limited to the philosophical aspect of the phenetic-phyletic controversy, but also, of the various issues which have been raised, it will deal with only two – the relation of phylogeny to classification and the species problem. Many of the objections raised against evolutionary taxonomy are actually criticisms of the synthetic theory of evolution, rather than of the classifications built upon it. Nor are these criticisms of recent origin. Every objection raised by the pheneticists to evolutionary theory and evolutionary taxonomy can be found in the work of earlier biologists, usually in the writings of the evolutionists themselves. The difference is that the evolutionists are optimistic about the eventual resolution of these difficulties, whereas the pheneticists, in the early years of the school, believed that they were insoluble. When viewed in the context of the development of bioloy during the past thirty years, phenetic taxonomy does not appear so much a recent insurrection as the culmination of long-standing grievances.

Soon after the turn of the century, both taxonomy and evolutionary theory had reached a low ebb in the esteem of the rest of the scientific community. Taxonomists seemed to be engaged in a frenzy of splitting and were viewed as nit-picking, skin-sorters, more as quarrelsome old librarians than scientists. Evolutionists had indulged themselves in reconstructing phylogenies in far greater detail and scope than the data and theory warranted and were looked upon as uncritical speculators, more authors of science fiction than science. Among evolutionists themselves, there were controversies. Were the laws of macroevolution different from those of microevolution? Was there such a thing as orthogenesis and, if so, what were the mechanisms for it? At this critical period, Mendel's laws were rediscovered, but instead of clarifying the situation, the birth of modern genetics confused it even further. A whole series of prejudices, conceptual confusions, and peculiarly pernicious terminologies made it seem as if the new genetics conflicted with evolutionary theory. Adding to the intensity of the controversy was the fact that evolutionists tended to be museum and field workers, whereas geneticists were, by and large, experimentalists at home in the laboratory. It was in this setting that the synthetic theory of evolution and the New Systematics had their inception.

The initial impetus for the rebirth of evolutionary theory was Fisher's *The Genetical Theory of Natural Selection* [49], followed by similar works by Haldane [58] and Wright [130, 131]. In these works it was shown that a mathematical model of evolutionary theory could be constructed in which the genetic mechanisms of Mendelian genetics meshed perfectly with the selective mechanisms of evolutionary theory. Evolutionary theory and, hence, evolutionary taxonomy had become respectable again.

However, the models supplied by Fisher, Haldane, and Wright were highly restrictive and very far removed from any situation a naturalist was likely to encounter in nature. Using the techniques of idealization which had proved so successful in physics, they showed that in certain overly simple, ideal cases, natural selection working on mutations which obeyed the laws of Mendelian genetics could result in the gradual evolution and splitting of species. The task still remained of showing how the insights gained in these idealizations could be applied to real situations in nature.[4] The classic works on this are those by Dobzhansky [37], Mayr [80, 85, 86], Huxley [68, 69], Simpson [99, 100, 102, 103], Rensch [94, 95], Stebbins [120], Hennig [60, 16], and Remane [93]. In the following discussion the earliest works of these authorities will be cited as freely as their later works because the basic features of the synthetic theory of evolution and evolutionary taxonomy have changed very little during this period.

I. PHYLOGENY AND CLASSIFICATION

One of the most persistent problems in biology has been the quest for a natural classification. Prior to Darwin a natural classification was one based on the essential natures of the organisms under study. Of the possible patterns that could be recognized in nature, a taxonomist would settle on one, partly because of his own peculiar psychological make-up and partly because of the scientific theories he held. Of course, another taxonomist with a different psychological make-up, perhaps holding different theoretical views, frequently recognized a different pattern. The controversies that ensued were usually settled by force of authority. The case of Cuvier and his disciples Owen and Agassiz is typical in this respect [83]. There are four basic plans in the animal kingdom, no more, no less!

Evolutionary theory promised to put an end to all this dogmatic haggling. After Darwin a natural classification would be one that was genealogical. No longer would biologists have to search fruitlessly for some ideal plan but would need only to discover the genealogical relationships among the organisms being studied and record this information in their classifications. The alacrity with which many biologists adopted Darwin's suggestion stemmed in part from two illegitimate sources – an inherent vagueness in the proposal and a misconception of the relation which any system of indented, discontinuous words can have to something as continuous and complex as phylogeny. As Darwin [34] observed of Naudin's simile of a tree and classification, "He cannot, I think, have reflected much on the subject, otherwise he would see that genealogy by itself does not give classification." Nearly a century later Gilmour [56] was still forced to remark that he doubted "whether the real significance of the term 'phylogenetic relationship' is yet fully understood."

The purpose of this section will be to investigate the relationships which phylogeny can have to biological classification – assuming that phylogeny can be known with sufficient certainty. The major criticism of evolutionary taxonomy by pheneticists has been that such reconstructions are too often impossible to make. Discussion of this criticism will be postponed until the next section.

No term in taxonomy seems immune to ambiguity and misunderstanding; this includes the term *classification*. Mayr [86] has already pointed out the process-product distinction between the process of classifying and the end product of this enterprise – a classification. But even the words *a classification* are open to misunderstanding. At one extreme, a classification is nothing but a list of taxa names indented to indicate category levels. Others would also include all the characters and the taxonomic principles used to construct a classification as part of the classification. At the other extreme, some authors use the words *a classification* to refer to the entire taxonomic monograph. Unless otherwise stipulated, *a classification* in the following pages will be used in the first, restricted sense.

The simplest view of the relation of a biological classification to phylogeny is that, given a classification, one can infer the phylogeny from which it was derived. One source of this misconception is a naive yet pervasive misconstrual of the relation between a hierarchical classification and a dendritic representation of phylogeny. According to this mistaken view, the classification of Order I sketched below

Order I
  Family A
    Genus 1
      species a

Genus 2
- species b
- species c

Genus 3
- species d
- species e
- species f

Family B

Genus 4
- species g
- species h

Genus 5
- species i
- species j

corresponds to the phylogenetic tree in Figure 1. However, Figure 1 is not a dendritic representation of a possible phylogeny. Rather it is merely a representation of the hierarchic indentations of the classification in a dendritic form. A true dendrogram of the possible phylogenetic development of the organisms involved would consist only of the species listed in the classification. One possible phylogeny from which the classification of Order I could have been derived is shown in Figure 2.

In this section we assume the phylogenetic development of the groups under discussion to be completely known. Hence, all the ancestral species are included in the classification along with extant species. In actual classifications, of course, not all ancestral species are known, but at least some are. At least sometimes, biological classifications contain reference to extinct forms. Hence, the interpretation of I in Figure 1 as the unknown stem species which gave rise to Order I, of *A* and *B* as the unknown stem species which gave rise to Families *A* and *B* respectively, and so on, cannot be carried through consistently. On occasion, at least, ancestral species will be known and will be included in the classification. The mistake is to confuse the inclusion relations in the taxonomic hierarchy with species splitting [103]. An order does not split into genera nor genera into species.

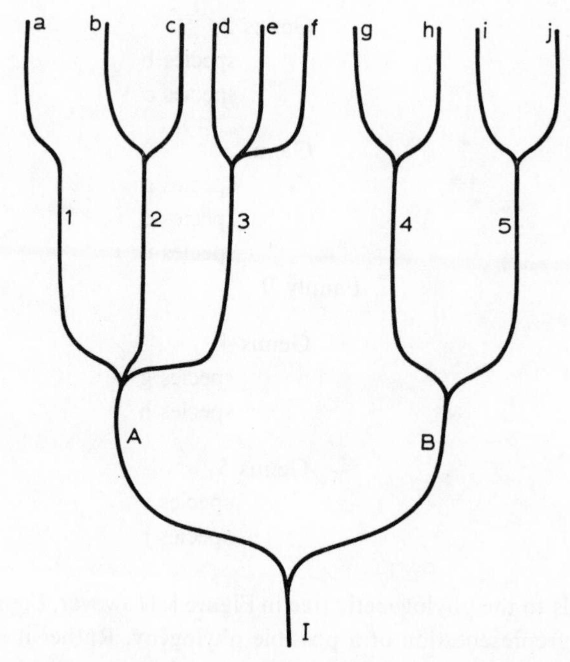

Fig. 1 A dendritic representation of a hierarchical classification.

A second impediment to seeing clearly the relation between phylogeny and classification has been a failure to distinguish cladistic from patristic relations [2, 5, 23, 75, 76, 84, 89, 115, 116]. The primary difference between the phylogenetic school of Hennig, Kiriakoff, and Brundin and the evolutionary school of Dobzhansky, Mayr, and Simpson is that the former want classification to reflect only cladistic affinity, whereas the latter feel that classification should also reflect such factors as degree of divergence, amount of diversification, or in general, patristic affinity.

Hennig's principles of classification are extremely straightforward [60, 61]. The stem species of every single higher taxon must be included in that taxon and must be indicated as the stem species by not being included in any of the other subgroups of that taxon. Splitting is the only mechanism of species formation that is recognized. Even though a group may evolve progressively until later members are extremely divergent from their ancestors, if no splitting has taken place, all the individuals are considered members of the same species. Upon splitting, the parental species is al-

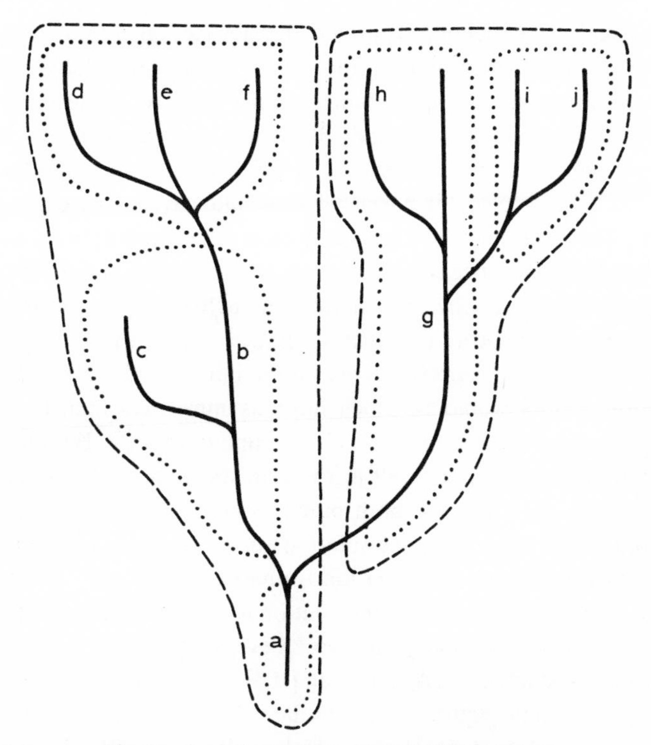

Fig. 2. A phylogenetic tree subdivided into taxa.

ways considered to be extinct, even though individuals may persist which are morphologically identical to members of the parent species. As far as ranking is concerned, sister groups must always be given coordinate ranks. In addition, Hennig is predisposed to Bigelow's [2–5] observation that in a truly phylogenetic classification recency of common ancestry must be considered a criterion for ranking. Taxa that evolved earlier should be given a higher taxonomic rank than those that evolved later.[5]

The major consequences of the adoption of these principles of classification is precisely the one intended by Hennig. Given a strictly phylogenetic classification, cladistic development can be read off directly. That is, given a classification and Hennig's principles, a dendrogram could be constructed which would accurately represent the cladistic relations of the groups classified. Hennig's principles of classification have something esthetically satisfying about them. They are straightforward and exceptionless. But this satisfaction is purchased at a price higher than many biologists are willing to pay. Early groups, even if they immediately became extinct without leaving descendants, would have to be recognized as separate phyla, equivalent to highly diversified, persistent groups. Hence, if it could be shown that a species split off in the Precambrian but gave rise to no other species, it nevertheless would have to be classed as a phylum. The resulting classification would be exceedingly monotypic. Our increasing ignorance as phylogeny is traced further back through the geological strata saves phylogeneticists from actually having to introduce such extreme asymmetries into their classification, but even so, enough is known so that classifications erected on the purely cladistic principles of the Hennig school would be much more asymmetrical than those now commonly accepted. Evolutionists also complain that the Hennig school is too narrow since it limits itself just to cladistic affinity. Patristic affinity is also important. Thus, for both practical and theoretical reasons, the evolutionists feel that Hennig's solution to the problem of the relation between phylogeny and classification is unacceptable.

When we turn our attention to evolutionary taxonomy, the situation is not so straightforward. The principles of evolutionary taxonomy are extremely fluid and intricate. As Simpson [103] has said, the practice of evolutionary taxonomy requires a certain flair. There is an art to taxonomy. Vagueness as to the actual relation which evolutionary classification is to have to phylogeny can be discerned in the earliest statements on the subject. The main purpose of Dobzhansky's *Genetics and the Origin of Species* [37] was to reconcile the differences between naturalists and geneticists: to convince the naturalist that the geneticists' experimental findings in the laboratory were relevant to their work in the field and museum, and to convince the geneticists that their understanding of evolutionary theory was grossly inadequate. Dobzhansky is not a systematist and is not especially interested in the problems of systematics. The little that Dobzhansky [37] had to say about classification can be quoted in its entirety.

A knowledge of the position of an organism in an ideal natural system would permit the formation of a sufficient number of deductive propositions for its complete description. Hence, a system based on the empirically existing discontinuities in the materials to be classified, and following the hierarchical order of the discontinuous arrays, approaches most closely to the ideal natural one. Every subdivision made in such a system conveys to the student the greatest possible amount of information pertaining to the objects before him. The modern classification of organisms uses the principles on which an ideal system could be built, although it would be an exaggeration to think that the two are consubstantial.

On the other hand, since the time of Darwin and his immediate followers the term 'natural classification' has meant in biology one based on the hypothetical common descent of organisms. The forms united together in a species, genus, or phylum were supposed to have descended from a single common ancestor, or from a group of very similar ancestors. The lines of separation between the systematic categories were, hence, adjusted, at least in theory, not so much to the discontinuities in the observed variations as to the branching of real or assumed phylogenetic trees. And yet the classification has continued to be based chiefly on morphological studies of the existing organisms rather than of the phylogenetic series of fossils. The logical difficulty thus incurred is circumvented with the aid of a hypothesis according to which the similarity between the organisms is a function of their descent. In other words, it is believed that one may safely base the classification on studies on the structures and functions of the organisms existing at one time level, in the assurance that if such studies are made complete enough, a picture of the phylogeny will emerge automatically. This comfortably complacent theory has received some rude shocks from certain palaeontological data that cast a grave doubt on the proposition that similarity is always a function of descent. Now, if similar organisms may, however rarely, develop from dissimilar ancestors, a phylogenetic classification must sometimes unite dissimilar, and separate similar, forms. The resulting system will be, at least in some of its parts, neither natural in the sense defined above nor convenient for practical purposes.

Fortunately, the difficulty just stated is more abstract than real. The fact is that the classification of organisms that existed before the advent of evolutionary theory has undergone surprisingly little change in the times following it, and whatever changes have been made depended only to a trifling extent on the elucidation of the actual phylogenetic relationships through palaeontological evidence. The phylogenetic interpretation has been simply superimposed on the existing classifications; a rejection of the former fails to do any violence to the latter. The subdivisions of the animal and plant kingdoms established by Linnaeus are, with few exceptions, retained in the modern classification, and this despite the enormous number of new forms discovered since then. These new forms were either included in the Linnaean groups, or else new groups were created to accommodate them. There has been no necessity for a basic change in the classification. This fact is taken for granted by most systematists, and all too frequently overlooked by the representatives of other biological disciplines. Its connotations are worth considering. For the only inference that can be drawn from it is that the classification now adopted is not an arbitrary but a natural one, reflecting the objective state of things.[6]

To begin with, the position of an organism in a hierarchical classification permits the inference of numerous propositions about it only if the characters used to classify the organisms are also listed. For example, knowledge that an organism is a chordate in conjunction with the defining characters of Chordata permits the inference that at some time in its ontogenetic development it has gill slits, a dorsal, hollow nerve cord, and probably a notochord. Knowledge that it is a vertebrate in conjunction with the defining characters of Vertebrata and the fact that Vertebrata is included in Chordata permits additional inferences and so on. A claim frequently made in the recent literature is that the best classification is the one with the highest information content; that is, the one which permits the greatest number of inferences. Colless [31], for example, says:

The current conflict between the 'phenetic' and 'phylogenetic' approaches to taxonomy thus boils down to whether a classification should in some fashion act as a storage-and-retrieval system for information about the distribution of attributes over organisms, and thus as a theory that predicts unexamined parts of that distribution: or whether it should reflect, as closely as possible, the historical course of evolution of the organisms concerned.

What is being blurred in the preceding quotation is that a biological classification as such (whether phenetic of phyletic) permits little in the way of inferences. Only a classification in conjunction with the principles and characters used to construct it is sufficient to permit any extensive inferences about the organisms being classified. For example, given the phenon levels of a phenetic classification, it is possible to infer that members of two taxa at a particular phenon level share a certain percentage of their characteristics, but it is not possible to infer which these may be. Similarly, from an evolutionary point of view, it would be reasonable to infer that two organisms classed together at the 40 phenon level are likely to have a more recent common ancestor than two organisms which are not classed together until the 10 phenon level – if one were given just this information.[7] Only when the characteristics used to partition the organisms into taxa are included can specific predictions be made about which organisms are likely to have which characters. But with the addition of such information, we are rapidly approaching the point at which the word *classification* has become expanded to include the entire monograph. Classifications in the narrow sense are incapable of storing much in the way of specific information. Rather than being storage-and-retrieval systems themselves, they serve as indexes to such storage-and-retrieval systems. The information resides in the monograph, not in the classification [128]. The classification merely provides a nested set of names which can be used to refer to the relevant taxa in as felicitous a manner as possible.

A second basic misunderstanding concerning the relation between a classification and a phylogeny has contributed to the belief that phylogeny can be inferred from an evolutionary classification. One commonly meets the assertion that proximity of names in a classification implies propinguity of descent. A glance back at Figure 2 shows that this belief is mistaken. If our knowledge of phylogeny were reasonably complete, every single higher taxon would contain at least one species which would be as closely related to a species in another taxon of the same rank as it is to its closest relative in its own taxon. For example, in the classification sketched on p. 403, species *a* is twice removed from species *b* (species *b* appears two lines below species *a*) and nine places removed from species *g* (species *g* appears nine lines below species *a*) – and yet both of these species are directly descended from *a* (see Figure 2).

When stated so baldly, the claim that inferences concerning propinquity of descent can actually be made from an evolutionary classification seems incredible; yet such a view is implicit in the writings of many phyleticists. From an evolutionary classification, even in conjunction with the stated criteria of classification, implications of cladistic relations are not possible. With a reconstructed phylogeny, indefinitely many classifications are possible. With any one of these classifications, an indefinite number of phylogenies are compatible. As Mayr has observed, "Even if we had a perfect understanding of phylogeny, it would be possible to convert it into many different classifications" [86; see also 14, 62, 103]. Of course, one way to falsify these claims is to expand the meaning of *a classification* to include phylogenetic dendrograms. Then, in a trivial sense, phylogeny can be inferred from an evolutionary classification.

Note that not all classifications are acceptable to an evolutionist. For example, all taxa must be 'monophyletic'. Each taxon can 'contain only the descendants of a common ancestor'. Early in the history of evolutionary theory, this meant that all the members of a taxon had to be descended from at most a single individual or pair of individuals in an immediately ancestral taxon. As the emphasis in evolutionary theory shifted from individuals to populations and species, the principle of monophyly was expanded so that descent from a single immediately ancestral species was all that was necessary for a taxon to be monophyletic. Hennig and the phylogenetic school still retain this rather stringent notion of monophyly. Unfortunately, if this principle is adhered to, many well-known and easily recognizable taxa such as the mammals (à la Simpson) become polyphyletic. The compromise suggested by Simpson [101, 103], and Gilmour [56] before him, is that all the members of a taxon may be descended, not from a single, immediately ancestral species, but from a single, immediately ancestral taxon of the same or lower rank. (For opposing views, see [32].) Thus, since all the species which contributed to the class Mammalia were in all likelihood therapsid reptiles, Mammalia is minimally monophyletic. As reasonable as this decision seems from the point of view of retaining well-marked groups and reflecting degree of divergence, its adoption further weakens the relation between classification and phylogeny. Not all classifications are compatible with a given phylogeny, but too many to permit any precise inferences.

Numerous authors before and after Dobzhansky [37] have observed that, from their classifications alone, "it is practically impossible to tell whether zoologists of the middle decades of the nineteenth century were evolutionists or not." Evolutionists have taken this fact to imply that pre-evolutionary taxonomists had been reflecting evolution in their classifications all along, though unwittingly. Pheneticists have argued for an additional factor. Classifications, before and after the introduction of evolutionary theory, are basically phenetic. Evolutionary theory, for all intents and purposes, is irrelevant to biological classification. It has been argued in this section that a third factor is actually responsible for the similarity between pre- and post-evolutionary classifications. Hierarchical classifications, in the absence of a rigid adherence to principles of classification like those of Hennig, do not permit any extensive inferences – whether phyletic or phenetic. Hennig says that hierarchic classifications are completely adequate to indicate phylogeny because he has incorporated the requirements of hierarchic classification into his principles of classification. From a strictly phylogenetic classification (just a list of indented names of taxa) and Hennig's principles, cladistic relations can be deduced. To the extent that this is not done, to that extent the number and variety of phylogenetic inferences which can be drawn from a classification will be diminished [28, 62, 125, 126].

Thus, biologists who maintain that biological classifications should be genealogical are presented with a dilemma. If they adopted a system like Hennig's, in which cladistic development is inferable from a classification, they would have to put up with the loss of information about patristic affinities and the cumbersome classifications that would result. If they retained the more tractable classifications that result from the more pliant principles of evolutionary taxonomy, they would have to abandon the ideal that classifications imply anything very precise about phylogenetic development. Evolution and evolutionary theory would still influence evolutionary classifications, but mainly in decisions as to homologies and the basic units of classification. The way in which evolutionary theory influences estimations of homologies will be discussed in the next section. The relation between evolutionary theory and the basic units of classification will be treated in the last section.

II. PHYLETIC INFERENCES AND PHENETIC TAXONOMY

In the preceding section, certain formal difficulties inherent in any attempt to establish a systematic relationship between classification and phylogeny were pointed out. The main thrust of the pheneticists' objections to evolutionary taxonomy, however, has been against permitting phylogeny to influence biological classification in the first place. The chief reasons that the pheneticists have given for excluding evolutionary considerations from biological classification are as follows: (1) We cannot make use of phylogeny in classification since, in the vast majority of cases, phylogenies are unknown [3, 4, 6, 43, 119]. (2) The methods which evolutionists use to reconstruct phylogeny, when not blatantly fallacious, are not sufficiently explicit and quantitative [6, 45, 115, 119]. (3) With the help of techniques being developed by the pheneticists, it eventually may be possible to reconstruct reasonably accurate phylogenies for certain groups of organisms, but since phylogeny cannot be known with sufficient certainty for all groups, it should not be used in those few cases in which we do have good reconstructions [6, 24, 115, 119]. (4) Even if the necessary evidence were available for all groups and the methods of reconstructing phylogeny were reformulated to make them completely acceptable, the resulting evolutionary classification would still be a special purpose classification and inadequate for biology as a whole; a general purpose classification would still be needed [55–57; 108–110, 119].

Like other criticisms of evolutionary taxonomy, these are not new. As early as 1874, Huxley [70], hardly an enemy of evolution, can be found saying, "Valuable and important as phylogenetic speculations are, as guides to, and suggestions of, investigation, they are pure hypotheses incapable of any objective test; and there is no little danger of introducing confusion into science by mixing up such hypotheses with Taxonomy, which should be a precise and logical arrangement of verifiable facts."

There is little that a philosopher can say about the first two objections to evolutionary taxonomy. After obvious inconsistencies have been removed and warnings about the type of certainty possible in empirical science duly entoned, the controversy becomes largely an empirical matter to be decided by scientists, not philosophers [64, 66]. If extensive fossil evidence for a group is necessary for reconstructing the phylogeny of that group, then the phylogenetic development of a majority of plants and animals will never be known, but many biologists think that various laws (or rules of thumb, if you prefer) can be used to reconstruct tentative phylogenies even in the absence of more direct evidence [85, 103, 124].

An interesting development in the phyletic-phenetic controversy is that some numerically minded biologists are beginning to set out formalisms for inferring phylogeny which they feel fulfill the various criteria of objectivity, etc., which more traditional methods are reputed to lack [22–24, 27, 40, 47, 48, 77, 91, 112, 118, 126]. Implicit in this endeavor is the conviction that attempts to reconstruct phylogeny even in the absence of fossil evidence are not inherently fallacious. Perhaps the practice of some evolutionists has been slipshod and certain reconstructions of the methods by which phylogenies are inferred have been mistaken, but the phyletic enterprise as such is not hopeless. For example, Colless [30] says, "I must stress at the outset that I am *not* denying that we can, and do, have available a body of reasonably credible phylogenies, which are probably fair reproductions of historical fact. I do, however, assert that some influential taxonomists have an erroneous view of the process by which such phylogenies are inferred; and, if my view is correct, such a situation clearly invites faulty inferences and sterile controversy."

Initially, phenetic and evolutionary taxonomy were treated by all those concerned as if they were in opposition to each other [84, 90, 108, 115, 116]. Pheneticists argued that evolutionary classifications, based on a priori weighting, were limited in their uses because they were biased toward a single scientific theory. Phenetic classifications, on the other hand, were general purpose classifications, based on the total number of unweighted or equally weighted characters, and were equally useful to all scientists because they were biased toward no scientific theory whatsoever. Pheneticists like Cain [18–21] attributed the mistakes which early taxonomists like Aristotle, Linnaeus, and Cuvier made to their letting theoretical and philosophical beliefs affect their classifications. Evolutionists had carried on in this misbegotten tradition. To eliminate such errors, pheneticists argued that no theoretical considerations should enter into the initial stages of a purely phenetic classification. A pheneticist must classify as if he were completely ignorant of all the scientific achievements (and failures) which preceded him. Characters must be delineated, homologies established, and clusters derived without recourse to any preconceived ideas whatsoever. No character could be weighted more heavily than another because it proved to be a 'good' character in previous studies (unless those studies themselves were phenetic) or because the studies were theoretically important according to current scientific theories. There must be no a priori weighting! Later, after several such purely phenetic studies had been run, certain characters would be found that tended to covary. They then could be weighted a posteriori. This a posteriori weighting would be, however, purely a function of the observed covariations of the characters being studied, nor of any theoretical considerations. Finally, evolutionary interpretations could be placed on these purely phenetic classifications which would transform them into special purpose evolutionary classifications. In short, phenetic taxonomy was just look, see, code, cluster.

This initial sharp contrast between evolutionary and phenetic classification has been modified considerably in recent years. In their latest utterances, pheneticists tend to view phenetic taxonomy somewhat differently. Doubts are raised as to whether any pheneticist ever held the views described above. Purely phenetic studies are still considered necessary preliminaries to scientific endeavors of any kind, including the construction of evolutionary classifications, but these phenetic studies are no longer thought of as being performed in isolation from all scientific theories – just from evolutionary theory. Homologies are not established just by observation, but are inferred via relevant genetic, embryological, physiological, and other scientific theories. Prior to any phenetic study, decisions are made as to which characters are to be considered the same, and in what sense they are to be so considered. For example, two organs which are structurally very similar in adult forms might be considered different organs because they have decidedly different embryological developments. Phenetic taxonomy is a matter of look, see, infer, code, and cluster. The resulting phenetic classifications are general purpose classifications because they have been constructed using all available knowledge, including all well-established scientific theories – except evolutionary theory. Finally, evolutionary interpretations can be placed on these phenetic classifications, but if the phenetic classification is properly constructed to begin with, it will actually be an evolutionary classification. Hence, phenetic and evolutionary classifications, when properly constructed, are equivalent to each other and are equally general purpose classifications.

It will be the purpose of this section to trace the change in phenetic taxonomy from its early, antitheory stage to its current state and to point out the fallacies in the early phenetic position which made it seem attractive and the reasons for changing it. It will be argued that purely phenetic classifications, as they were originally explicated, are impossible and that even if they were possible, they would be undesirable. To the question 'Theory now or theory later?' only one answer is possible. The two processes of constructing classifications and of discovering scientific laws and formulating scientific theories must be carried on together. Neither can outstrip the other very far without engendering mutually injurious effects. The idea that an extensive and elaborate classification can be constructed in isolation from all scientific theories and then transformed only later into a theoretically significant classification is purely illusory. A priori weighting of the theoretical kind is not only desirable in taxonomy, it is necessary. The price one pays for theoretical significance is, obviously, that any change or abandonment of the theories which gave rise to the classification will necessitate corresponding changes in the classification [52, 53, 67, 68, 69].

There is less to criticize in the latest versions of the phenetic position. One still must question why, of all scientific theories, evolutionary theory must be scrupulously excluded from the process of biological classification. There may be reasons for such a rejection, but the pheneticists have not been very articulate in stating them. Most of the objections which they have raised against evolutionary theory would count equally against any scientific theory and must be interpreted as utterances stemming from their early, antitheory stage of development. Now that pheneticists are willing to accept the role of theory in science, it would be helpful if they were to spell out exactly what faults they still find with evolutionary theory. A final question must also be asked before we turn to a detailed analysis of the evolution of phenetic taxonomy. What was all the controversy about? Except for a greater emphasis on making taxonomic practice explicit and perhaps even quantitative, how does phenetic taxonomy differ from evolutionary taxonomy? If patristic affinity is equivalent to some function of phenetics, chronistics, and cladistics, why all the acrimony? Has the phenetic-phyletic controversy been just one extended terminological confusion?

That the pheneticists actually held the early views attributed to them can easily be demonstrated. For example, as late as 1965, Sokal *et al.* [116] can be found saying, "Numerical taxonomists *do not disparage* interpretation or speculation or the inductive-deductive method in science. They simply feel that the process of constructing classification should be as free from such inferences as possible...." (See also [18–23, 29, 30, 39, 115, 116]) According to Colless [30], phenetic taxonomy makes reference "only to the observed properties of such entities, without any reference to inferences that may be drawn *a posteriori* from the patterns displayed. Such a classification can, and, to be strictly phenetic *must*, provide nothing more than a summary of observed facts." Even in their most recent publications, pheneticists can still be found making such extremely empirical claims; for example, Sokal [114] says that taxonomy is "the grouping of like organisms based on direct observation."

The key notion in the empiricist philosophy is the claim that, ideally, a priori weighting is to be completely expunged from taxonomic practice. What pheneticists have intended by such interdictions has been extremely equivocal. At one extreme they claim that homologies must be established on the basis of pure observation (as if there were such a thing). Two instances of a character are instances of the same character if they look, smell, taste, sound, and feel the same; otherwise not. Systematics "is a pure science of relation, unconcerned with time, space, or cause" [15]. All operational homologies are observational homologies.

So far no pheneticist has produced anything like a strict phenetic classification as described above. Pheneticists make reference to things like wings, antennae, anal gills, dorsal nerve cords, enzymes, and nucleotides. These are hardly pure observation terms. They presuppose all sorts of previous knowledge of a highly theoretical kind. For example, a taxonomist working on brachiopods today describes his specimens and forgets that at one time considerable effort was expended to decide whether brachiopod valves were front and back, dorsal and ventral, or right and left and that the eventual decision reached was based on various theoretical beliefs concerning their ontogenetic and phylogenetic development [35].

As Sneath [108] has observed, "Many taxonomic problems start part of the way along the classificatory process, and one is apt to forget what previous knowledge is assumed."

Pheneticists take this to be a fault with traditional taxonomy rather than a characteristic of all scientific undertakings, including their own. They think that, ideally, a purely descriptive, non-theoretical classification must be possible. The source of the persuasiveness of this view can be found in empiricist epistemology, according to which all empirical knowledge stems from sense impressions. Hence, all knowledge must be reducible to pure observation statements. Empiricists themselves have shown that such a reduction is impossible and, specifically, that scientific theories are not replaceable by sets of observation statements [59]. There remains the metaphysical compulsion to believe that such a reduction must be possible, and with it, the notion of a purely phenetic classification.

At times pheneticists are a little more liberal in their interpretation of what is to count as a priori weighting. For example, Colless [29] says, "Of course, the simple act of observation of 'existing' entities involves inferences, but they are of a primitive nature and, I believe, can be clearly distinguished from those which I am concerned to exclude." But how primitive is primitive enough? What criteria does Colless have for making this distinction? And why are primitive a priori weightings acceptable but sophisticated ones illegitimate?

There is a continuum between terms that are largely observational, like white precipitate, flammable fluids, and red appendage, and those that are more theoretical, like inertia, unit charge, and selection pressure. The reason why pheneticists want classifications to be constructed using those terms nearer the observational end of the scale is all too apparent. Time and again Cain [18–21] has argued that the greatest source of error in early classifications is their reliance on scientific theories which we now know to be erroneous. Wouldn't the safest procedure be to classify neutrally? That way theories could come and go and the classification, nevertheless, remain unchanged. Such a procedure would assuredly be safe, but in the extreme it is impossible to accomplish and in moderation undesirable.

The basic fallacy underlying the phenetic position on a priori weighting is the confusion of the logical order of epistemological reconstructions with the temporal order in actual scientific investigations [50]. Perhaps an analogous example from a different discipline will help to bring this fallacy into sharper focus. In the epistemological approach advocated by Sneath [108], a classification of inorganic substances must begin with purely phenetic studies in which samples are collected of a wide variety of inorganic substances, purely observational homologies established, and various clustering techniques used to group these substances into OTUs. Certain characters might then turn out to be good indicators of certain clusters and weighted more heavily for future runs. Eventually, a classification would emerge which would be equally useful for all purposes. Later, if one wished, this general purpose classification could be transformed into a special purpose classification by introducing atomic theory and weighting atomic number more heavily than all other characters put together.

The actual history of the construction of the periodic table does not, of course, read anything like this epistemological reconstruction. For example, gold was originally recognized and defined in terms of its color, malleability, weight, and so on – a characterization inadequate to distinguish gold from various alloys. Thus, Archimedes was presented with the problem of discovering a more important characteristic of gold. He hit upon specific density. What we tend to forget is that his selection of specific density rather than a host of other characters was his acceptance of the physics of his day in which the four elements were fire, air, earth, and water! Later, as physical theory developed, atomic weight replaced specific density as the key character in distinguishing inorganic substances. In the interim a new concept of element had evolved in the context of atomic theory. Not until atomic number replaced atomic weight could elements, in this new sense, be distinguished from each other and from compounds.

The analogy to the development of evolutionary theory and the species concept is obvious. The point is that a priori considerations were, not after-the-fact interpretations, but necessary factors in every step of the formation of the periodic table. Inorganic elements are distinguished from compounds and from each other on largely theoretical grounds. Incidentally, some very rough clusters of observable characters also accompany this theoretically significant classification. Atomic number, even if considered a phenetic character, was not treated as of equal weight to all other characters. Nor was its weight established a posteriori by discovering that numerous other characters tended to covary with it. The correlation between atomic number and the overall similarity of physical elements is about on the same order of magnitude as that observed by Dobzhansky between breeding habits and the overall similarity of living organisms.

Pheneticists might reply that perhaps this is how the periodic table was constructed, but it shouldn't have been. It should have been constructed by purely phenetic means, and to be justified it must be. This contention has yet to be proven. To do so, pheneticists would have to sample all inorganic substances. They could not limit themselves to just the elements, because that would presuppose that they knew which inorganic substances were elements, a blatant instance of a priori weighting. After establishing homologies purely on the basis of observation, pheneticists would then have to erect various alternative phenetic classifications. Atomic number could hardly appear as one of these phenetic characters, since electrons are observable in only the widest sense of the word. If electrons are observable, so is evolutionary development! If one of these phenetic classifications can distinguish between elements and compounds and can order the elements as they are ordered on the periodic table, then the pheneticists will have proved their case. If recourse to atomic theory is permitted in the early stages of the investigation and atomic number weighted more heavily than all other phenetic characters put together, then phenetic taxonomy, as it was originally explicated and as it is still propounded by many, has been abandoned. If it is to be abandoned, then the original criticisms of evolutionary classifications need to be re-evaluated.

Pheneticists seem to have gradually come to realize that the notion of a theoretically neutral phenetic classification is an illusion and have modified their position accordingly. Operational homologies are established utilizing any respectable scientific theory except evolutionary theory. The reasons given for permitting morphological, behavioral, physiological, serological, and DNA homologies, but forbidding evolutionary homologies, have all depended on repeated equivocations on the terms *phenetic character* and *operational homology*. Pheneticists claim that operational homologies are observed, whereas evolutionary homologies must be inferred. In the first place, only characters are observed. That two instances of a character are instances of the same character (i.e., that they are operationally homologous) must be inferred. Only if operational homologies are limited to observational homologies (i.e., if they both look blue then they are blue) will these inferences be made solely on the basis of observation. All other types of inferences to operational homologies will make essential reference to a particular scientific theory, and with the introduction of theory the overly simplistic notion of observational homology must be abandoned. One cannot observe that two nucleotides are operationally homologous. Both the existence of the nucleotides and which of the nucleotides are homologous must be inferred from extremely indirect evidence in the context of current biochemical theories.[8]

Colless (30) claims that there is a 'phylogenetic fallacy' – the view that "in reconstructing phylogenies, we can employ something more than the observed attributes of individual specimens, plus some concept of 'overall resemblance' and some concept of 'attribute' of a set or class of such specimens." Scientists in general, not just evolutionists, do employ something more than observed attributes and some concept of overall resemblance. This something more is scientific theory. As Colless [30] himself says, "The codon elements thus employed as attributes must, surely, be the ultimate approximation to our notion of 'unit attributes'..." What is or is not a codon is determined in large measure by biochemical theory. Codons are certainly not observable. In this instance, pheneticists and not phylogeneticists are guilty of reasoning fallaciously. The phenetic fallacy is the belief that in reconstructing phylogenies, we employ anything less than all the data and all the scientific theories at our disposal. For example, even a theory as far removed from biology as quantum theory is used in the process of carbon dating.

Each of the various kinds of homology has its own special problems. For example, behavioral homologies cannot be obtained very readily for extinct species, nor are the results obtained for extant species by controlled experiments in the laboratory very reliable. Thus, the argument that evolutionary homologies should not be used for any group because we cannot obtain them for all groups cannot be cogent, since, if it were, it would count against all types of homologies. Even morphological homology, the most pervasive type of homology used in classification, has limited applicability. For example, individual viruses and bacteria have few morphological characters which can be used in classifying them. The likelihood of obtaining extensive information about DNA homologies for more than an infinitesimally small percentage of species (and these all extant) is very slim, and yet no one would want to argue that this information should not be used when we do have it. Sokal and Camin [115] say, "Because phenetic classifications require only description, they are possible for all groups and are more likely to be obtained as a first stage in the taxonomic process." The preceding claim is true only if operational homologies are limited to observational homologies. If not, then phenetic classifications require more than description. They require the establishment of theoretically significant operational homologies.

However, the abandonment of the distinction between a priori and a posteriori weighting has certain ramifications for the notions of overall similarity and a general purpose classification. If it is admitted that the establishment of homologies presupposes various scientific theories, then the idea of a single parameter which might be termed *overall similarity* loses much of its plausibility and all classifications become special purpose classifications. As Edwards and Cavalli-Sforza [40] observed, "To say that the purpose of a classification is 'general' is, in our view, too vague to be of use in its construction." The idea of a general purpose classification is still another phenetic illusion. Pheneticists themselves have come come to realize that too many parameters exist which have equal right to be termed measures of overall similarity and, hence, that there is no such thing as a general purpose classification. As the Ehrlichs [44] have said recently, "Theoretical considerations make it seem unlikely that the idea of 'overall similarity' has any validity.... *All* classifications are inherently special." They quickly add, however, that "no special classification is any more or less 'correct' than any other." (See also [45, 50–53, 73, 110, 115, 116]).

All actual biological classifications are mixed classifications; that is to say, they are affected to a greater or lesser degree by all current biological theories. No classification is purely evolutionary, purely embryological, and certainly, none is purely phenetic. The justifications for this irregular mixing of these various considerations in a single classification are both practical and theoretical. In the current state of these theories, evolutionary considerations could no more be untwined from all other considerations and excluded from classification than could embryological or physiological considerations. They are too interconnected. They are interconnected because the theories from which they are partially derived are themselves partially interdependent. Of course, this situation need not be permanent. These various theories may gradually become more carefully and completely formulated, and the relevant derivations more distinct. When this happens, the ideal of providing a straightforward reconstruction of the inferences involved in biological classification can be more closely approximated. We must resist at all costs the tendency to superimpose a false simplicity on the exterior of science to hide incompletely formulated theoretical foundations.[9]

3 III. THE BIOLOGICAL SPECIES CONCEPT

Although Dobzhansky [36–38] first emphasized the biological species concept, it has received its most extensive development at the hands of Ernst Mayr. From his earliest to his most recent writings, Mayr [80–83, 85–87] has set himself the task of demolishing the typological species concept and replacing it with a species concept adequate for its role in evolutionary theory. According to the typological species concept, each species is distinguished by one set of essential characteristics. The possession of each essential character is necessary for membership in the species, and the possession of all the essential characters sufficient. On this view, either a character is essential or it is not. There is nothing intermediate. If a character is essential, it is all-important. If it is accidental, then it is of no importance [63, 65].

In taxonomy, the essentialist position is known as *typology*, a word with decidedly bad connotations. In the recent literature, every school of taxonomy has been called typological at one time or another. The phylogeneticists term the evolutionists typologists because they let degree of divergence take precedence over recency of common ancestry in their classifications [75, 76]. The pheneticists call both evolutionists and phylogeneticists typologists because they claim to use criteria which are rarely tested and may not actually obtain [113, 116, 117]. The pheneticists in turn are called typologists because their classifications are intended to reflect overall similarity [71, 98, 103]. The pheneticists reply that they are typologists but without types and of a statistical variety [113]. Their opponents reply that this is not typology but nominalism [84–86]! To put a nice edge on the dispute, some taxonomists openly claim the honor of being called typologists. "Now the great object of classification everywhere is the same. It is to group the objects of study in accordance with their essential natures." (See also [13, 15, 97, 121–123, 127]).

The key feature of essentialism is the claim that natural kinds have real essences which can be defined by a set of properties which are severally necessary and jointly sufficient for membership. Hence, strictly speaking, there can be no such thing as statistical typology. Biologists were always aware that the characters which they used to distinguish species did not always universally covary, as the essentialist metaphysics which they tacitly assumed entailed, but not until evolutionary theory were they forced to admit that such variation was not an accidental feature of the organic world, but intrinsic to it. After evolutionary theory was accepted, variation was acknowledged as the rule, not the exception [63, 65]. Instead of ignoring it, taxonomists had to take variation into account by describing it statistically. No one specimen could possibly be typical in any but a statistical sense. Species could no longer be viewed as homogeneous groups of individuals, but as polytypic groups, often with significant subdivisions. Polythetic definitions, in terms of statistically covarying properties, replaced essentialist definitions in terms of a single character or several universally covarying characters [1, 26, 33, 63, 92].

One of the accompanying characteristics of essentialism was the gradual insinuation of metaphysical properties and entities into taxonomy. Whenever naturalists attempted to define natural kinds in terms of observable attributes of the organisms being studied, exceptions always turned up. One way to reconcile this apparent contradiction was to dismiss all exceptions as monsters. Another way was to define the names of natural kinds in terms of unobservable attributes. However, two kinds of unobservables must be distinguished at this juncture – metaphysical entities and theoretical entities which, in the context of a particular scientific theory, are indirectly observable. The entities and attributes postulated by classical essentialists tended to be of the former type. The genetic criteria of the biological definition of species may be tested very rarely, but they are testable and, hence, are not metaphysical. What the pheneticists have in common with typologists is a belief in the existence of natural units of overall similarity. They differ in that these units can be defined only polythetically.

Recognizing the existence of variation among contemporary forms as a necessary consequence of the synthetic theory of evolution is one thing; formulating a methodology in taxonomy sufficient to handle such variation is another. The history of the biological species concept is a story of successive attempts to define species so that the resulting groups are significant units in evolution, or in Simpson's [101, 103] words, an evolutionary species is an "ancestral-descendant sequence of populations... evolving separately from others and with its own unitary evolutionary role and tendencies." Dobzhansky [36, 37] began by defining a species as that stage of the evolutionary process "at which the once actually or potentially interbreeding array of forms becomes segregated in two or more separate arrays which are physiologically incapable of interbreeding," and he emphasized the necessity of geographic isolation in species formation. "Species formation without isolation is impossible." Mayr concurred with Dobzhansky and distinguished with him between various isolating mechanisms, as such, and geographic and ecological isolation, since these latter are temporary and are readily removed. The species level is reached "when the process of speciation has become irreversible, even if some of the (component) isolating mechanisms have not yet reached perfection" [85]. The classic formulation of the biological species definition is as follows:

A species consists of a group of populations which replace each other geographically or ecologically and of which the neighboring ones intergrade or interbreed wherever they are in contact or which are potentially capable of doing so (with one or more of the populations) in those cases where contact is prevented by geographical or ecological barriers.

Or it may be defined more briefly:

Species are groups of actually or potentially interbreeding natural populations, which are reproductively isolated from other such groups [80].

Special attention in the preceding definition must be paid to the fact that it is populations which are said to be actually or potentially interbreeding, reproductively isolated, and so on; not individuals. In ordinary discourse, the same terms are applied both to individuals and to groups of individuals – like populations. For example, both individuals and populations are frequently said to interbreed. In most cases, the use of two distinct senses of interbreed causes no confusion, especially since the notion of populations interbreeding is defined in terms of individuals interbreeding. Similarly, Mayr [85] says of isolating mechanisms that they are "biological properties of individuals that prevent the interbreeding of populations that are actually or potentially sympatric." By their very nature, claims about populations interbreeding, etc., are statistical notions derived from the corresponding actions and properties of individuals. Thus, complaints that evolutionists continue to consider two groups as separate species even though members of these groups occasionally cross and produce fertile offspring are misplaced. It is the amount of crossing and the degree of viability and fertility of the offspring that matter. Complaints that values for these variables are too often difficult or impossible to specify are obviously relevant.

Since there is a definitional interdependence between species and population, charges of circularity must be allayed before we proceed further. *Species* is defined in terms of interbreeding, potential interbreeding, and reproductive isolation. Populations are included in species. Hence, populations must at least fulfill all the requirements for species. Additional requirements are added for populations. *Populations* are defined in terms of geographic distribution, ecological continuity, and genetic exchange. A population is "the total sum of conspecific individuals of a particular locality comprising a single potential interbreeding unit" [85]. The members of a population must not be separated from each other by ecological or geographic barriers. They must be actually interbreeding among themselves. As a unit, they are potentially interbreeding with other such units.

Throughout his long career, Mayr has continually opposed the typological species concept and essentialism, and yet on some interpretations, the biological species concept has itself been treated typologically, as if it provided both necessary and sufficient conditions for species status. Dobzhansky [37, 82], for example, has argued that individuals which never reproduce by interbreeding can form neither populations nor species because potential interbreeding is a necessary condition for the correct application of these terms. He even goes so far as to say that the terminal populations of a *Rassenkreis,* if intersterile, are to be included in separate species, even though these populations are exchanging genes through intermediary populations! Dobzhansky seems to be confusing the importance of a particular species criterion with the importance of the species concept. The crucial issue is not whether some one character is possessed, but whether the units function in evolution as species. As Mayr (87) has said, "*Species are the real units of evolution*, they are the entities which specialize, which become adapted, or which shift their adaption." Do asexual 'species' specialize, become adapted, split, diverge, become extinct, invade new ecological niches, complete, etc.? If so, then from the point of view of evolutionary theory, they form species and criteria must be found to delimit them.

The three elements in the biological species definition are actual interbreeding, potential interbreeding, and reproductive isolation. As succinct as Mayr's shorter version of the biological species definition is, it nevertheless contains redundancies. Two or more populations are reproductively isolated from each other if, and only if, they are neither actually nor potentially interbreeding with each other. Thus, one or the other side of the equivalence could be omitted with no loss of assertive content. Species are groups of natural populations which are not reproductively isolated from each other but which are reproductively isolated from other such groups. In his most recent publication, Mayr himself omits reference to potential interbreeding in his revised version of the biological species definition: "Species are groups of interbreeding natural populations that are reproductively isolated from other such groups" [86].

In his new biological species definition Mayr still retains reference to interbreeding to indicate that the definition is applicable only to populations whose members reproduce by interbreeding and because successful interbreeding is the most directly observable criterion for species status. Reference to potential interbreeding is omitted because anything that can be said in terms of potential interbreeding can be said in terms of reproductive isolation. Neither morphological similarity nor time is mentioned in any of the formulations of the biological species definition. Among synchronous populations, morphological similarity and difference are of no significance, as far as species status is concerned. Questions of inferring species status aside, they function only in distinguishing phena of the same population, subspecies, sibling species, etc. (See Mayr's [86] discrimination grid.)

The omission of any temporal dimension from the biological species definition is of greater significance. The application of the biological species definition successively in time would lead to the recognition of a series of biological species with minimal temporal dimensions. What is to integrate these successive time slices into temporally extended species? The answer, as Simpson [103] pointed out earlier, is descent. If species are to be significant evolutionary units, some reference to descent eventually must be made. It is also implicit in any definition of population, since males, females, young and adults, workers and asexual castes are all to be included in the same populations. Morphological similarity won't do, because the types of individuals listed are often morphologically quite dissimilar. However, once a temporal dimension is introduced into the species concept and speciation without splitting is permitted (contra Hennig), an additional criterion must be introduced to divide gradually evolving phyletic lineages into species. The only candidate for such a criterion is degree of divergence, as indicated by morphological and physiological similarity and difference. Thus, in the discernment of biological species, morphological similarity and difference play a dual role, in most cases as the evidence by which the fulfillment of the other criteria is inferred and in some instances as criteria themselves. By now it should be readily apparent that any adequate definition of species as evolutionary units can no more be typological in form than can any definition of any theoretically significant term in science. As Julian Huxley [68] observed quite early in the development of the synthetic theory of evolution, "Species and other taxonomic categories may be of very different types and significance in different groups; and also that there is no single criterion of species". [10]

The objections, however, which have been made most frequently by the pheneticists against the biological species concept are not those just enumerated, but the following: (1) As important as biological species may be in evolutionary theory, such theoretical considerations should not be allowed to intrude into biological classification, both because they are theoretical and because the presence or absence of reproductive isolation can seldom be inferred with sufficient certainty. (2) There may be fairly pervasive evolutionary units in nature, but reproductive isolation does not mark them. (3) There are no pervasive evolutionary units in nature, regardless of the criteria used to discern them.

As in the case of inferring phylogeny, the commonest complaint raised by extreme empiricists in general, and the pheneticists in particular, against the biological species concept is that too often reproductive isolation cannot be inferred with sufficient certainty to warrant its intrusion into classification. As early as the *New Systematics* [68] Hogben objected that biological species could not be determined often enough, and recently Mayr [85] has said that to "determine whether or not an incipient species has reached the point of irreversibility is often impossible." The problem is not distinguishing one taxon from another but deciding when one or more taxa have reached the level of evolutionary unity and distinctness required of species. If two groups are reproductively isolated from each other, then they are included in separate species; but how often and with what degree of certainty can the presence or absence of reproductive isolation be determined?

If just the two factors space and time are taken into account, four possible situations confront the taxonomist: In the ideal case, two populations, for a while at least, are synchronous and partially overlap. Here, in principle, it is possible to confirm species status by observation. In practice, the situation is not so ideal because the making of such observations is expensive, time consuming, and difficult – not to mention that decisions have to be made regarding the frequency of crossing, the degree of viability and fertility of the offspring, etc. In most cases, even under such optimal conditions, taxonomists depend heavily on inferences from morphological similarity to aid them in their decisions. In cases of synchronic but allopatric populations, the presence or absence of reproductive isolation must be inferred. The advantage here is that on occasion such inferences can be checked, both indirectly by fertility tests in the laboratory and directly, if the populations happen to meet in nature. Usually of course, species status is inferred via morphological similarity and difference. When two populations are separated by appreciable durations of time, inferences of species status are even more circumstantial and can never be checked by any of the more direct means. "Hence, while the definition of the BSC [biological species concept] does not involve phenetics, the actual determination of a biological species always will do so, even in the optimal case" [117].

Pheneticists have objected both to the failure of evolutionists to give phenetics its just due in the application of the biological species concept and to the deficiencies of phenetic similarity as an indicator of reproductive isolation. Since phenetics plays such a predominant role in species determination anyway and since inferences from phenetic similarity to interbreeding status are very shaky at best, they ask why one should not abandon oneself to phenetic taxonomy right from the start. The problem in replying to this question is in deciding precisely what phenetic taxonomy is. By a rigid interpretation, phenetic taxonomy, as it was originally set out, is something radically new, but by this inetrpretation it can be shown that there can be no such thing as phenetic taxonomy. By a more reasonable interpretation, phenetic taxonomy loses its originality, since it becomes by and large what traditional taxonomists have been doing all along. The jargon of phenetic taxonomy is different, and greater emphasis is placed on mathematical techniques of evaluation, but with such an interpretation phenetic taxonomy is not very revolutionary.

Sokal and Crovello (117) complain that since the words *potential interbreeding* have "never really been defined, let alone defined operationally... it appears to us that the only possible answer one could get from the question whether or not two samples are potentially interbreeding is 'don't know'." In the first place, potential interbreeding has been defined. If two populations are kept from interbreeding only by geographical or ecological barriers, then they are potentially interbreeding; otherwise not. It is another story, of course, whether or not ecologists and population biologists are in a position to make reasonable inferences on these matters. Sometimes, however, detailed analyses of particular situations have been provided and biologists are in a position to say more than 'don't know'. With equal justification, an evolutionist could say that since the words *phenetic similarity* have never really been defined, let alone defined operationally, the only possible answer one could get from the question whether or not two samples are phenetically similar is 'don't know'. Of course, for specific studies, when the OTUs, characters, and clustering method are specified, more specific decisions can be made, but the same is true for potential interbreeding claims. In both disciplines loose and specific questions can be asked.

As unflattering as the appellation may sound, *phenetic* has been a weasel word in phenetic taxonomy. Its meaning changes as the occasion demands. When the principles of other schools of taxonomy are being criticized, it is given a strict interpretation. Phenetic taxonomy is look, see, code, cluster. A methodologically sophisticated ignoramus could do it. But when the pheneticists turn to the elaboration of the methods and procedures of phenetic taxonomy, it takes on a whole spectrum of more significant meanings, heedless of the fact that under these various interpretations the original criticisms of other taxonomic schools lose much of their decisiveness.

For example, in the flow chart designed by Sokal and Crovello [117] for the recognition of biological species, they begin by grouping individuals into rough-and-ready samples. "In the initial stages of the study it may be that sufficient estimations of phenetic similarity can be determined by visual inspection of the specimens." But they go on to admit that such groupings are not the result of mindless look-see. "Knowledge of the biology of the organisms involved may be invoked". Throughout this flow chart, *phenetically homogenous sets* must include all stages in the life cycle of the organism, various castes in social insects, males and females, etc., regardless of the polymorphisms involved [11, 16, 85, 88, 95, 96]. They see this as a practical difficulty, when it is plainly a theoretical difficulty. The admission of such theoretical considerations in the initial stages of a phenetic study means that the pheneticists themselves are practicing a priori weighting, a practice which they have roundly condemned in others. Decisions to include males and females in the same taxon do not stem from earlier phenetic clustering but from previously accepted biological theories. Evolutionists emphasize reproductive isolation because they feel that it is of extreme importance in the phylogenetic development of species. They don't want to see evolutionary units broken up and scattered throughout the nomenclatural system. Similarly, biologists emphasize cellular continuity as a criterion for individuality because they feel that it is of extreme importance in the embryological development of the individual. They don't want to see embryological units broken up and scattered throughout the nomenclatural system. The theory of the individual, as Hennig calls it, may be so fundamental that it has become commonplace, but a biological theory does not cease to be a theory just because it has been around for a long time. As was argued earlier in the section on inferring phylogeny, pheneticists themselves admit theoretical (i.e., a priori) considerations in the initial stages of their studies – as well they should. The point in making this observation is not that pheneticists should be more rigorous in purging their procedures of such theoretical considerations – which are absolutely necessary – but that pheneticists should recognize them for what they are and modify their criticisms of evolutionary taxonomy accordingly.

What is a phenetic property, a phenetic classification, phenetic similarity? If a phenetic property is to be some minimal attribute analyzed in the absence of all scientific theories, regardless of how rudimentary, such characters will certainly be useless in any attempt to construct a scientifically meaningful classification. Arguments have even been set out that, in principle, such an analysis is impossible. If a phenetic property is to be some minimal unit analyzed in the context of some but not all scientific theories (and, in particular, not of evolutionary theory), then the criteria for deciding which scientific theories are legitimate and which illegitimate must be stated explicitly and defended. If some scientific theories are to be admitted even at the initial stages of a phenetic study, then the criticisms of comparable admissions of evolutionary theory must be re-evaluated. The establishment of evolutionary homologies on the basis of evolutionary theory may still be illegitimate, but not just because it is a scientific theory entering into the initial stages of a taxonomic study.

Sokal and Crovello [117] say that phenetic taxonomy is closely related to what Blackwelder calls practical taxonomy – "the straight-forward description of the patterns of variation in nature for the purpose of ordering knowledge." As efforts of the pheneticists have ably proved, there are indefinitely many ways of describing the patterns of variation in nature, and in each way there are indefinitely many patterns to be recognized. The problem is not so much that there is nothing which might be called overall phenetic similarity, but that there are too many things which might answer to this title. The question is whether or not some of these possible ways of ordering knowledge are perhaps more significant than others. The whole course of science attests to the reply that there are some preferable orderings – those which are most compatible with current scientific theories.

Evolutionists claim that their classifications, though they may be constructed in part by intuitive means, are objective, real, nonarbitrary, and so on, because they reflect something which really exists in nature. Pheneticists reply that character covariation also really exists in nature. As might be expected, this sort of exchange has done little to clarify the issues. The difference between evolutionary and phenetic taxonomy in this respect is that evolutionists have biologically significant reasons for making one decision rather than another while, by a strict interpretation, pheneticists do not. On purely phenetic criteria, any group of organisms can be arranged in indefinitely many OTUs with coefficients of similarity ranging from zero to unity. In contrast, evolutionists contend that biological species are important units in nature, more important than numerous other units which might be discernible. They are functioning as evolutionary units in evolution. Hence, from the point of view of evolutionary theory, there is good reason to pay special attention to these units and not to others.

If science were a theoretically neutral exercise, all dicisions would be on a par. There would be no difference between the claims that it rains a lot in San Francisco and that all bodies attract each other with a force equal to the product of their masses divided by the square of their distances. As soon as scientific theory is allowed to intrude, certain alternatives are closed, certain decisions are preferable. This is the important sense of natural which has lurked behind the distinction between natural and artificial classifications from the beginning.

In the absence of any scientific theory, the only difference between a natural and an artificial classification is the number of characters used. A natural classification is constructed using a large number of characters, while an artificial classification is constructed using only a few [22, 55–57, 78, 115, 128]. Biologists have tended to object to this characterization because it seemed to leave something out, but they have not been too articulate in describing this something. They have argued that a natural biological classification is one based on biologically relevant attributes – as many as possible. An artificial classification is one based on biologically irrelevant attributes – regardless of how many. The controversy has surrounded the sense in which attributes can be biologically relevant or irrelevant.

Taxonomists have tended to term an attribute relevant or taxonomically useful if it has served to cluster organisms into reasonably discrete groups. Thus, for future runs on a group, it would be given greater weight a posteriori. Pheneticists are in full agreement with this usage. But taxonomists also wish to extend their taxonomically useful attributes to cover additional, unstudied groups. This is the a priori weighting to which the pheneticists raised such vocal objections. The justification for such an extension, when it is justified, rests on the second and more important sense of biologically relevant. Certain concepts are central to biological theories; others are not. For example, canalization, geographic isolation, crossing over, epistatic interaction, and gene flow are important concepts in contemporary biological theory. Hence, a classification in which they were central would be natural in the above sense. Of course, gene flow is not used to define the name of a particular taxon, but it does serve two other functions. It plays an important part in the definition of species, and this definition, in turn, determines which taxa are classed at the species level and which are not. In addition, it might play a part in justifying the claim that an attribute which was taxonomically useful in group *A* should also prove to be taxonomically useful in group *B*. To the extent that such claims are justified, they must be backed up by appropriate scientific laws.

An empiricist might object that all attributes of organisms are equally real. This is certainly true. The broken setae of an insect are as real as a mutation which permits it to produce double the number of offspring, but they hardly are equally important. Just as physical elements are classified on the basis of their atomic number – an attribute selected because of its theoretical significance – evolutionary elements are classified on the basis of their reproductive habits and for the same reasons. Evolutionists contend that if all the data were available, a high percentage of organisms which reproduce by interbreeding could be grouped for long periods of their duration into phylogenetically significant units by the biological species definition.

The pheneticists have attacked this contention on two fronts. First, they have argued that biological species, like phenetic species, are arbitrary units and, second, that biological species, even if they could be determined, would not form pervasive, significant units in evolution. At the heart of the first criticism is the evaluative term *arbitrary*. Claiming to use *arbitrary* in Simpson's sense, Sokal and Crovello [117] say, "Our study of the operations necessary to delimit a biological species revealed considerable arbitrariness in the application of the concept. This is in direct conflict with the claim of nonarbitrariness by proponents of the BSC.... The degree of sterility required in any given cross, the number of fertile crosses between members of populations, not to mention the necessarily arbitrary decisions proper to the hidden phenetic components of the BSC, make this concept no less arbitrary than a purely phenetic species concept, and perhaps even more so, since phenetics is one of its components."

Simpson's definition of *arbitrary* is hardly relevant to the issues at hand. According to Simpson [101, 103], when there is a criterion of classification and a classification, groups in this classification are nonarbitrary to the extent that they have actually been classified according to the criterion. For example, if species *A* is defined in terms of property *f*, then the species is nonarbitrary if all of its members have *f*; otherwise, it is not. Simpson's definition is extraneous to this discussion since it assumes precisely what is at issue.

What then do Sokal and Crovello mean by *arbitrary*? Since they repeatedly designate decisions in phenetic taxonomy as arbitrary and since they are advocates of phenetic taxonomy, one might reasonably infer that they do not take it to be a term of condemnation. Yet in one place they talk of arbitrariness as being a drawback to various species definitions. *Arbitrary* is used in ordinary discourse in a host of different senses, and the pheneticists, in a manner not confined to themselves, seem to switch casually from one to another in their criticisms of evolutionary taxonomy. At one extreme, a decision is arbitrary if more than one choice is possible. This is unfortunate because in science more than one reasonable decision is always possible. Hence, all scientific decisions become arbitrary, and the term ceases to make a distinction. For example, should physicists retain Euclidean geometry and complicate their physical laws, or should they retain the simplicity of their laws and treat space as non-Euclidean? Either choice is possible, but physicists' decision for the latter is hardly arbitrary.

A more reasonable use of *arbitrary* is in the division of continua into segments. Biologists of all persuasions commonly admit that whenever an even gradation exists, any classificatory decision automatically becomes arbitrary [17, 103, 119]. Here there are not just two or a few possible choices, but many, perhaps infinitely many. Hidden in this line of reasoning is the essentialist prejudice that the only distinctions that exist are sharp distinctions. Unless there is a complete, abrupt break in the distribution of the characters being used for classification, no meaningful decisions can be made. This prejudice was one of the primary motives for philosophers' refusing to countenance even the possibility of evolution by gradual variation and for many philosophers' and biologists' opting for evolution by saltation [65]. But this prejudice runs counter to both the very nature of modern science ahd the methods being introduced by the pheneticists. Various statistical means exist for clustering elements, even when at least one element ex ists at every point in the distributional space. For example, there are reasons for dividing a bimodal curve at some points rather than at others. Darwin argued that species as well as varieties integraded insensibly. He concluded, therefore, that they were equally arbitrary. Owing to the mathematical and philosophical prejudices of his day, Darwin's conclusion is understandable. There is no excuse for similar prejudices still persisting [25, 72].

All decisions in phenetic taxonomy are hardly arbitrary in any meaningful sense. If they were, then all the techniques of phenetic taxonomy could be replaced by the single expedient of flipping a coin. Similarly, all decisions as to the degree of crossing, the number of fertile offspring and their viability, etc., sufficient to assure the presence or absence of reproductive isolation are hardly arbitrary in any meaningful sense of the term. From all indications, various thresholds exist in the empirical world. The temperature of water can be varied continuously, but it does not follow thereby that the attendant physical phenomena also vary continuously. At the boiling point, at the freezing point, and near absolute zero, a change of a single degree is accompanied by extremely discontinuous changes in the attendant physical phenomena. Similarly, for example, Simpson refers to quantum evolution, the burst of proliferation that follows a population managing to make its way through an adaptive valley to invade a new ecological niche [88, 101, 103].

There seems to be no question that such significant thresholds exist in evolution. Recently, however, pheneticists have contended that the biological species concept does not mark such a threshold [42, 45, 115]. Of all the criticisms leveled at evolutionary taxonomy in the last ten years, this is the most serious. Most of the other criticisms have been largely methodological, resting uneasily on certain dubious philosophical positions, but this criticism is empirical. In a recent study by Ehrlich and Raven (46), evidence was adduced to show that selection is so overwhelmingly important in speciation that the occasional effects of gene flow can safely be ignored in the general evolutionary picture. If this contention is borne out by additional investigation, then the role of the biological definition of species will have been fatally undermined and the synthetic theory of evolution will have to be modified accordingly.

Sokal and Crovello (117), concurring with the position of Ehrlich and Raven, observe that "possibly concepts such as the BSC are more of a burden than a help in understanding evolution." They go on to conclude, however, that "the phenetic species as normally described and whose definition may be improved by numerical taxonomy is the desirable appropriate concept to be associated with the category, species, while the local population may be the most useful unit for evolutionary study." If it can be shown that biological species are not significant units in evolution, then from the point of view of evolutionary taxonomy, the role of the biological species has been fatally undermined. It does not follow, therefore, that the phenetic species, as normally described, should automatically replace them in biological classification, if for no other reason than that no description has been provided yet for the phenetic species. Instead, there are literally an infinite number of phenetic units, all of which have an equal right, on the principles of numerical taxonomy, to be called species.

IV. CONCLUSION

Numerous distinctions have been drawn in the preceding pages, but little notice has been taken of the most important distinction underlying the phenetic-phyletic controversy – the difference between explicit and implicit or intuitive taxonomy. Simpson [103] has argued that taxonomy, like many other sciences, is a combination of science and art. For example, tempering vertical with horizontal classification, dividing a gradually evolving lineage into species, deciding how much interbreeding is permissible before two populations are included in the same species, the assignment of category rank above the species level, choices between alternative ways of classifying the same phylogeny, balancing splitting and lumping tendencies, and the inductive inferences by which phylogenies are inferred are all to some extent part of the art of taxonomy. The question is whether the intuitive element in taxonomy should be decreased and, if so, at what cost.

It has been assumed in this paper that decreasing the amount of art in taxonomy is desirable. Taxonomists can be trained to produce quite excellent classifications without being able to enunciate the principles by which they are classifying, just as pigeons can be trained to use the first-order functional calculus in logic. Human beings can be trained to be quite efficient classifying machines. They can scan complex and subtle data and produce estimates of similarity with an accuracy which far exceeds the capacity of current techniques of multivariate analysis. Taxonomists as classifying machines, however, have several undesirable qualities. Although taxonomists, once trained, tend to produce consistent, accurate classifications, the programs by which they are producing these classifications are unknown to other taxonomists and vary from worker to worker. In addition, just when a taxonomist is reaching the peak of his abilities, he tends to die. Only recently one of the most accomplished taxonomists passed away and with her, all the experience which she had accumulated during decades of doing taxonomy.

The resistance to making taxonomic practice and procedures explicit seems to have stemmed from two sources: one, an obscurantist obsession with the ultimate mystery of the human intellect; the other, a concern over how much theoretical significance one must sacrifice in order to make biological classification explicit. With respect to the first reservation. Kaplan [74] has distinguished between reconstructed logic and logic-in-use. Frequently, during the course of development of formal and empirical science, empirical scientists use certain modes of inference which are beyond the current formal reconstructions. There is the tendency to dismiss these modes of inference by attributing them to genius, imagination, and unanalyzable, fortuitous guesswork. Kaplan [74] views the intuition of great scientists, not as lucky guesswork, but as currently unreconstructed logic-in-use. Intuition is any logic-in-use which is preconscious and outside the inference schemata for which we have readily available reconstructions. "We speak of intuition, in short, when neither we nor the discoverer himself knows quite how he arrived at his discoveries, while the frequency or pattern of their occurrence makes us reluctant to ascribe them merely to chance."

The second reservation which taxonomists have had about making taxonomy less intuitive and more explicit is less subtle, but equally important. In the early days of phenetic taxonomy, pheneticists seemed willing to dismiss the theoretical side of biological classification, since it seemed to make straightforward reconstructions extremely difficult, if not impossible. They tended to conflate the complexity of taxonomic inferences with taxonomists being muddle-headed. Certainly some of the complexity of traditional taxonomy may well have been due just to sloppy thinking, but instead of this evaluation being the immediate, initial response, it should have been the last resort. Traditional taxonomists and computer taxonomists are going to have to adapt to each other, but this adaptation cannot be purchased at the expense of the purposes of scientific investigation. These ends are better characterized by the words *theoretical significance* than by *usefulness*. An extremely accurate scientific theory of great scope will certainly be useful, but there are many things which are useful, though of little theoretical significance.

ACKNOWLEDGMENTS

I wish to thank Donald H. Colless, Theodore J. Crovello, Michael T. Ghiselin, Ernst Mayr, and Robert R. Sokal for reading and criticizing this paper. The preparation of this paper was supported in part by NSF grant GS-1971.

University of Wisconsin, Milwaukee

NOTES

[1] For the most recent statement of the pheneticists' philosophy of classification, see Robert R. Sokal, 'Classification: Purposes, Principles, Progress, Prospects', *Science* **185** (1974) 1115–1123; for recent criticism, see Michael Ruse, *Philosophy of Biology*, London, 1973.

[2] In this paper, I lumped the views of Ernst Mayr and G. G. Simpson together, but as Gareth Nelson has argued, the two are different in several important respects; see Gareth Nelson, 'The Influence of Hennig's 'Phylogenetic Systematics' upon Ichthyology', *Systematic Zoology* **21** (1972), 364–372.

[3] For those interested in a review of the numerical aspects of the phenetic-phyletic controversy, I recommend Johnson (73).

[4] For a more realistic, formal axiomatization of evolutionary theory, see Williams (129).

[5] Since this paper was written, Hennig's principles have been adopted and expanded upon by increasing numbers of American taxonomists; see for example the papers presented in a symposium on contemporary systematic philosophies in *Systematic Zoology* **22**, 4 (1973). Ernst Mayr's criticisms of cladistic taxonomy can be found in his 'Cladistic Analysis or Cladistic Classification?', *Sonderdruck aus Z. f. zool. Systematik u. Evolutionsforschung* **12** (1974), 94–128. For opposing views, see G. Nelson, 'Darwin-Hennig Classification: A Reply to Ernst Mayr', *Systematic Zoology* **23** (1974) 452–458 and D. E. Rosen, 'Cladism or Gradism: A Reply to Ernst Mayr', *Systematic Zoology* **23** (1974) 446–451.

[6] Dobzhansky condenses the preceding discussion to about half its length in the 3rd edition of his work and Mayr quotes the final paragraph in his *Systematics and the Origin of Species* (80).

[7] One of the most persistent problems in taxonomy has been the explication of the notion of 'similarity' which is to be some function of descent. An analysis of this concept must be postponed until the next section.

[8] For a more detailed discussion of the term 'phenetic' as it is used in the taxonomic literature, see W. J. van der Steen and W. Boontje, 'Phylogenetic versus Phenetic Taxonomy: A Reappraisal', *Systematic Zoology* **22** (1973) 55–63.

[9] If I had it to write over again, I would emphasize that even though these particular pheneticists' criticisms of evolutionary taxonomy are in principle unwarranted, in practice they have some point.

[10] If species are taken to be classes, then the preceding discussion is still appropriate to them. However, I now think that species are better characterized as 'individuals', See Michael Ghiselin, *The Triumph of the Darwinian Method*, Berkeley, 1969; David L. Hull, *Philosophy of Biological Science*, Englewood Cliffs, N. J., 1974; M. Ghiselin, 'A Radical Solution to the Species Problem', *Systematic Zoology* **23** (1974).

BIBLIOGRAPHY

[1] Beckner, M.: 1959, *The Biological Way of Thought*, New York: Columbia Univ. Press, 200 pp.
[2] Bigelow, R. S.: 1956, 'Monophyletic Classification and Evolution', *Syst. Zool.* **5**, 145–46.
[3] Bigelow, R S.: 1958, 'Classification and P ylogeny', *Syst. Zool.* **7**, 49–59.
[4] Bigelow, R. S.: 1959, 'Similarity, Ancestry, and Scientific Principles', *Syst. Zool.* **8**, 165–68.
[5] Bigelow, R. S.: 1961, 'Higher Categories and Phylogeny', *Syst. Zool.* **10**, 86–91.
[6] Birch, L. C. and Ehrlich, P. R.: 1967, 'Evolutionary History and Population Biology', *Nature* **214**, 349–52.
[7] Blackwelder, R. E.: 1959, 'The Present Status of Systematic Zoology', *Syst. Zool.* **8**, 69–75.
[8] Blackwelder, R. E.: 1959, 'The Functions and Limitations of Classification', *Syst. Zool.* **8**, 202–11.
[9] Blackwelder, R. E.: 1962, 'Animal Taxonomy and the New Systematics', *Surv. Biol. Progr.* **4**, 1–57.
[10] Blackwelder, R. E.: 1964, 'Phyletic and Phenetic *versus* Omnispective Classification., in *Phenetic and Phylogenetic Classification* (ed. by V. H. Heywood and J. McNeill), 17–28, London: Systematics Assoc. 164 pp.
[11] Blackwelder, R. E.: 1967, 'A Critique of Numerical Taxonomy', *Syst. Zool.* **16**, 64–72.
[12] Blackwelder, R. E.: 1967, *Taxonomy*, New York: Wiley, 698 pp.
[13] Blackwelder, R. E. and Boyden, A.: 1952, 'The Nature of Systematics', *Syst. Zool.* **1**, 26–33.
[14] Bock, W. J.: 1963, 'Evolution and Phylogeny in Morphologically Uniform Groups', *Am. Natur.* **97**, 265–85.
[15] Borgmeier, T.: 1957, 'Basic Questions of Systematics', *Syst. Zool.* **6**, 53–69.
[16] Boyce, A. J.: 'The Value of Some Methods of Numerical Taxonomy with Reference to Hominoid Classification'. See Ref. [10, 47–65].
[17] Burma, B. H.: 1949, 'The Species Concept: a Semantic Review', *Evolution* **3**, 369–70.
[18] Cain, A. J.: 1968, 'Logic and Memory, in Linnaeus's System of Taxonomy', *Proc. Linn. Soc. London* **169**, 144–63.
[19] Cain, A. J.: 1959, 'Deductive and Inductive, Methods in Post-Linnaen Taxonomy, *Proc. Linn. Soc. London* **170**, 185–217.
[20] Cain, A. J.: 1959, 'Taxonomic Concepts', *Ibis* **101**, 302–18.
[21] Cain, A. J.: 1962, 'Zoological Classification', *Aslib Proc.* **14**, 226–30.
[22] Cain, A. J. and Harrison, G. A.: 1958, 'An Analysis of the Taxonomist's Judgment of Affinity', *Proc. Zool. Soc. London* **131**, 85–98.
[23] Cain, A. J. and Harrison, G. A.: 1960, 'Phyletic Weighting', *Proc. Zool. Soc. London* **135**, 1–31.
[24] Camin, J. H. and Sokal, R. R.: 1965, 'A Method for Deducing Branching Sequences in Phylogeny', *Evolution* **19**, 311–26.
[25] Cargile, J.: 1969, 'The Sorites Paradox', *Brit. J. Phil. Sci.* **20**, 193–202.
[26] Carmichael, J. W., George, J. A., and Julius, R. S.: 1968, 'Finding Natural Clusters', *Syst. Zool.* **17**, 144–50.
[27] Cavalli-Sforza, L. L. and Edwards, A. W. F.: 1967, 'Phylogenetic Analysis:

Models and Estimation Procedures', *Evolution* **21**, 550–70.
[28] Clark, R. B.: 1956, 'Species and Systematics', *Syst. Zool.* **5**, 1–10.
[29] Colless, D. H.: 1967, 'An Examination of Certain Concepts in Phenetic Taxonomy', *Syst. Zool.* **16**, 6–27.
[30] Colless, D. H.: 1967, 'The Phylogenetic Fallacy', *Syst. Zool.* **16**, 289–95.
[31] Colless, D. H.: 1970, 'The Relationship of Evolutionary Theory to Phenetic Taxonomy', *Evolution* **24**.
[32] Crowson, R. A.: 1965, 'Classification, Statistics and Phylogeny', *Syst. Zool.* **14**, 144–48.
[33] Daly, H. V.: 1961, 'Phenetic Classification and Typology', *Syst. Zool.* **10**, 176–79.
[34] Darwin, F.: 1959, *The Life and Letters of Charles Darwin.* New York: Basic Books, 2 vols, 558 pp. & 562 pp.
[35] Dexter, R. W.: 1966, 'Historical Aspects of Studies on the Brachipoda by E. E. Morse', *Syst. Zool.* **15**, 241–43.
[36] Dobzhansky, T.: 1935, 'A Critique of the Species Concept in Biology', *Phil. Sci.* **2**, 344–55.
[37] Dobzhansky, T.: 1937, *Genetics and the Origin of Species*, New York: Columbia Univ. Press, 364 pp.
[38] Dobzhansky, T.: 1940, 'Speciation as a Stage in Evolutionary Divergence', *Am. Natur.* **74**, 312–21.
[39] DuPraw, E. J.: 1964, 'Non-Linnaean Taxonomy', *Nature* **202**, 849–52.
[40] Edwards, A. W. F. and Cavalli-Sforza, L. L.: 'Reconstruction of Evolutionary Trees'. [See Ref. 10, 67–76]
[41] Ehrlich, P. R.: 1958, 'Problems of Higher Classification', *Syst. Zool.* **7**, 180–84.
[42] Ehrlich, P. R.: 1961, 'Has the Biological Species Concept Outlived Its Usefulness?', *Syst. Zool.* **10**, 167–76.
[43] Ehrlich, P. R.: 1964, 'Some Axioms of Taxonomy', *Syst. Zool.* **13**, 109–23.
[44] Ehrlich, P. R. and Ehrlich, A. H.: 1967, 'The Phenetic Relationships of the Butterflies', *Syst. Zool.* **16**, 301–27.
[45] Ehrlich, P. R. and Holm, R. W.: 1962, 'Patterns and Populations', *Science* **137**, 652–57.
[46] Ehrlich, P. R. and Raven, P. H.: 1969, 'Differentiation of Populations', *Science* **165**, 1228–31.
[47] Farris, J. S.: 1967, 'The Meaning of Relationship and Taxonomic Procedure', *Syst. Zool.* **16**, 44–51.
[48] Farris, J. S.: 1968, 'Categorical Rank and Evolutionary Taxa in Numerical Taxonomy', *Syst. Zool.* **17**, 151–59.
[49] Fisher, R. A.: 1930, *The Genetical Theory of Natural Selection*, Oxford: Clarendon, 272 pp.
[50] Ghiselin, M. T.: 1966, 'On Psychologism in the Logic of Taxonomic Principles', *Syst. Zool.* **15**, 207–15.
[51] Ghiselin, M. T.: 1967, 'Further Remarks on Logical Errors in Systematic Theory', *Syst. Zool.* **16**, 347–48.
[52] Ghiselin, M. T.: 1969, *The Triumph of the Darwinian Method*, Berkeley: Univ. California Press, 287 pp.
[53] Ghiselin, M. T.: 1969, 'The Principles and Concepts of Systematic Biology', in *Systematic Biology. Publ. 1962 Nat. Acad. Sci.* (ed. by C. G. Sibley), 45–55, 632 pp.

[54] Gilmartein, A. J.: 1967, 'Numerical Taxonomy – an Eclectic Viewpoint', *Taxon* **16**, 8–12.
[55] Gilmour, J. S. L.: 1937, 'A Taxonomic Problem', *Nature* **139**, 1040–47.
[56] Gilmour, J. S. L.: 1940, 'Taxonomy and Philosophy', in *The New Systematics* (ed. by J. Huxley), 461–74, London: Oxford Univ. Press, 583 pp.
[57] Gilmour, J. S. L. and Walters, S. M.: 1964, 'Philosophy and Classification', *Vistas Bot.* **4**, 1–22.
[58] Haldane, J. B. S.: 1932, *The Causes of Evolution*, London: Harpers, 234 pp.
[59] Hempel, C. G.: 1965, *Aspects of Scientific Explanation*, New York: Free Press, 505 pp.
[60] Hennig, W.: 1950, *Grundzüge einer Theorie der phylogenetischen Systematik*, Berlin: Deut. Zentralverlag, 370 pp.
[61] Hennig, W.: 1966, *Phylogenetic Systematics*, Urbana, Univ. Illinois Press, 263 pp.
[62] Hull, D. L.: 1964, 'Consistency and Monophyly', *Syst. Zool.* **13**, 1–11.
[63] Hull, D. L.: 1965, 'The Effect of Essentialism on Taxonomy', *Brit. J. Phil. Sci.* **15**, 314–26, **16**, 1–18.
[64] Hull, D. L.: 1967, 'Certainty and Circularity in Evolutionary Taxonomy', *Evolution* **2**, 174–89.
[65] Hull, D. L.: 1967, 'The Metaphysics of Evolution', *Brit. J. Hist. Sci.* **3**, 309–37.
[66] Hull, D. L.: 1968, 'The Operational Imperative – Sense and Nonsense in Operationism', *Syst. Zool.* **16**, 438–57.
[67] Hull, D. L.: 1969, 'The Natural System and the Species Problem', in *Systematic Biology. Publ. 1962 Nat. Acad. Sci.* (ed. by C. G. Sibley), 56–61, 632 pp.
[68] Huxley, J. (ed.): 1940, *The New Systematics*, London: Oxford Univ. Press, 583 pp.
[69] Huxley, J.: 1942, *Evolution: the Modern Synthesis*, London: Allen & Unwin, 645 pp.
[70] Huxley, T. H.: 1874, 'On the Classification of the Animal Kingdom', *Nature* **11**, 101–2.
[71] Inger, R. R.: 1958, 'Comments on the Definition of Genera', *Evolution* **12**, 370–84.
[72] James, M. T.: 1963, 'Numerical vs. Phylogenetic Taxonomy', *Syst. Zool.* **12**, 91–93.
[73] Johnson, L. A. S.: 1968, 'Rainbow's End: the Quest for an Optimal Taxonomy', *Proc. Linn. Soc. N.S.W.* **93**, 8–45.
[74] Kaplan, A.: 1964, *The Conduct of Inquiry*, San Francisco: Chandler, 428 pp.
[75] Kiriakoff, S. G.: 1959, 'Phylogenetic Systematics versus Typology', *Syst. Zool.* **8**, 117–18.
[76] Kiriakoff, S. G.: 1965, 'Cladism and Phylogeny', *Syst. Zool.* **15**, 91–93.
[77] Kluge, A. G. and Farris, J. S.: 1969, 'Quantitative Phyletics and the Evolution of Anurans', *Syst. Zool.* **18**, 1–32.
[78] Lorch, J.: 1961, 'The Natural System in Biology', *Phil. Sci.* **28**, 282–95.
[79] Mackin, J. H.: 1963, 'Rational and Empirical Methods of Investigation in Geology', in *The Fabric of Geology* (ed. by C. C. Albritton), 135–63, New York: Addison-Wesley, 372 pp.
[80] Mayr, E.: 1942, *Systematics and the Origin of Species*, New York: Columbia Univ. Press, 334 pp.
[81] Mayr, E.: 1949, 'The Species Concept', *Evolution* **3**, 371–72.
[82] Mayr, E.: 1957, *The Species Problem. AAAS Publ. N. 50*, Washington, 338 pp.

[83] Mayr, E.: 1959, 'Agassiz, Darwin, and Evolution', *Harvard Libr. Bull.* **13**, 165–94.
[84] Mayr, E.: 1965, 'Numerical Phenetics and Taxonomic Theory', *Syst. Zool.* **14**, 73–97.
[85] Mayr, E.: *Animal Species and Evolution*, Cambridge: Harvard Univ. Press, 797 pp.
[86] Mayr, E.: 1969, *Principles of Systematic Zoology*, New York: McGraw-Hill, 428 pp.
[87] Mayr, E.: 1969, 'The Biological Meaning of Species', *Biol. J. Linn. Soc.* **1**, 311–20.
[88] Megletsch, P. A.: 1954, 'On the Nature of the Species', *Syst. Zool.* **3**, 49–65.
[89] Michener, C. D.: 1957, 'Some Bases for Higher Categories in Classification', *Syst. Zool.* **6**, 160–73.
[90] Michener, C. D.: 1963, 'Some Future Developments in Taxonomy', *Syst. Zool.* **12**, 151–72.
[91] Michener, C. D. and Sokal, R. R.: 1957, 'A Quantitative Approach to a Problem in Classification', *Evolution* **11**, 130–62.
[92] Minkoff, E. C.: 1964, 'The Present State of Numerical Taxonomy', *Syst. Zool.* **13**, 98–100.
[93] Remane, A.: 1952, *Die Grundlagen des natürlichen Systems, der vergleichenden Anatomie und der Phylogenetik*, Leipzig: Geest and Portig, 364 pp.
[94] Rensch, B.: 1929, *Das Prinzip geographischer Rassenkreise und das Problem der Artbildung*, Berlin: Borntraeger.
[95] Rensch, B.: 1947, *Neure Probleme der Abstammungslehre*, Stuttgart: Enke, 407 pp.
[96] Rohlf, F. J.: 1963, 'The Consequence of Larval and Adult Classification in Aedes', *Syst. Zool.* **12**, 97–117.
[97] Sattler, R.: 1963, 'Methodological Problems in Taxonomy', *Syst. Zool.* **13**, 19–27.
[98] Sattler, R.: 1963, 'Phenetic Contra Phyletic Systems', *Syst. Zool.* **12**, 94–95.
[99] Simpson, G. G.: 1944, *Tempo and Mode in Evolution*, New York: Columbia Univ. Press, 434 pp.
[100] Simpson, G. G.: 1945, 'The Principles of Classification and a Classification of Mammals', *Bull. Am. Mus. Natur. Hist.* **85**, 1–350.
[101] Simpson, G. G.: 1951, 'The Species Concept', *Evolution* **5**, 285–98.
[102] Simpson, G. G.: 1953, *The Major Features of Evolution*, New York: Columbia Univ. Press, 434 pp.
[103] Simpson, G. G.: 1961, *Principles of Animal Taxonomy*, New York: Columbia Univ. Press, 247 pp.
[104] Simpson, G. G.: 1964, 'Numerical Taxonomy and Biological Classification', *Science* **144**, 712–13.
[105] Simpson, G. G., Roe, A., and Lewontin, R. C.: 1960, *Quantitative Zoology*, New York: Harcourt, Brace & World, 440 pp.
[106] Sneath, P. H. A.: 1957, 'The Application of Computors to Taxonomy', *J. Gen. Microbiol.* **17**, 201–26.
[107] Sneath, P. H. A.: 1958, 'Some Aspects of Adansonian Classification and of the Taxonomic Theory of Correlated Features', *Ann. Microbiol. Enzimol.* **8**, 261–68.
[108] Sneath, P. H. A.: 1961, 'Recent Developments in Theoretical and Quantitative Taxonomy', *Syst. Zool.* **10**, 118–39.
[109] Sneath, P. H. A.: 'Introduction'. [See Ref. 10, 43–45]
[110] Sneath, P. H. A.: 1968, 'International Conference on Numerical Taxonomy',

Syst. Zool. **17**, 88–92.
[111] Sokal, R. R.: 1959, 'Comments on Quantitative Systematics', *Evolution* **13**, 420–23.
[112] Sokal, R. R.: 1961, 'Distance as a Measure of Taxonomic Similarity', *Syst. Zool.* **10**, 70–79.
[113] Sokal, R. R.: 1962, 'Typology and Empiricism in Taxonomy', *J. Theor. Biol.* **3**, 230–67.
[114] Sokal, R. R.: 1969, 'Review of Mayr's *Principles of Systematic Zoology*, *Quart. Rev. Biol.* **44**, 209–11.
[115] Sokal, R. R. and Camin, J. H.: 1965, 'The Two Taxonomies: Areas of Agreement and Conflict', *Syst. Zool.* **14**, 176–95.
[116] Sokal, R. R., Camin, J. H., Rohlf, F. J., and Sneath, P. H. A.: 1965, 'Numerical Taxonomy: Some Points of View, *Syst. Zool.* **14**, 237–43.
[117] Sokal, R. R. and Crovello, T. J.: 1970, 'The Biological Species Concept: a Critical Evaluation', *Am. Natur.* **104**, 127–53.
[118] Sokal, R. R. and Michener, C. D.: 1958, 'A Statistical Method for Evaluating Systematic Relationships', *Univ. Kansas Sci. Bull.* **38**, 1409–38.
[119] Sokal, R. R. and Sneath, P. H. A.: 1963, *The Principles of Numerical Taxonomy*, San Francisco: Freeman, 359 pp.
[120] Stebbins, G. L.: 1950, *Variation and Evolution in Plants*, New York: Columbia Univ. Press, 643 pp.
[121] Thompson, W. R.: 1952, 'The Philosophical Foundations of Systematics', *Can. Entomol.* **84**, 1–16.
[122] Thompson, W. R.: 1960, 'Systematics: the Ideal and the Reality', *Studio Entomol.* **3**, 493–99.
[123] Thompson, W. R.: 1962, 'Evolution and Taxonomy', *Studio Entomol.* **5**, 549–70.
[124] Thorne, R. F.: 1963, 'Some Problems and Guiding Principles of Angiosperm Phylogeny', *Am. Natur.* **97**, 287–305.
[125] Throckmorton, L. H.: 1965, 'Similarity *versus* Relationship in *Drosophila*', *Syst. Zool.* **14**, 221–36.
[126] Throckmorton, L. H.: 1968, 'Concordance and Discordance of Taxonomic Characters in *Drosophila* classification', *Syst. Zool.* **17**, 355–87.
[127] Troll, W.: 1944, 'Urbild und Ursache in der Biologie', *Bot. Arch.* **45**, 396-416.
[128] Warburton, F. E.: 1967, 'The Purposes of Classification', *Syst. Zool.* **16**, 241–45.
[129] Williams, M. B.: 1970, 'Deducing the Consequences of Evolution: a Mathematical Model', *J. Theor. Biol.*
[130] Wright, S.: 1931, 'Evolution in Mendelian Populations', *Genetics* **16**, 97–159.
[131] Wright, S.: 1931, 'Statistical Theory of Evolution', *Am. Statist. J.* March suppl., 201–8.

FURTHER READING

General, Anthologies and Texts

Ayala, Francisco J. and Dobzhansky, Theodosius (eds.): 1974, *Studies in the Philosophy of Biology*, Berkeley, University of California Press.

Beckner, Morton: 1968, *The Biological Way of Thought*, 2nd edn., Berkeley, University of California Press.

Bertalanffy, Ludwig von: 1960, *Problems of Life*, 2nd edn., New York, Harper and Row.

Blackburn, Robert T. (ed.): 1966, *Interrelations: The Biological and Physical Sciences*, Chicago, Scott, Foresmann.

Carlson, Elof A. (ed.): 1967, *Modern Biology: Its Conceptual Foundations*, New York, Braziller.

Commoner, Barry: 1961, 'In Defense of Biology', *Science* **133**, 1745–8.

Delbrück, Max: 1949, 'A Physicist Looks at Biology', *Transactions of the Connecticut Academy of Arts and Sciences*, **38**, 173–190.

Goodfield, June: 1969, 'Theories and Hypotheses in Biology', *Boston Studies in the Philosophy of Science*, V, 421–449.

Gregg, John R. and Harris, F. T. C. (eds.): 1964, *Form and Strategy in Science*, D. Reidel, Dordrecht.

Grene, Marjorie: 1974, 'The Understanding of Nature: Essays in the Philosophy of Biology', *Boston Studies in the Philosophy of Science* **23**, Dordrecht, Boston.

Hawkins, D.: 1964, *The Language of Nature*, London, W. H. Freeman.

Hull, David L.: 1974, *Philosophy of Biological Science*, Englewood Cliffs, N.J., Prentice Hall.

Hull, David: 1969, 'What Philosophy of Biology is Not', *Synthese* **20**, 157–184.

Munson, Ronald (ed.): 1971, *Man and Nature. Philosophical Issues in Biology*, New York, Dell.

Ruse, Michael: 1973, *Philosophy of Biology*, London, Hutchinson.

Shapere, Dudley: 1969, 'Biology and the Unity of Science', *J. History Biology* **2**, 3–18.

Simon, Michael A.: 1971, *The Matter of Life*, New Haven, Yale Univ. Press.

Simpson, George Gaylord: 1969, *This View of Life*, New York, Harcourt, Brace and World.

Smart, J. J. C.: 1959, 'Can Biology Be An Exact Science?', *Synthese* **11**, 359–368.

Woodger, J. H.: 1952, *Biology and Language*, Cambridge, Cambridge Univ. Press.

Section One

Benton, E.: 1974, 'Vitalism in Nineteenth-Century Scientific Thought: A Typology and Reassessment', *Stud. Hist. Phil. Sci.* **5**, 17–48.

Bowler, Peter J.: 1974, 'Evolutionism in the Englightenment', *Hist. Sci.* **12**, 159–183.

Goodfield, June: 1960, *The Growth of Scientific Physiology*, London, Hutchinson.

Mendelsohn, Everett: 1965, 'Physical Models and Physiological Concepts: Explanation in Nineteenth-Century Biology', *Boston Studies in the Philosophy of Science* **II**, 127–150.

Provine, William B.: 1971, *The Origins of Theoretical Population Genetics*, Chicago, Univ. Chicago Press.
Rudwick, Martin: 1974, 'Darwin and Glen Roy: A "Great Failure" in Scientific Method?', *Stud. Hist. Phil. Sci.* **5**, 97–185.
Rudwick, Martin: 1972, *The Meaning of Fossils*, London, Macdonald.

Sections Two and Three

Ackermann, Robert: 1969, 'Mechanism, Methodology and Biological Theory', *Synthese* **20**, 219–229.
Beckner, Morton: 1967, 'Aspects of Explanation in Biological Theory', in S. Morgenbesser (ed.), *Philosophy of Science Today*, New York, Basic Books.
Berlinski, David: 1972, 'Philosophical Aspects of Molecular Biology', *J. Phil.* **69**, 319–335.
Canfield, J. V. (ed.): 1966, *Purpose in Nature*, Englewood Cliffs, N.J., Prentice-Hall.
Causey, Robert L.: 1969. 'Polanyi on Structure and Reduction', *Synthese* **20**, 230–237.
Elsasser, W. M.: 1966, *Atom and Organism*, Princeton, Princeton Univ. Press.
Glass, Bentley: 1963, 'The Relation of the Physical Sciences to Biology – Indeterminary and Causality', in B. Baumrin (ed.), *Philosophy of Science: The Delaware Seminar*, Vol. 1, New York, Wiley, pp. 223–249.
Gregory, R. L.: 1969, 'On How So Little Information Controls So Much Behaviour', in D. H. Waddington (ed.), *Towards A Theoretical Biology*, II, Edinburgh, Edinburgh Univ. Press, pp. 236–47.
Grene, M. (ed.): 1971, *Interpretations of Life and Mind*, New York, Humanities Press, including Selected Bibliography.
Hein, Hilde: 1969, 'Molecular Biology vs Organicism', *Synthese* **20**, 238–254.
Hirschmann, D. and Manser, A. R.: 1973, 'Function and Explanation', *Aristotelian Soc.* **47**, 19–52.
Hull, David: 1972, 'Reduction in Genetics – Biology or Philosophy?', *Phil. Sci.* **39**, 491–499.
Johnson, H. A.: 1970, 'Information Theory in Biology after 18 Years', *Science* **168**, 1545–1550.
Koestler, A. and Smythies, J. R. (eds.): 1969, *Beyond Reductionism*, London, Hutchinson.
Lehman, Hugh: 1965, 'Functional Explanation in Biology', *Phil. Sci.* **32**, 1–20.
Lewontin, Richard C.: 1969, 'The Bases of Conflict in Biological Explanation', *J. History Biology* **2**, 35–47.
Pattee, H. H. (ed.): 1973, *Hierarchy Theory: The Challenge of Complex Systems*, New York, G. Braziller.
Pollard, E. C.: 1963, 'Are Life Processes Governed by Physical Laws?', in Bernard Baumrin (ed.), *Philosophy of Science: The Delaware Seminar*, Vol. 2, New York, Wiley, pp. 295–410.
Pratt, Vernon: 1974, 'Explaining the Properties of Organisms', *Stud. Hist. Phil. Sci.* **5**, 1–15.
Prigogine, I.: 1969, 'Structure, Dissipation and Life', in M. Marois (ed.), *Theoretical Physics and Biology*, Amsterdam, North Holland, pp. 23–52.
Roll-Hansen, Nils: 1969, 'On the Reduction of Biology to Physical Science', *Synthese* **20**, 277–289.
Ruse, Michael: 1971, 'Functional Statements in Biology', *Phil. Sci.* **38**, 87–95.

Schaffner, Kenneth: 1974, 'Logic of Discovery and Justification in Regulatory Genetics', *Stud. Hist. Phil. Sci* **4**, 349–385.
Schaffner, Kenneth: 1969, 'Theories and Explanation in Biology', *J. History Biology* **2**, 19–33.
Stent, Gunther: 1968, 'That Was the Molecular Biology That Was', *Science* **160**, 390–395.
Wimsatt, William C.: 1972, 'Teleology and the Logical Structure of Function Statements', *Stud. Hist. Phil. Sci.* **3**, 1–80.
Wright, Larry: 1972, 'Explanation and Teleology', *Phil. Sci.* **39**, 204–218.
Wright, Larry: 1973, 'Functions', *Phil. Rev.* **82**, 139–168.
Wright, Larry: 1974, 'Mechanisms and Purposive Behavior IV', *Phil. Sci.* **41**, 345–360.

Section IV

Erlich, P. R. and Holm, R. W.: 1962, 'Patterns and Population', *Science* **137**, 652–7.
Ghiselin, Michael T.: 1969, *The Triumph of the Darwinian Method*, Berkeley, Univ. California Press.
Gruner, R.: 1969, 'Uniqueness in Nature and History', *Perspect. Biol. Med.* **19**, 145–154.
Levins, R.: 1968, *Evolution in Changing Environments*, Princeton, Princeton University Press.
Lewontin, R. C.: 1974, 'The Analysis of Variance and the Analysis of Causes', *Am. J. Human Genetics* **26**, 400–411.
Lewontin, Richard: 1970, 'The Units of Selection', *Ann. Rev. Ecol. Syst.* **1**, 1–18.
Mayr, E.: 1962, 'Accident or Design, the Paradox of Evolution', in G. W. Leeper (ed.), *The Evolution of Living Organisms*, Parkville, Melbourne University Press, pp. 1–14.
Mayr, Ernst: 1963, *Animal Species and Evolution*, Cambridge, Harvard University Press.
Muller, H. J.: 1947–48, 'Evidence of the Precision of Genetic Adaptation', *The Harvey Society of New York: The Harvey Lectures*, XLIII, pp. 165–229.
Ruse, Michael: 1975, 'Darwin's Debt to Philosophy: An Examination of the Influence of the Philosophical Ideas of John R. W. Herschel and William Whewell on the Development of Charles Darwin's Theory of Evolution', *Stud. Hist. Phil. Sci.* **6**, 159–181.
Scriven, Michael: 1959, 'Explanation and Prediction in Evolutionary Theory', *Science* **130**, 477–482.
Slobodokin, L. B. and Rapoport, A.: 1974, 'An Optimal Strategy of Evolution', *Quart. Rev. Biol.* **49**, 181–200.
Stebbins, G. L.: 1969, *The Basis of Progressive Evolution*, Chapel Hill, University of North Carolina Press.
Stebbins, G. L.: 1959, 'The Synthetic Approach to Problems of Organic Evolution', *Cold Spring Harbor Symposia on Quantitative Biology*, XXIV, pp. 305–311.
Stebbins, G. L. and Lewontin, R. C.: 1970–1971, 'Comparative Evolution at the Levels of Molecules, Organisms and Populations', *Berkeley Symposium on Mathematical Statistics and Probability, Proceedings of the Sixth*, V, pp. 23–42.
Thoday, J. M.: 1953, 'Components of Fitness', *Society for Experimental Biology Symposia*, VII, pp. 96–113.
Williams, G. C.: 1966, *Adaptation and Natural Selection*, Princeton, Princeton University Press.

Williams, G. C.: 1971, *Group Selection: A Controversy in Biology*, Chicago, Aldine-Atherton.

Williams, M. B.: 1970, 'Deducing the Consequences of Evolution: A Mathematical Model', *J. Theor. Biol.* **29**, 343–385.

Young, Robert M.: 1971, 'Darwin's Metaphor: Does Nature Select?', *Monist* **55**, 442–503.

Section Five

Ehrlich, P. R.: 1961, 'Has the Biological Species Concept Outlived Its Usefulness?', *Syst. Zool.* **10**, 167–176.

Gilmour, J. S. L.: 1951, 'The Development of Taxonomic Theory Since 1851', *Nature* **168**, 400–402.

Hull, David: 1964–65, 'The Effect of Essentialism on Taxonomy – Two Thousand Years of Stasis', *British J. Phil. Sci.* **15**, 1–18, 314–326.

Mayr, E. (ed.): 1957, *The Species Problem*, (A Symposium presented at the Atlanta meeting of the American Association for the Advancement of Science, December 28–29, 1955), Washington.

Mayr, E.: 1963, *Animal Species and Evolution*, Cambridge, The Belknap Press of the Harvard University Press, 1963; London, Oxford University Press.

Ruse, Michael: 1969, 'Definitions of Species in Biology', *British J. Phil. Sci.* **20**, 97–119.

Simpson, G. G.: 1961, *Principles of Animal Taxonomy*, New York, Columbia University Press.

INDEX

SYNTHESE LIBRARY

Monographs on Epistemology, Logic, Methodology,
Philosophy of Science, Sociology of Science and of Knowledge, and on the
Mathematical Methods of Social and Behavioral Sciences

1. J. M. BOCHEŃSKI, *A Precis of Mathematical Logic*. 1959, X + 100 pp.
2. P. L. GUIRAUD, *Problèmes et méthodes de la statistique linguistique*. 1960, VI + 146 pp.
3. HANS FREUDENTHAL (ed.), *The Concept and the Role of the Model in Mathematics and Natural and Social Sciences, Proceedings of a Colloquium held at Utrecht, The Netherlands, January 1960*. 1961, VI + 194 pp.
4. EVERT W. BETH, *Formal Methods. An Introduction to Symbolic Logic and the Study of Effective Operations in Arithmetic and Logic:* 1962, XIV + 170 pp.
5. B. H. KAZEMIER and D. VUYSJE (eds.), *Logic and Language. Studies dedicated to Professor Rudolf Carnap on the Occasion of his Seventieth Birthday*. 1962, VI + 256 pp.
6. MARX W. WARTOFSKY (ed.), *Proceedings of the Boston Colloquium for the Philosophy of Science, 1961–1962*, Boston Studies in the Philosophy of Science (ed. by Robert S. Cohen and Marx W. Wartofsky), Volume I. 1973, VIII + 212 pp.
7. A. A. ZINOV'EV, *Philosophical Problems of Many-Valued Logic*. 1963, XIV + 155 pp.
8. GEORGES GURVITCH, *The Spectrum of Social Time*. 1964, XXVI + 152 pp.
9. PAUL LORENZEN, *Formal Logic*. 1965, VIII + 123 pp.
10. ROBERT S. COHEN and MARX W. WARTOFSKY (eds.), *In Honor of Philipp Frank*, Boston Studies in the Philosophy of Science (ed. by Robert S. Cohen and Marx W. Wartofsky), Volume II. 1965, XXXIV + 475 pp.
11. EVERT W. BETH, *Mathematical Thought. An Introduction to the Philosophy of Mathematics*. 1965, XII + 208 pp.
12. EVERT W. BETH and JEAN PIAGET, *Mathematical Epistemology and Psychology*. 1966, XII + 326 pp.
13. GUIDO KÜNG, *Ontology and the Logistic Analysis of Language. An Enquiry into the Contemporary Views on Universals*. 1967, XI + 210 pp.
14. ROBERT S. COHEN and MARX W. WARTOFSKY (eds.), *Proceedings of the Boston Colloquium for the Philosophy of Science 1964–1966, in Memory of Norwood Russell Hanson*, Boston Studies in the Philosophy of Science (ed. by Robert S. Cohen and Marx W. Wartofsky), Volume III. 1967, XLIX + 489 pp.

15. C. D. Broad, *Induction, Probability, and Causation. Selected Papers.* 1968, XI + 296 pp.
16. Günther Patzig, *Aristotle's Theory of the Syllogism. A Logical-Philosophical Study of Book A of the Prior Analytics.* 1968, XVII + 215 pp.
17. Nicholas Rescher, *Topics in Philosophical Logic.* 1968, XIV + 347 pp.
18. Robert S. Cohen and Marx W. Wartofsky (eds.), *Proceedings of the Boston Colloquium for the Philosophy of Science 1966–1968*, Boston Studies in the Philosophy of Science (ed. by Robert S. Cohen and Marx W. Wartofsky), Volume IV. 1969, VIII + 537 pp.
19. Robert S. Cohen and Marx W. Wartofsky (eds.), *Proceedings of the Boston Colloquium for the Philosophy of Science 1966–1968*, Boston Studies in the Philosophy of Science (ed. by Robert S. Cohen and Marx W. Wartofsky), Volume V. 1969, VIII + 482 pp.
20. J. W. Davis, D. J. Hockney, and W. K. Wilson (eds.), *Philosophical Logic.* 1969, VIII + 277 pp.
21. D. Davidson and J. Hintikka (eds.), *Words and Objections: Essays on the Work of W. V. Quine.* 1969, VIII + 366 pp.
22. Patrick Suppes, *Studies in the Methodology and Foundations of Science. Selected Papers from 1911 to 1969*, XII + 473 pp.
23. Jaakko Hintikka, *Models for Modalities. Selected Essays.* 1969, IX + 220 pp.
24. Nicholas Rescher *et al.* (eds.). *Essay in Honor of Carl G. Hempel. A Tribute on the Occasion of his Sixty-Fifth Birthday.* 1969, VII + 272 pp.
25. P. V. Tavanec (ed.), *Problems of the Logic of Scientific Knowledge.* 1969, XII + 429 pp.
26. Marshall Swain (ed.), *Induction, Acceptance, and Rational Belief.* 1970. VII + 232 pp.
27. Robert S. Cohen and Raymond J. Seeger (eds.), *Ernst Mach; Physicist and Philosopher*, Boston Studies in the Philosophy of Science (ed. by Robert S. Cohen and Marx W. Wartofsky), Volume VI. 1970, VIII + 295 pp.
28. Jaakko Hintikka and Patrick Suppes, *Information and Inference.* 1970, X + 366 pp.
29. Karel Lambert, *Philosophical Problems in Logic. Some Recent Developments.* 1970, VII + 176 pp.
30. Rolf A. Eberle, *Nominalistic Systems.* 1970, IX + 217 pp.
31. Paul Weingartner and Gerhard Zecha (eds.), *Induction, Physics, and Ethics, Proceedings and Discussions of the 1968 Salzburg Colloquium in the Philosophy of Science.* 1970, X + 382 pp.
32. Evert W. Beth, *Aspects of Modern Logic.* 1970, XI + 176 pp.
33. Risto Hilpinen (ed.), *Deontic Logic: Introductory and Systematic Readings.* 1971, VII + 182 pp.
34. Jean-Louis Krivine, *Introduction to Axiomatic Set Theory.* 1971, VII + 98 pp.
35. Joseph D. Sneed, *The Logical Stricture of Mathematical Physics.* 1971, XV + 311 pp.
36. Carl R. Kordig, *The Justification of Scientific Change.* 1971, XIV + 119 pp.
37. Milič Čapek, *Bergson and Modern Physics*, Boston Studies in the Philosophy of Science (ed. by Robert S. Cohen and Marx W. Wartofsky), Volume VII, 1971, XV + 414 pp.
38. Norwood Russell Hanson, *What I do not Believe, and other Essays*, ed. by Stephen Toulmin and Harry Woolf, 1971, XII + 390 pp.

39. Roger C. Buck and Robert S. Cohen (eds.), *PSA 1970. In Memory of Rudolf Carnap*, Boston Studies in the Philosophy of Science (ed. by Robert S. Cohen and Marx W. Wartofsky, Volume VIII. 1971, LXVI + 615 pp. Also available as a paperback.
40. Donald Davidson and Gilbert Harman (eds.), *Semantics of Natural Language*. 1972, X + 769 pp. Also available as a paperback.
41. Yehoshua Bar-Hillel (ed)., *Pragmatics of Natural Languages*. 1971, VII + 231 pp.
42. Sören Stenlund, *Combinators, λ-Terms and Proof Theory*. 1972, 184 pp.
43. Martin Strauss, *Modern Physics and Its Philosophy. Selected Papers in the Logic, History, and Philosophy of Science*. 1972, X + 297 pp.
44. Mario Bunge, *Method, Model and Matter*. 1973, VII + 196 pp.
45. Mario Bunge, *Philosophy of Physics*. 1973, IX + 248 pp.
46. A. A. Zinov'ev, *Foundations of the Logical Theory of Scientific Knowledge (Complex Logic)*, Boston Studies in the Philosophy of Science (ed. by Robert S. Cohen and Marx W. Wartofsky), Volume IX. Revised and enlarged English edition with an appendix, by G. A. Smirnov, E. A. Sidorenko, A. M. Fedina, and L. A. Bobrova 1973, XXII + 301 pp. Also available as a paperback.
47. Ladislov Tondl, *Scientific Procedures*, Boston Studies in the Philosophy of Science (ed. by Robert S. Cohen and Marx W. Wartofsky), Volume X. 1973, XII + 268 pp. Also available as a paperback.
48. Norwood Russell Hanson, *Constellations and Conjectures*, ed. by Willard C. Humphreys, Jr. 1973, X + 282 pp.
49. K. J. J. Hintikka, J. M. E. Moravcsik, and P. Suppes (eds.), *Approaches to Natural Language. Proceedings of the 1970 Stanford Workshop on Grammar and Semantics*. 1973, VIII + 526 pp. Also available as a paperback.
50. Mario Bunge (ed.), *Exact Philosophy – Problems, Tools, and Goals*. 1973, X + 214 pp.
51. Radu J. Bogdan and Ilkka Niiniluoto (eds.), *Logic, Language, and Probability*. A selection of papers contributed to Sections IV, VI, and XI of the Fourth International Congress for Logic, Methodology, and Philosophy of Science, Bucharest, September 1971. 1973, X + 323 pp.
52. Glenn Pearce and Patrick Maynard (eds.), *Conceptual Chance*. 1973, XII + 282 pp.
53. Ilkka Niiniluoto and Raimo Tuomela, *Theoretical Concepts and Hypothetico-Inductive Inference*. 1973, VII + 264 pp.
54. Roland Fraïssé, *Course of Mathematical Logic* – Volume I: *Relation and Logical Formula*. 1973, XVI + 186 pp. Also available as a paperback.
55. Adolf Grünbaum, *Philosophical Problems of Space and Time*. Second, enlarged edition, Boston Studies in the Philosophy of Science (ed. by Robert S. Cohen and Marx W. Wartofsky), Volume XII. 1973, XXIII + 884 pp. Also available as a paperback.
56. Patrick Suppes (ed.), *Space, Time, and Geometry*. 1973, XI + 424 pp.
57. Hans Kelsen, *Essays in Legal and Moral Philosophy*, selected and introduced by Ota Weinberger. 1973, XXVIII + 300 pp.
58. R. J. Seeger and Robert S. Cohen (eds.), *Philosophical Foundations of Science. Proceedings of an AAAS Program, 1969*. Boston Studies in the Philosophy of Science (ed. by Robert S. Cohen and Marx W. Wartofsky), Volume XI. 1974, X + 545 pp. Also available as paperback.
59. Robert S. Cohen and Marx W. Wartofsky (eds.), *Logical and Epistemological*

Studies in Contemporary Physics, Boston Studies in the Philosophy of Science (ed. by Robert S. Cohen and Marx W. Wartofsky), Volume XIII. 1973, VIII + 462 pp. Also available as paperback.

60. Robert S. Cohen and Marx W. Wartofsky (eds.), *Methodological and Historical Essays in the Natural and Social Sciences. Proceedings of the Boston Colloquium for the Philosophy of Science, 1969–1972*, Boston Studies in the Philosophy of Science (ed. by Robert S. Cohen and Marx W. Wartofsky), Volume XIV. 1974, VIII + 405 pp. Also available as paperback.
61. Robert S. Cohen, J. J. Stachel, and Marx W. Wartofsky (eds.), *For Dirk Struik. Scientific, Historical and Political Essays in Honor of Dirk J. Struik*, Boston Studies in the Philosophy of Science (ed. by Robert S. Cohen and Marx W. Wartofsky), Volume XV. 1974, XXVII + 652 pp. Also available as paperback.
62. Kazimierz Ajdukiewicz, *Pragmatic Logic*, transl. from the Polish by Olgierd Wojtasiewicz. 1974, XV + 460 pp.
63. Sören Stenlund (ed.), *Logical Theory and Semantic Analysis. Essays Dedicated to Stig Kanger on His Fiftieth Birthday*. 1974, V + 217 pp.
64. Kenneth F. Schaffner and Robert S. Cohen (eds.), *Proceedings of the 1972 Biennial Meeting, Philosophy of Science Association*, Boston Studies in the Philosophy of Science (ed. by Robert S. Cohen and Marx W. Wartofsky), Volume XX. 1974, IX + 444 pp. Also available as paperback.
65. Henry E. Kyburg, Jr., *The Logical Foundations of Statistical Inference*. 1974, IX + 421 pp.
66. Marjorie Grene, *The Understanding of Nature: Essays in the Philosophy of Biology*, Boston Studies in the Philosophy of Science (ed. by Robert S. Cohen and Marx W. Wartofsky), Volume XXIII. 1974, XII + 360 pp. Also available as paperback.
67. Jan M. Broekman, *Structuralism: Moscow, Prague, Paris*. 1974, IX + 117 pp.
68. Norman Geschwind, *Selected Papers on Language and the Brain*, Boston Studies in the Philosophy of Science (ed. by Robert S. Cohen and Marx W. Wartofsky), Volume XVI. 1974, XII + 549 pp. Also available as paperback.
69. Roland Fraïssé. *Course of Mathematical Logic* – Volume II: *Model Theory*. 1974, XIX + 192 pp.
70. Andrzej Grzegorczyk, *An Outline of Mathematical Logic. Fundamental Results and Notions Explained with all Details*. 1974, X + 596 pp.
71. Franz von Kutschera, *Philosophy of Language*. 1975, VII + 305 pp.
75. Jaakko Hintikka and Unto Remes, *The Method of Analysis. Its Geometrical Origin and Its General Significance*. 1974, XVIII + 144 pp.
76. John Emery Murdoch and Edith Dudley Sylla, *The Cultural Context of Medieval Learning. Proceedings of the First International Colloquium on Philosophy, Science, and Theology in the Middle Ages – September 1973*. Boston Studies in the Philosophy of Science (ed. by Robert S. Cohen and Marx. W. Wartofsky), Volume XXVI. 1975, X + 566 pp. Also available as paperback.
77. Stefan Amsterdamski, *Between Experience and Metaphysics. Philosophical Problems of the Evolution of Science*. Boston Studies in the Philosophy of Science (ed. by Robert S. Cohen and Marx W. Wartofsky), Volume XXXV. 1975, XVIII + 193 pp. Also available as paperback.

SYNTHESE HISTORICAL LIBRARY

Texts and Studies
in the History of Logic and Philosophy

Editors:

1. M. T. BEONIO-BROCCHIERI FUMAGALLI, *The Logic of Abelard.* Translated from the Italian. 1969, IX + 101 pp.
2. GOTTFRIED WILHELM LEIBNITZ, *Philosophical Papers and Letters.* A selection translated and edited with an introduction, by Leroy E. Loemker. 1969, XII + 736 pp.
3. ERNST MALLY, *Logische Schriften,* ed. by Karl Wolf and Paul Weingartner. 1971, X + 340 pp.
4. LEWIS WHITE BECK (ed.), *Proceedings of the Third International Kant Congress.* 1972, XI + 718 pp.
5. BERNARD BOLZANO, *Theory of Science,* ed. by Jan Berg. 1973, XV + 398 pp.
6. J. M. E. MORAVCSIK (ed.), *Patterns in Plato's Thought. Papers arising out of the 1971 West Coast Greek Philosophy Conference.* 1973, VIII + 212 pp.
7. NABIL SHEHABY, *The Propositional Logic of Avicenna: A Translation from al-Shifā': al-Qiyās,* with Introduction, Commentary and Glossary. 1973, XIII + 296 pp.
8. DESMOND PAUL HENRY, *Commentary on De Grammatico: The Historical-Logical Dimensions of a Dialogue of St. Anselm's.* 1974, IX + 345 pp.
9. JOHN CORCORAN, *Ancient Logic and Its Modern Interpretations.* 1974. X + 208 pp.
10. E. M. BARTH, *The Logic of the Articles in Traditional Philosophy.* 1974, XXVII + 533 pp.
11. JAAKKO HINTIKKA, *Knowledge and the Known. Historical Perspectives in Epistemology.* 1974, XII + 243 pp.
12. E. J. ASHWORTH, *Language and Logic in the Post-Medieval Period.* 1974, XIII + 304 pp.
13. ARISTOTLE, *The Nicomachean Ethics.* Translated with Commentaries and Glossary by Hyppocrates G. Apostle. 1975, XXI+372 pp.
14. R. M. DANCY, *Sense and Contradiction: A Study in Aristotle.* 1975, XII+184 pp.
15. WILBUR RICHARD KNORR, *The Evolution of the Euclidean Elements. A Study of the Theory of Incommensurable Magnitudes and Its Significance for Early Greek Geometry.* 1975, IX+374 pp.
16. AUGUSTINE, *De Dialectica.* Translated with the Introduction and Notes by B. Darrell Jackson. 1975, XI+151 pp.

Kevin Stultz

Nihil sub sole novum
nothing under sun new

You should ask "the reason why it is that way" in 4 ways (Aristotle)

1. final - that for the sake of which
2. efficient - of which - cause for motion
3. material cause - That that is able to take on 1 form or another
4. formal - principle of organization

KENT
KRESGE

G. C. Williams
Adaptation and Nat. Selection

"Succession" because of entropy

"Science is the elaboration of common sense" but sometimes common sense is incorrect

body size in successional series

11 12 13 14 15
16 17 18 19 20 21 22
23 24 25 26 27 28 29
30 1 2 3 4 5 6
7 8 9 10 11 12 13

33 days
− 7
26

26
− 3
23